Normal Families of Meromorphic Functions

NORMAL FAMILIES OF MEROMORPHIC FUNCTIONS

CHI-TAI CHUANG

Department of Mathematics
Peking University, China

World Scientific
Singapore • New Jersey • London • Hong Kong

Published by

World Scientific Publishing Co. Pte. Ltd.
5 Toh Tuck Link, Singapore 596224
USA office: 27 Warren Street, Suite 401-402, Hackensack, NJ 07601
UK office: 57 Shelton Street, Covent Garden, London WC2H 9HE

British Library Cataloguing-in-Publication Data
A catalogue record for this book is available from the British Library.

NORMAL FAMILIES OF MEROMORPHIC FUNCTIONS

ISBN-13 978-981-02-1257-5
ISBN-10 981-02-1257-7

PREFACE

The theory of normal families of holomorphic functions or more generally of meromorphic functions was created by Montel about eighty years ago. His definition of normal families of holomorphic functions may be stated as follows:

Definition 1. A family of holomorphic functions in a domain D is said to be normal in D, if from every sequence of functions $f_n(z)$ $(n=1,2,\cdots)$ of the family, we can extract a subsequence $f_{n_k}(z)$ $(k=1,2,\cdots)$ which converges locally uniformly in D to a holomorphic function or the constant ∞.

In order to extend this definition to the case of families of meromorphic functions, it is necessary to give a definition of the locally uniform convergence of a sequence of meromorphic functions in a domain. Landau and Carathéodory first gave a such definition. Later Ostrowski pointed out that it can be equivalently defined by means of the spherical distance $|w_1, w_2|$ between two points w_1, w_2 of the extended complex plane $\hat{\mathbf{C}}=\mathbf{C}\bigcup(\infty)$. A locally uniformly convergent sequence of meromorphic functions in a domain with respect to the spherical distance, has a limit function which is a meromorphic function in the domain or the constant ∞. So the notion of normal families of meromorphic functions can be simply defined as follows:

Definition 2. A family of meromorphic functions in a domain D is said to be normal in D, if from every sequence of functions $f_n(z)$ $(n=1,2,\cdots)$ of the family we can extract a subsequence $f_{n_k}(z)$ $(k=1,2,\cdots)$ which is locally uniformly convergent in D with respect to the spherical distance.

In particular for a family of holomorphic functions in a domain, this definition is equivalent to Definition 1.

The theorey of normal families of meromorphic functions has various applications such as uniform convergence of sequences of holomorphic functions or meromorphic functions, extensions of Picard's theorems, Schottky's theorem and Landau's theorem, iteration of rational functions or entire functions and conformal mapping.

Next to the notion of normal families of meromorphic functions, Montel also introduced the notion of quasi-normal families of meromorphic functions which may be defined as follows:

Definition 3. A family of meromorphic functions in a domain D is said to be quasi-normal in D, if from every sequence of functions $f_n(z)$ $(n=1,2,\cdots)$ of the family, we can extract a subsequence $f_{n_k}(z)$ $(k=1,2,\cdots)$ which is locally uniformly convergent in D-σ with respect to the spherical distance, where σ is a set of points of D having no point of accumulation in D and depending on the sequence of functions $f_n(z)$ $(n=1,2,\cdots)$.

Chuang observed that if we denote by E the set of points of D in the neighborhood of each of which the extracted sequence $f_{n_k}(z)$ $(k=1,2,\cdots)$ is not uniformly convergent with respect to the spherical distance, then normality and quasi-normality of a family of meromorphic functions in D correspond respectively to the conditions $E=\emptyset$ and $E'=\emptyset$, where E' is the derived set in D of E. It is natural to go still further and introduce notions corresponding respectively to the conditions $E''=\emptyset, E'''=\emptyset, \cdots$. In doing so, for each integer $m\geqslant 0$, Chuang defined a corresponding notion of Q_m-normality corresponding to the condition $E^{(m)}=\emptyset$, such that in particular for $m=0$ and $m=1$, it coincides respectively with the notions of normality and quasi-normality defined respectively by Definitions 2 and 3.

The theory of quasi-normal families of meromorphic functions and more generally that of Q_m-normal families of meromorphic functions also have various applications.

This book consists of nine chapters and two appendices. They as a whole constitute a systematic general theory of normal families, quasi-normal families and Q_m-normal families of meromorphic functions with many applications. Many materials in this book are research works of the author unpublished before.

Chapters 1, 2 and 3 are concerned with normal families of meromorphic functions. They contain mainly a precise definition and three necessary and sufficient conditions of the normality of a family of meromorphic functions in a domain, some standard criterions of normality and their applications.

Chapter 4 deals with closed families of meromorphic functions. A closed family of meromorphic functions in a domain is a normal family in that domain, but the converse is untrue. By means of continuous functionals and continuous operators defined on a closed family of meromorphic functions, for many extremal problems, the existence of the solution can be deduced. A covering theorem concerning the existence of an extremal domain covered at least p times by the values of each function of a closed family of meromorphic functions satisfying a certain condition, is proved.

Chapter 5 is concerned with quasi-normal families of meromorphic functions. Besides a precise definition and some properties of such a family, this chapter consists mainly of two parts. The first part deals with quasi-normal families of order not exceeding an integer ν. Correspondingly, criterions and a necessary and sufficient condition of quasi-normality of order not exceeding an integer ν are obtained, with applications. The second part deals with quasi-normal families of finite total order. After some preliminary preperations, the notion of such families is defined. This is then followed by some examples and applications.

The main purpose of Chapters 6 and 7 is to give applications of the theory of normal families or quasi-normal families of meromorphic functions, which are not contained in the preceding chapters. However the criterions of normality or quasi-normality of families of holomorphic or meromorphic functions proved in Chapter 7 are also of importance.

Chapter 8 is concerned with the theory of Q_m-normal families of meromorphic functions. It consists mainly of the definition and necessary and sufficient conditions of Q_m-normality, the definition and necessary and sufficient conditions of Q_m-normality of order not exceeding an integer ν, criterions of Q_m-normality or Q_m-normality of order not exceeding an integer ν and an existence theorem which justifies the introduction of the notion of Q_m-normal families of meromorphic functions. The theory of Q_m-normal families of meromorphic functions has various applications which are given in Chapter 9.

In Appendices A and B are proved some theorems which are used in the proofs of some criterions of normality or quasi-normality in Chapters 2,3 and 5.

The author has tried to write this book as simply and as clearly as possible. For instance, in this book the definitions of normality and quasi-normality of a family of meromorphic functions are in substance equivalent to those given by Montel, but in a different form which is much clearer and easier to handle. However the proofs of some theorems are long. In order to facilitate the reading, they are given in detail. The author also has tried to keep this book at an elementary level. It is self-contained, except only that in Appendix B some elementary knowledge of Nevanlinna's theory of meromorphic functions is needed, which can be found from any one of the following three books:

Nevanlinna, R., Le Théorème de Picard-Borel et la Théorie des Fonctions Méromorphes, Gauthier-Villars, Paris, 1929.

Hayman, W. K., Meromorphic Functions, Clarendon Press, Oxford, 1964.

Chuang, C. T. and Yang, C. C., Fix-Points and Factorization of Meromorphic Functions, World Scientific Publishing Co., Singapore, 1990.

It is assumed that the reader is familiar with the basic theory of functions of a complex variable and has some elementary knowledge of Nevanlinna's theory of meromorphic functions. For reference, the reader may consult the following books:

Montel, P., Lecons sur les familles normales de fonctions analytiques et leurs applications, Gauthier-Villars, Paris, 1927.

Valiron, G., Familles Normales et Quasi-normales de Fonctions Méromorphes, Mémorial des Sc. Math., Fasc. 38, 1929.

Gu, Y. X., Normal Families of Meromorphic Functions (in Chinese), Educational Press of Szechuan, 1991.

Unlike the first two books, the third book deals only with normal families of meromorphic functions and the emphasis is placed on generalizations of a theorem of Miranda.

The References at the end of this book comprise only those papers and books quoted in the text, with few exceptions. For other references the reader may consult the above three books.

Finally I express my sincere thanks to Mrs. Hu Yang and Mrs. Wang Hui Li for their efforts to arrange to transform my manuscript into a camera-ready manuscript by the Computer Centre of Peking University. I thank especially Ms. Yan Ling and Mr. Yan An of that centre who undertake this laborious task. I am also very grateful to Dr. K. K. Phua, Mr. J. G. Xu and the staff of the World Scientific Publishing Company for the publication of this book.

Chi-tai Chuang

CONTENTS

1

BASIC NOTIONS AND THEOREMS

1. 1. SPHERICAL DISTANCE

In the figure, the equation of the sphere S is

$$x^2 + y^2 + (u - \frac{1}{2})^2 = \frac{1}{4}. \qquad (1.1)$$

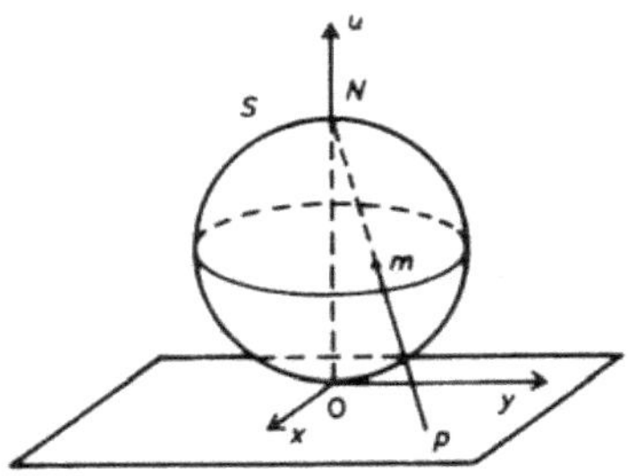

Consider a complex number $z=x+iy$. Let p be the point of the xy plane corresponding to z, whose coordinates are (x, y). The straight line joining the two points N and p intersects S at a point m distinct from N. We call m the point of S corresponding to z. Let us find out the coordinates (X, Y, u) of m. We have

$$X = hx, \quad Y = hy, \quad u - 1 = - h,$$

where h is a positive number. Substituting into (1. 1) we get

$$h = \frac{1}{x^2 + y^2 + 1} = \frac{1}{1 + |z|^2}$$

and

$$X = \frac{x}{1 + |z|^2}, \quad Y = \frac{y}{1 + |z|^2}, \quad u = \frac{|z|^2}{1 + |z|^2}. \qquad (1.2)$$

The point of S corresponding to ∞ is the point N whose coordinates are (0, 0, 1).

Definition 1. 1. Let z_1, z_2 be two points of the extended complex plane $\hat{\mathbf{C}}=\mathbf{C}\cup\infty$ and m_1, m_2 the two points of S corresponding respectively to z_1, z_2. The length of the line segment $\overline{m_1m_2}$ is defined to be the spherical distance between z_1, z_2 and is denoted by $|z_1, z_2|$.

Let us find out an expression of $|z_1, z_2|$. Distinguish three cases:

1) z_1, z_2 are both finite. Let $z_j=x_j+iy_j (j=1,2)$ and set $k_j=1+|z_j|^2 (j=1,2)$. By (1.2), we have

$$(k_1k_2|z_1,z_2|)^2 = (k_2x_1 - k_1x_2)^2 + (k_2y_1 - k_1y_2)^2 + (k_1 - k_2)^2$$
$$= k_2^2k_1 + k_1^2k_2 - 2k_1k_2(x_1x_2 + y_1y_2 + 1)$$

and hence

$$k_1k_2|z_1,z_2|^2 = |z_1|^2 + |z_2|^2 - 2(x_1x_2 + y_1y_2). \tag{1.3}$$

Next using the relations

$$2x_j = z_j + \bar{z}_j, \quad 2iy_j = z_j - \bar{z}_j \quad (j = 1,2),$$

we find that the right member of (1. 3) is equal to $|z_1-z_2|^2$. So we have the formula

$$|z_1,z_2| = \frac{|z_1 - z_2|}{(1 + |z_1|^2)^{\frac{1}{2}}(1 + |z_2|^2)^{\frac{1}{2}}}. \tag{1.4}$$

2) One of z_1, z_2 is finite and the other is infinite. For instance $z_1=x_1+iy_1$ is finite and $z_2=\infty$. Then

$$|z_1,z_2|^2 = \frac{x_1^2}{(1 + |z_1|^2)^2} + \frac{y_1^2}{(1 + |z_1|^2)^2} + \frac{1}{(1 + |z_1|^2)^2}$$
$$= \frac{1}{1 + |z_1|^2}$$

and hence

$$|z_1, z_2| = \frac{1}{(1 + |z_1|^2)^{\frac{1}{2}}}. \tag{1.5}$$

3) z_1, z_2 are both infinite. Evidently $|z_1, z_2| = 0$.

By Definition 1.1 the triangular inequality

$$|z_1, z_3| \leqslant |z_1, z_2| + |z_2, z_3| \tag{1.6}$$

obviously holds for any three points $z_j (j=1,2,3)$ of $\hat{\mathbf{C}}$. Also it is easily verified that the formula

$$|z_1, z_2| = \left|\frac{1}{z_1}, \frac{1}{z_2}\right| \tag{1.7}$$

holds for any two points $z_j (j=1,2)$ of $\hat{\mathbf{C}}$.

Finally we prove two lemmas which are sometimes useful.

Lemma 1.1. Let z_1, z_2 and $a \neq \infty$ be three points of $\hat{\mathbf{C}}$. Then

$$|z_1 - a, z_2 - a| \geqslant \frac{1}{2} |a, \infty|^2 |z_1, z_2|. \tag{1.8}$$

Proof. Assume first that $z_j \neq \infty (j=1,2)$. In this case, (1.8) follows from the formula

$$|z_1 - a, z_2 - a| = \frac{|z_1 - z_2|}{(1 + |z_1 - a|^2)^{\frac{1}{2}} (1 + |z_2 - a|^2)^{\frac{1}{2}}}$$

and the inequality

$$\begin{aligned} 1 + |\zeta_1 - \zeta_2|^2 &= 1 + |\zeta_1|^2 + |\zeta_2|^2 - 2Re(\zeta_1 \bar{\zeta}_2) \\ &< 2(1 + |\zeta_1|^2 + |\zeta_2|^2 + |\zeta_1|^2 |\zeta_2|^2) \end{aligned}$$

$$= 2(1 + |\zeta_1|^2)(1 + |\zeta_2|^2).$$

If one of z_1, z_2 is finite and the other is infinite, for instance $z_1 \neq \infty, z_2 = \infty$, we apply (1.8) to z_1 and $z_2' \neq \infty$, and then let $z_2' \to \infty$.

Lemma 1.2. Let $A, B (A < B)$ be two positive numbers. Then there is a positive number $\mu = \mu(A, B)$ depending only on A and B such that for $|z_1| \leqslant A$, $|z_2| \geqslant B$, we have $|z_1, z_2| \geqslant \mu$.

Proof. If $|z_1| \leqslant A$, $|z_2| \geqslant B$, $z_2 \neq \infty$, then

$$|z_1, z_2| = \frac{|z_1 - z_2|}{(1 + |z_1|^2)^{\frac{1}{2}}(1 + |z_2|^2)^{\frac{1}{2}}} \geqslant \frac{1 - \left|\frac{z_1}{z_2}\right|}{(1 + |z_1|^2)^{\frac{1}{2}}\left(1 + \frac{1}{|z_2|^2}\right)^{\frac{1}{2}}}$$

$$\geqslant \frac{1 - \frac{A}{B}}{(1 + A^2)^{\frac{1}{2}}\left(1 + \frac{1}{B^2}\right)^{\frac{1}{2}}},$$

which is also true, when $z_2 = \infty$.

1.2. SEQUENCES OF MEROMORPHIC FUNCTIONS

Definition 1.2. A sequence of points $z_n (n = 1, 2, \cdots)$ of $\hat{C}$ is said to be convergent with respect to the spherical distance, if to each positive number ε corresponds a positive integer N such that, for $n \geqslant N, m \geqslant N$, we have

$$|z_n, z_m| < \varepsilon. \tag{1.9}$$

Lemma 1.3. If a sequence of points $z_n (n = 1, 2, \cdots)$ of $\hat{C}$ is convergent with respect to the spherical distance, then there exists an unique point Z of $\hat{C}$ such that

$$\lim_{n\to+\infty} |z_n, Z| = 0. \tag{1.10}$$

Z is called the limit of the sequence $z_n (n=1,2,\cdots)$ with respect to the spherical distance.

Proof. We first show that such a point Z exists. In fact, if

$$\lim_{n\to\infty} |z_n, \infty| = 0,$$

then $Z=\infty$ is such a point. Otherwise we can find a positive number ε_0 and an increasing sequence of positive integers $n_k (k=1,2,\cdots)$ such that

$$|z_{n_k}, \infty| \geqslant \varepsilon_0 \qquad (k = 1,2,\cdots)$$

which implies that z_{n_k} is finite and

$$|z_{n_k}| < (1 + |z_{n_k}|^2)^{\frac{1}{2}} = \frac{1}{|z_{n_k}, \infty|} \leqslant \frac{1}{\varepsilon_0}.$$

The sequence $z_{n_k} (k=1,2,\cdots)$ is then bounded. Let $Z\neq\infty$ be a limiting point of the sequence $z_{n_k} (k=1,2,\cdots)$. Then to any positive number η and any positive integer K, corresponds a positive integer k such that

$$k \geqslant K, \quad |z_{n_k} - Z| < \eta.$$

Since

$$n_k \geqslant k, \quad |z_{n_k}, Z| \leqslant |z_{n_k} - Z|,$$

we have, à fortiori,

$$n' \geqslant K, \quad |z_{n'}, Z| < \eta \qquad (n' = n_k).$$

Now given a positive number ε, let N be a positive integer such that

$$|z_n, z_m| < \frac{\varepsilon}{2}$$

for $n \geqslant N$, $m \geqslant N$, and then let n_0 be a positive integer such that

$$n_0 \geqslant N, |z_{n_0}, Z| < \frac{\varepsilon}{2}.$$

Consequently for $n \geqslant N$, we have

$$|z_n, Z| \leqslant |z_n, z_{n_0}| + |z_{n_0}, Z| < \frac{\varepsilon}{2} + \frac{\varepsilon}{2} = \varepsilon.$$

Hence the point Z satisfies the condition (1.10). Finally the uniqueness of such a point Z follows immediately from the inequality

$$|Z, Z'| \leqslant |Z, z_n| + |z_n, Z'|.$$

Definition 1.3. Let $S: f_n(z)(n=1,2,\cdots)$ be a sequence of meromorphic functions in a domain D. Let E be a set of points of D. The sequence S is said to be uniformly convergent in E with respect to the spherical distance, if to each positive number ε corresponds a positive integer N such that, when $n \geqslant N$, $m \geqslant N$, we have

$$|f_n(z), f_m(z)| < \varepsilon \tag{1.11}$$

in E.

Assume that the condition in Deflnition 1.3 is satisfied. Then for each point z_0 of E, the sequence $f_n(z_0)(n=1,2,\cdots)$ is convergent with respect to the spherical distance, hence it has a limit $F(z_0)$ with respect to the spherical distance, by Lemma 1.3. $F(z)$ is a function defined in E. We are going to show that to each positive number ε corresponds a positive integer N such that, when $n \geqslant N$, we have

$$|f_n(z), F(z)| < \varepsilon \tag{1.12}$$

in E.

In fact, given a positive number ε, by hypothesis, there is a positive integer N such that, when $n \geqslant N$, $m \geqslant N$, we have

$$|f_n(z), f_m(z)| < \frac{\varepsilon}{2}$$

in E. Consequently when $n \geqslant N$, $m \geqslant N$ and $z \in E$, we have

$$|f_n(z), F(z)| \leqslant |f_n(z), f_m(z)| + |f_m(z), F(z)|$$
$$< \frac{\varepsilon}{2} + |f_m(z), F(z)|.$$

In this inequality keeping $n \geqslant N$ and $z \in E$ fixed and letting $m \to +\infty$, we get

$$|f_n(z), F(z)| \leqslant \frac{\varepsilon}{2} < \varepsilon,$$

as asserted. We say that as $n \to +\infty$, $f_n(z)$ converges uniformly to $F(z)$ in E with respect to the spherical distance.

Theorem 1. 1. Let $S: f_n(z)(n=1,2,\cdots)$ be a sequence of meromorphic functions in a circle $\Gamma: |z-z_0|<r$. Assume that the sequence S is uniformly convergent in Γ with respect to the spherical distance and let $F(z)$ be the limit function of S defined in Γ with respect to the spherical distance. Then the following assertions are true.

1^0 If $F(z_0) \neq \infty$, then we can find a circle $\Gamma_0: |z-z_0|<r_0 (0<r_0 \leqslant r)$ and a positive integer n_0 such that the functions $f_n(z)(n \geqslant n_0)$ and $F(z)$ are holomorphic in Γ_0, and

$$\lim_{\substack{n \to +\infty \\ n \geqslant n_0}} |f_n(z) - F(z)| = 0 \tag{1.13}$$

uniformly in Γ_0.

2^0 If $F(z_0)=\infty$, then we can find a circle $\Gamma_0: |z-z_0|<r_0 (0<r_0\leqslant r)$ and a positive integer n_0 such that the functions $1/f_n(z) (n\geqslant n_0)$ and $1/F(z)$ are holomorphic in Γ_0, and

$$\lim_{\substack{n\to+\infty \\ n\geqslant n_0}} \left|\frac{1}{f_n(z)} - \frac{1}{F(z)}\right| = 0 \tag{1.14}$$

uniformly in Γ_0.

For the proof of this theorem, we need the following lemma.

Lemma 1.4. If $f(z)$ is a meromorphic function in a domain D, then $f(z)$ is continuous in D with respect to the spherical distance. In other words, for each point z_0 of D, we have

$$\lim_{z\to z_0} |f(z), f(z_0)| = 0. \tag{1.15}$$

Proof. Consider a point z_0 of D and distinguish two cases. If $f(z_0)\neq\infty$, then there is a circle $c: |z-z_0|<r$ belonging to D, such that $f(z)$ is holomorphic in c. Consequently by (1.4), we have

$$|f(z), f(z_0)| \leqslant |f(z) - f(z_0)| \tag{1.16}$$

in c. Evidently (1.16) implies (1.15). If $f(z_0)=\infty$, then the result just obtained is applicable to the function $1/f(z)$. So we have

$$\lim_{z\to z_0} \left|\frac{1}{f(z)}, \frac{1}{f(z_0)}\right| = 0$$

and then by the identity

$$|f(z), f(z_0)| = \left|\frac{1}{f(z)}, \frac{1}{f(z_0)}\right|$$

we get again (1.15).

Now let us return to the proof of Theorem 1.1. Consider first the case $F(z_0)\neq\infty$. Then $d=|f(z_0),\infty|>0$. Set

$$A=\frac{2}{d},\quad B=\frac{2}{d}+1$$

and let $\mu(A,B)$ be the positive number defined in Lemma 1.2. Set

$$\varepsilon_0=min(\frac{d}{6},\mu(A,B)). \tag{1.17}$$

By the result (1.12) obtained above, we can find a positive integer n_0 such that, when $n\geqslant n_0$, we have

$$|f_n(z),F(z)|<\varepsilon_0 \tag{1.18}$$

in the circle Γ. Then by lemma 1.4, we can find a circle Γ_0: $|z-z_0|<r_0(0<r_0\leqslant r)$ such that in Γ_0 we have

$$|f_{n_0}(z),f_{n_0}(z_0)|<\frac{d}{6}.$$

Consequently for $z\in\Gamma_0$, we have

$$\begin{aligned}|F(z_0),\infty| &\leqslant |F(z_0),f_{n_0}(z_0)|+|f_{n_0}(z_0),f_{n_0}(z)|\\ &\quad+|f_{n_0}(z),F(z)|+|F(z),\infty|\\ &<3\frac{d}{6}+|F(z),\infty|,\\ |F(z),\infty| &>\frac{d}{2}\end{aligned}$$

and hence

$$|F(z),\infty|=\frac{1}{(1+|F(z)|^2)^{\frac{1}{2}}},\quad |F(z)|<A. \tag{1.19}$$

Then from (1.17), (1.18) and (1.19), we see, by Lemma 1.2, that for $n \geqslant n_0$ and $z \in \Gamma_0$, we have

$$|f_n(z)| < B. \tag{1.20}$$

Next from (1.19) and (1.20), we have, for $n \geqslant n_0$ and $z \in \Gamma_0$,

$$\begin{aligned}|f_n(z) - F(z)| &= \{1 + |f_n(z)|^2\}^{\frac{1}{2}}\{1 + |F(z)|^2\}^{\frac{1}{2}}|f_n(z), F(z)| \\ &\leqslant (1 + B^2)^{\frac{1}{2}}(1 + A^2)^{\frac{1}{2}}|f_n(z), F(z)|.\end{aligned}$$

This inequality and the result (1.12) evidently implies the assertion 1^0 in Theorem 1.1.

Consider now the case $F(z_0) = \infty$. Then the above argument is applicable to $1/f_n(z)$ $(n = 1, 2, \cdots)$ and $1/F(z)$, hence the assertion 2^0 in Theorem 1.1 is also true.

Definition 1.4. Let S: $f_n(z)(n = 1, 2, \cdots)$ be a sequence of meromorphic functions in a domain D. A point z_0 of D is called a C_0-point of the sequence S, if there is a circle $\Gamma: |z - z_0| < r$ belonging to D such that the sequence S is uniformly convergent in Γ with respect to the spherical distance. The sequence S is said to be a C_0-sequence in D, if each point of D is a C_0-point of S.

Assume that the sequence S is a C_0-sequence in D. Let z_0 be a point of D. Then by hypothesis, the sequence S is uniformly convergent in a circle $\Gamma: |z - z_0| < r$ with respect to the spherical distance. Consequently, by the result (1.12), the sequence S has a limit function F(z) defined in Γ with respect to the spherical distance and as $n \to +\infty$, $f_n(z)$ converges uniformly to F(z) in Γ with respect to the spherical distance. Hence we have the following theorem.

Theorem 1.2. Let $S: f_n(z)(n = 1, 2, \cdots)$ be a C_0-sequence of meromorphic functions in a domain D. Then S has a limit function F(z) defined in D with respect to the spherical distance, such that as $n \to +\infty$, $f_n(z)$ converges locally uniformly to F(z) in D with respet to the spherical distance.

Furthermore we have the following theorem.

Theorem 1.3. Let $S: f_n(z)(n=1,2,\cdots)$ be a C_0-sequence of meromorphic functions in a domain D. Then the limit function $F(z)$ of S with respect to the spherical distance is a meromorphic function in D or the constant ∞.

Proof. Denote by σ the set of the points z' of D such that $F(z')=\infty$. Distinguish two cases.

1) Assume that σ has a point of accumulation z_0 in D. By hypothesis, there is a circle $\Gamma: |z-z_0|<r$ belonging to D such that the sequence S is uniformly convergent in Γ with respect to the spherical distance. We can not have $F(z_0)\neq\infty$, because then by Theorem 1.1, $F(z)$ would be finite in a circle $\Gamma_0: |z-z_0|<r_0$. Hence $F(z_0)=\infty$ and by Theorem 1.1, the function $1/F(z)$ is holomorphic in a circle $\Gamma_0: |z-z_0|<r_0$. Since the points of the set σ are zeros of $1/F(z)$, so the function $1/F(z)$ is identically equal to zero in Γ_0 and $F(z)=\infty$ in Γ_0. Consider a point $z_1 (z_1\neq z_0)$ of D. Join z_0 to z_1 by a polygonal line $z=p(t)$ $(0\leqslant t\leqslant b)$ lying in D such that $p(0)=z_0, p(b)=z_1$. Denote by T the set of the numbers t' $(0<t'<b)$ such that

$$F\{p(t)\}=\infty \tag{1.21}$$

for $0\leqslant t\leqslant t'$. Evidently the set T is non-empty. Let $\beta (0<\beta\leqslant b)$ be the least upper bound of T. Then (1.21) holds for $0\leqslant t<\beta$. It follows that the point $z_*=p(\beta)$ is a point of accumulation of the set σ, and hence there is a circle $\Gamma_*: |z-z_*|<r_*$ in which $F(z)=\infty$. So we must have $\beta=b$ and $F(z_1)=\infty$. Consequently $F(z)$ is the constant ∞.

2) Now assume that the set σ has no point of accumulation in D. Consider a point z_0 of D. If $F(z_0)\neq\infty$, then by Theorem 1.1, the function $F(z)$ is holomorphic in a circle $\Gamma_0: |z-z_0|<r_0$. If $F(z_0)=\infty$, then by Theorem 1.1, the function $1/F(z)$ is holomorphic in a circle $\Gamma_0: |z-z_0|<r_0$. Let r_0' be a number such that $0<r_0'\leqslant r_0$ and $F(z)\neq\infty$ for $0<|z-z_0|<r_0'$. Then in the circle $\Gamma_0': |z-z_0|<r_0'$ we may write $F(z)=1/G(z)$, where the function $G(z)$ is holomorphic in Γ_0', $G(z_0)=0$ and $G(z)\neq 0$ for $0<|z-z_0|<r_0'$. Hence z_0 is a pole of $F(z)$. Consequently $F(z)$ is a meromorphic function in D.

In particular for sequences of holomorphic functions we have the following corollary.

Corollary 1. 1. Let S: $f_n(z)(n=1,2,\cdots)$ be a sequence of holomorphic functions in a domain D. In order that S is a C_0-sequence in D, it is necessary and sufficient that one of the following two conditions is satisfied.

1^0 As $n\to+\infty$, $f_n(z)$ converges locally uniformly to a holomorphic function in D.

2^0 As $n\to+\infty$, $f_n(z)$ converges locally uniformly to the constant ∞ in D.

Proof. Assume that S is a C_0-sequence in D and let $F(z)$ be the limit function of S with respect to the spherical distance. Consider first the case that $F(z)$ is a meromorphic function in D. We are going to show that $F(z)$ has no pole in D. In fact, if $F(z_0)=\infty$ at a point z_0 of D, then by Theorem 1. 1, we can find a circle $\Gamma_0: |z-z_0|<r_0$ and a positive integer n_0 such that the functions $1/f_n(z)(n\geqslant n_0)$ and $1/F(z)$ are holomorphic in Γ_0, and as $n\to+\infty$, $1/f_n(z)$ converges uniformly to $1/F(z)$ in Γ_0. This implies that when n is sufficiently large, the function $1/f_n(z)$ would have zero in Γ_0, incompatible to the fact that $f_n(z)$ is finite in D. Consequently the function $F(z)$ is holomorphic in D. Then again by Theorem 1. 1, we see that as $n\to+\infty$, $f_n(z)$ converges locally uniformly to $F(z)$ in D.

Next consider the case that $F(z)$ is the constant ∞. Then by Theorem 1. 1, for each point z_0 of D, we can find a circle $\Gamma_0: |z-z_0|<r_0$ and a positive integer n_0 such that the functions $1/f_n(z)(n\geqslant n_0)$are holomorphic in Γ_0, and as $n\to+\infty$, $1/f_n(z)$ converges uniformly to zero in Γ_0 and hence $f_n(z)$ converges uniformly to ∞ in Γ_0.

Conversely assume that, for instance, the condition 1^0 is satisfied and let $F(z)$ be the limit function. Then the inequality

$$\begin{aligned}|f_n(z),f_m(z)| &\leqslant |f_n(z),F(z)| + |F(z),f_m(z)| \\ &\leqslant |f_n(z) - F(z)| + |F(z) - f_m(z)|\end{aligned}$$

iniplies clearly that the sequence S is a C_0-sequence in D. Similarly if the condition 2^0 is satisfied, then the inequality

$$\begin{aligned}|f_n(z),f_m(z)| &\leqslant |f_n(z),\infty| + |\infty,f_m(z)| \\ &= \frac{1}{(1+|f_n(z)|^2)^{\frac{1}{2}}} + \frac{1}{(1+|f_m(z)|^2)^{\frac{1}{2}}}\end{aligned}$$

implies that the sequence S is a C_0-sequence in D.

1. 3. FAMILIES OF MEROMORPHIC FUNCTIONS

Definition 1. 5. Let $\mathscr{F}$ be a family of meromorphic functions in a domain D. We say that the family $\mathscr{F}$ is normal in D, if from every sequence of functions $f_n(z)(n=1,2,\cdots)$ of the family $\mathscr{F}$, we can extract a subsequence $f_{n_k}(z)(k=1,2,\cdots,n_k<n_{k+1})$ which is a C_0-sequence in D.

In particular for families of holomorphic functions, we have the following corollary which results immediately from Definition 1. 5 and Corollary 1. 1.

Corollary 1. 2. Let $\mathscr{F}$ be a family of holomorphic functions in a domain D. In order that the family $\mathscr{F}$ be normal in D, it is necessary and sufficient that from every sequence of functions $f_n(z)(n=1,2,\cdots)$ of $\mathscr{F}$, we can extract a subsequence $f_{n_k}(z)(k=1,2,\cdots)$ satisfying one of the following two conditions:

1^0 As $k\to+\infty$, $f_{n_k}(z)$ converges locally uniformly to a holomorphic function in D.

2^0 As $k\to+\infty$, $f_{n_k}(z)$ converges locally uniformly to the constant ∞ in D.

Definition 1. 6. Let $\mathscr{F}$ be a family of meromorphic functions in a domain D and z_0 a point of D. We say that the family $\mathscr{F}$ is normal at z_0, if we can find a circle $\Gamma: |z-z_0|<r$ belonging to D, such tht the family $\mathscr{F}$ is normal in Γ.

Evidently if $\mathscr{F}$ is normal in D, then $\mathscr{F}$ is normal at each point of D. Conversely we have the following theorem.

Theorem 1. 4. Let $\mathscr{F}$ be a family of meromorphic functions in a domain D. If the family $\mathscr{F}$ is normal at each point of D, then $\mathscr{F}$ is normal in D.

Proof. First of all, we can find a sequence of points $z_j(j=1,2,\cdots)$ of D such that each point of D is a limiting point of the sequence $z_j(j=1,2,\cdots)$. One way to get such a sequence of points is to take the set of the rational points $a+ib$ (a,b rational numbers)of D. We know that this set is enumerable.

Consider a point z_j. By hypothesis there is a circle $|z-z_j|<r$ belonging to D, in which the family $\mathscr{F}$ is normal. Let R_j be the least upper bound of the set of the numbers r having this property. Let Γ_j be the circle $|z-z_j|<R_j/2$, if $R_j<$

$+\infty$, and the circle $|z-z_j|<1$, if $R_j=+\infty$. Γ_j belongs to D and the family $\mathscr{F}$ is normal in Γ_j. Now let $S: f_n(z)(n=1,2,\cdots)$ be a sequence of functions of the family $\mathscr{F}$. From S we can extract a subsequence

$S_1: f_{\alpha_1}(z), f_{\alpha_2}(z), \cdots$

which is a C_0-sequence in Γ_1. From S_1 we can extract a subsequence

$S_2: f_{\beta_1}(z), f_{\beta_2}(z), \cdots$

which is a C_0-sequence in Γ_2. From S_2 we can extract a subsequence

$S_3: f_{\gamma_1}(z), f_{\gamma_2}(z), \cdots$

which is a C_0-sequence in Γ_3. In this way, we get successively a sequence of sequences $S_p(p=1,2,\cdots)$ such that for each $p\geqslant 1$, S_p is a C_0-sequence in Γ_p and S_{p+1} is a subsequence of S_p. Consider the diagonal sequence

$S': f_{\alpha_1}(z), f_{\beta_2}(z), f_{\gamma_3}(z), \cdots, f_{\lambda_k}(z), \cdots.$

S' is a subsequence $f_{n_k}(z)(k=1,2,\cdots)$ of S. Since for each k, the terms $f_{n_k}(r)$, $f_{n_{k+1}}(z), \cdots$ all belong to the sequence S_k, hence S' is a C_0-sequence in each of the circles $\Gamma_j(j=1,2,\cdots)$.

Consider a point z' of D. By hypothesis, there is a circle $\Upsilon: |z-z'|<\rho(0<\rho<1)$ belonging to D, such that the family $\mathscr{F}$ is normal in Υ. Let z_j be such that $|z_j-z'|<\rho'$, $\rho'=\rho/4$. Then the circle $|z-z_j|<2\rho'$ belongs to Υ and hence $\mathscr{F}$ is normal in the circle $|z-z_j|<2\rho'$. Consequently by the definition of R_j, if $R_j<+\infty$, we have

$$2\rho'\leqslant R_j,\quad \rho'\leqslant R_j/2,$$

and the circle $|z-z_j|<\rho'$ belongs to Γ_j. On the other hand, if $R_j=+\infty$, then since $\rho'<1$, the circle $|z-z_j|<\rho'$ also belongs to Γ_j. But z' is a point of the circle $|z-z_j|<\rho'$, hence z' is a C_0-point of the sequence S'. Since $z'\in D$ is arbitrary, so S' is a C_0-sequence in D.

In what follows, we are going to prove some necessary and sufficient conditions for the normality of families of meromorphic functions.

Defintion 1. 7. Let $\mathscr{F}$ be a family of meromorphic functions in a domain D and z_0 a point of D. We say that the family $\mathscr{F}$ is equicontinuous at z_0 with respect to the spherical distance, if to each positive number ε corresponds a positive number δ such that the circle $\Upsilon: |z-z_0|<\delta$ belongs to D and that for each function $f(z)$ of the family $\mathscr{F}$, the inequality.

$$|f(z), f(z_0)| < \varepsilon$$

holds in γ. If $\mathscr{F}$ is equicontinuous at each point of D with respect to the spherical distance, we say that $\mathscr{F}$ is equicontinuous in D with respect to the spherical distance.

Lemma 1.5. Let $S: f_n(z)$ $(n=1,2,\cdots)$ be a sequence of meromorphic functions in a domain D and z_0 a point of D. If z_0 is a C_0-point of S, then S is equicontinuous at z_0 with respect to the spherical distance.

Proof. By hypothesis there is a circle $\Gamma: |z-z_0|<r$ belonging to D such that the sequence S is uniformly convergent in Γ with respect to the spherical distance. Given a positive number ε, let N be a positive integer such that, when $n \geqslant N$, $m \geqslant N$, we have

$$|f_n(z), f_m(z)| < \frac{\varepsilon}{3}$$

in Γ. Next let $\Gamma': |z-z_0|<r'$ $(0<r'<r)$ such that

$$|f_N(z), f_N(z_0)| < \frac{\varepsilon}{3}$$

in Γ'. Then for $n \geqslant N$, $z \in \Gamma'$, we have

$$\begin{aligned}|f_n(z), f_n(z_0)| &\leqslant |f_n(z), f_N(z)| + |f_N(z), f_N(z_0)| + |f_N(z_0), f_n(z_0)| \\ &< \frac{\varepsilon}{3} + \frac{\varepsilon}{3} + \frac{\varepsilon}{3} = \varepsilon.\end{aligned}$$

Since the functions $f_n(z)$ $(1 \leqslant n < N)$ are finite in number, we can find a positive number $\delta < r'$ such that for $1 \leqslant n < N$ and $|z-z_0|<\delta$, we have

$$|f_n(z), f_n(z_0)| < \varepsilon. \tag{1.22}$$

Consequently for each $n \geqslant 1$, the inequality (1.22) holds in the circle $|z-z_0|<\delta$.

Theorem 1.5. Let $\mathscr{F}$ be a family of meromorphic functions in a domain D. In order that the family $\mathscr{F}$ be normal in D, it is necessary and sufficient that $\mathscr{F}$ is equicontinuous in D with respect to the spherical distance.

This theorem is due to Ostrowski[30].

Proof. To prove the necessity, let z_0 be a point of D and assume that $\mathscr{F}$ is not equicontinuous at z_0 with respect to the spherical distance. Then, $\delta_n (n=1,2,\cdots)$ being a sequence of positive numbers such that

$$\lim_{n\to+\infty} \delta_n = 0, \tag{1.23}$$

for each n, there is a function $f_n(z)$ of $\mathscr{F}$ such that

$$\sup_{\Gamma_n} |f_n(z), f_n(z_0)| \geqslant \varepsilon_0, \tag{1.24}$$

where Γ_n denotes the circle $|z-z_0|<\delta_n$. Since $\mathscr{F}$ is normal in D, we can extract from the sequence $f_n(z)(n=1,2,\cdots)$ a subsequence $f_{n_k}(z)(k=1,2,\cdots)$ which is a C_0-sequence in D. So, in particular, z_0 is a C_0-point of the sequence $f_{n_k}(z)(k=1,2,\cdots)$. By Lemma 1,5, the sequence $f_{n_k}(z)(k=1,2,\cdots)$ is equicontinuous at z_0 with respect to the spherical distance. But this is incompatible with (1.23) and (1.24), so we get a contradiction.

Next we prove the sufficiency. Let $\zeta_p(p=1,2,\cdots)$ be a sequence of points of D, such that each point of D is a limiting point of the sequence $\zeta_p(p=1,2,\cdots)$. Let $S: f_n(z)(n=1,2,\cdots)$ be a sequence of the family $\mathscr{F}$. Consider the sequence of points $f_n(\zeta_1)(n=1,2,\cdots)$ of $\hat{\mathbf{C}}$. Evidently we can find a subsequence $f_{\alpha_j}(\zeta_1)(j=1,2,\cdots)$ of the sequence $f_n(\zeta_1)(n=1,2,\cdots)$ and a point $w_1 \in \hat{\mathbf{C}}$ such that

$$\lim_{j\to+\infty} |f_{\alpha_j}(\zeta_1), w_1| = 0.$$

Next we can find a subsequence $f_{\beta_l}(\zeta_2)(l=1,2,\cdots)$ of the sequence $f_{\alpha_j}(\zeta_2)(j=1,2,\cdots)$ and a point $w_2 \in \hat{\mathbf{C}}$ such that

$$\lim_{l\to+\infty}|f_{\beta_l}(\zeta_2),w_2|=0,$$

and so on. Finally in taking the diagonal sequence, we get a subsequence $f_{n_k}(z)$ $(k=1,2,\cdots)$ of the sequence S, such that, for each $p\geqslant 1$, we have

$$\lim_{k\to+\infty}|f_{n_k}(\zeta_p),w_p|=0, \tag{1.25}$$

where $w_p\in\hat{C}$. We are going to show that the sequence $F_k(z)=f_{n_k}(z)(k=1,2,\cdots)$ is a C_0-sequence in D. In fact, consider a circle $\bar{\Gamma}: |z-z_0|\leqslant r$ belonging to D. Evidently it is sufficient to show that given arbitrarily a positive number ε, there is a positive integer K such that when $k\geqslant K, k'\geqslant K$ and $z\in\bar{\Gamma}$, We have

$$|F_k(z),F_{k'}(z)|<\varepsilon.$$

For this, consider a point $z_*\in\bar{\Gamma}$. Since the family $\mathscr{F}$ is equicontinuous at z_* with respect to the spherical distance, there is a circle $\gamma_{z_*}: |z-z_*|<\rho$ belonging to D, such that for each function $f(z)\in\mathscr{F}$, the inequality

$$|f(z),f(z_*)|<\frac{\varepsilon}{6}$$

holds in γ_{z_*}. Then consider a point $\zeta_p\in\gamma_{z_*}$. By (1.25), there is a positive integer K_{z_*} such that for $k\geqslant K_{z_*}, k'\geqslant K_{z_*}$, we have

$$|F_k(\zeta_p),F_{k'}(\zeta_p)|<\frac{\varepsilon}{6}.$$

Then for $k\geqslant K_{z_*}, k'\geqslant K_{z_*}$ and $z\in\gamma_{z_*}$, we have

$$\begin{aligned}|F_k(z),F_{k'}(z)|&\leqslant|F_k(z),F_k(z_*)|+|F_k(z_*),F_k(\zeta_p)|+|F_k(\zeta_p),F_{k'}(\zeta_p)|\\&+|F_{k'}(\zeta_p),F_{k'}(z_*)|+|F_{k'}(z_*),F_{k'}(z)|<\frac{5}{6}\varepsilon<\varepsilon.\end{aligned}$$

Thus to each point $z_*\in\bar{\Gamma}$ correspond a circle γ_{z_*} and a positive integer K_{z_*}. By the finite covering theorem, we can find a finite number of points $z_j(j=1,2,\cdots,m)$ of $\bar{\Gamma}$, such that

$$\overline{\Gamma} \subset \bigcup_{j=1}^{m} \Gamma_{z_j}.$$

On the other hand, let

$$K = \max_{1 \leqslant j \leqslant m} K_{z_j}.$$

Then evidently this positive integer K has the required property mentioned above.

Definition 1.8. Let $\mathscr{F}$ be a family of holomorphic functions in a domain D. We say that $\mathscr{F}$ is locally uniformly bounded in D, if for each point z_0 of D, we can find a circle $\Gamma: |z-z_0|<r$ belonging to D and a positive number M such that for each function $f(z) \in \mathscr{F}$, the inequality

$$|f(z)| \leqslant M \tag{1.26}$$

holds in Γ.

Corollary 1.3. Let $\mathscr{F}$ be a family of holomorphic functions in a domain D. If $\mathscr{F}$ is locally uniformly bounded in D, then $\mathscr{F}$ is normal in D.

This theorem of Montel [26] is fundamental in his theory of normal families.

Proof. Consider a point z_0 of D. By hypothesis, we can find a circle $\overline{\Gamma}: |z-z_0| \leqslant r$ belonging to D and a positive number M such that for each function $f(z) \in \mathscr{F}$, the inequality (1.26) holds in $\overline{\Gamma}$. Consider a function $f(z) \in \mathscr{F}$, then in the circle $\Gamma: |z-z_0|<r$ we have

$$f(z) = \frac{1}{2\pi i} \int_c \frac{f(\zeta)}{\zeta - z} d\zeta,$$

where c is the circle $|z-z_0|=r$. From this formula, we deduce, for $z \in \Gamma$,

$$f(z) - f(z_0) = \frac{1}{2\pi i}(z - z_0)\int_{\Gamma} \frac{f(\zeta)}{(\zeta - z)(\zeta - z_0)} d\zeta.$$

In particular for $|z-z_0|<r/2$, we have

$$|f(z) - f(z_0)| < \frac{2M}{r}|z - z_0|.$$

Evidently this inequality implies that the family $\mathscr{F}$ is equicontinuous at z_0. Since

$$|f(z), f(z_0)| \leqslant |f(z) - f(z_0)|,$$

$\mathscr{F}$ is, à fortiori, equicontinuous at z_0 with respect to the spherical distance. So $\mathscr{F}$ is equicontinuous in D with respect to the spherical distance, and by Theorem 1.5, $\mathscr{F}$ is normal in D.

Corollary 1.4. Let $\mathscr{F}$ be a family of meromorphic functions in a domain D. In order that $\mathscr{F}$ be normal in D, it is necessary and sufficient that for each point z_0 of D, we can find a circle $\Gamma: |z-z_0|<r$ belonging to D and a positive number M such that each function f(z) of $\mathscr{F}$ satisfies in Γ one of the inequalities

$$|f(z)| < M, \quad \left|\frac{1}{f(z)}\right| < M. \tag{1.27}$$

Proof. Assume that the family $\mathscr{F}$ is normal in D. Let $\mu(1,2)$ be the positive number defined in Lemma 1.2 corresponding to $A=1, B=2$. Consider a point z_0 of D. By Theorem 1.5, $\mathscr{F}$ is equicontinuous at z_0 with respect to the spherical distance. Hence there is a circle $\Gamma: |z-z_0|<r$ belonging to D such that for each function f(z) of $\mathscr{F}$, the inequality

$$|f(z), f(z_0)| < \mu(1,2)$$

holds in Γ. Consequently by Lemma 1.2, if $|f(z_0)|\leqslant 1$, then in Γ, we have

$$|f(z)| < 2.$$

If $|f(z_0)|>1$, then since the inequality

$$\left|\frac{1}{f(z)},\frac{1}{f(z_0)}\right|<\mu(1,2)$$

also holds in Γ, we have

$$\left|\frac{1}{f(z)}\right|<2$$

in Γ.

Next assume that the condition in Corollary 1. 4 is satisfied. Let z_0 be a point of D. Then, by hypothesis, we can find a circle Γ: $|z-z_0|<r$ and a positive number M having the property stated in that corollary. By Theorem 1. 4, it is sufficient to show that the family $\mathscr{F}$ is normal in Γ. For this, denote by $\mathscr{F}_1$ and $\mathscr{F}_2$ respectively the subfamilies of $\mathscr{F}$ of functions $f(z)$ satisfying in Γ the first and the second of the two inequalities of (1. 27). Then $\mathscr{F}_1$ is a family of holomorphic functions uniformly bounded in Γ. By the proof of Corollary 1. 3, $\mathscr{F}_1$ is equicontinuous in Γ with respect to the spherical distance. Similarly we see that this is also true for $\mathscr{F}_2$. Consequently the family $\mathscr{F}$ is equicontinuous in Γ with respect to the spherical distance and hence normal in Γ by Theorem 1. 5.

Lemma 1. 6. Let $f(z)$ be a meromorphic function in a domain D and z_0 a point of D. Then the limit

$$\partial(z_0,f)=\lim_{z\to z_0}\frac{|f(z),f(z_0)|}{|z-z_0|} \tag{1. 28}$$

exists and is finite, and we have the formula

$$\partial(z_0,f)=\frac{|f'(z_0)|}{1+|f(z_0)|^2}, \tag{1. 29}$$

provided that, in the case $f(z_0)=\infty$, the right member of (1. 29) is understood to be the limit

$$\lim_{z\to z_0} \frac{|f'(z)|}{1+|f(z)|^2}.$$

$\partial(z_0, f)$ is called the spherical derivative of the function $f(z)$ at z_0.

Proof. Distinguish two cases:

1) $f(z_0)\neq\infty$. Then there is a circle $|z-z_0|<r$ belonging to D, in which the function $f(z)$ is holomorphic, and so by the formula

$$\frac{|f(z),f(z_0)|}{|z-z_0|} = \frac{|f(z)-f(z_0)|}{|z-z_0|}\,\frac{1}{(1+|f(z)|^2)^{\frac{1}{2}}(1+|f(z_0)|^2)^{\frac{1}{2}}}$$

for $0<|z-z_0|<r$, we have

$$\lim_{z\to z_0} \frac{|f(z),f(z_0)|}{|z-z_0|} = \frac{|f'(z_0)|}{1+|f(z_0)|^2}.$$

2) $f(z_0)=\infty$. Then the function $g(z)=1/f(z)$ is meromorphic in D and $g(z_0)=0$. Hence

$$\lim_{z\to z_0} \frac{|g(z),g(z_0)|}{|z-z_0|} = \frac{|g'(z_0)|}{1+|g(z_0)|^2}.$$

Next, since

$$|g(z),g(z_0)| = |f(z),f(z_0)|$$

and, when $z\neq z_0$ is in the neighborhood of z_0,

$$\frac{|g'(z)|}{1+|g(z)|^2} = \frac{|f'(z)|}{1+|f(z)|^2},$$

we have

$$\lim_{z\to z_0} \frac{|f(z),f(z_0)|}{|z-z_0|} = \frac{|g'(z_0)|}{1+|g(z_0)|^2} = \lim_{z\to z_0} \frac{|f'(z)|}{1+|f(z)|^2}.$$

$\partial(z,f)$ is a continuous function in D. In fact, let z_0 be a point of D. Then in a circle $|z-z_0|<r$, $f(z)$ is holomorphic and

$$\partial(z,f)=\frac{|f'(z)|}{1+|f(z)|^2},$$

hence $\partial(z,f)$ is continuous at z_0. If $f(z_0)=\infty$, this is also true, by the identity

$$\partial(z,f)=\partial(z,\frac{1}{f}). \tag{1.30}$$

Lemma 1.7. Let $f(z)$ be a meromorphic function in a domain D. Let z_1, z_2 ($z_1\neq z_2$) be two points of D, such that the segment $\sigma: z=z_1+t(z_2-z_1)$ $(0\leqslant t\leqslant 1)$ belongs to D. Set $m=\max_{z\in\sigma}\partial(z,f)$. Then we have

$$|f(z_1),f(z_2)|\leqslant m|z_1-z_2|. \tag{1.31}$$

Proof. Consider a positive number ε. We are going to show that

$$|f(z_1),f(z_2)|<(m+\varepsilon)|z_1-z_2|. \tag{1.32}$$

Assume that (1.32) is untrue. Let ζ be the middle point of σ, which divides σ into two segments $\sigma':\overline{z_1\zeta}$ and $\sigma'':\overline{\zeta z_2}$. We can not have at the same time,

$$|f(z_1),f(\zeta)|<(m+\varepsilon)|z_1-\zeta|, \tag{1.33}$$

$$|f(\zeta),f(z_2)|<(m+\varepsilon)|\zeta-z_2|, \tag{1.34}$$

because otherwise we would have

$$\begin{aligned}|f(z_1),f(z_2)|&\leqslant|f(z_1),f(\zeta)|+|f(\zeta),f(z_2)|\\&<(m+\varepsilon)(|z_1-\zeta|+|\zeta-z_2|)=(m+\varepsilon)|z_1-z_2|.\end{aligned}$$

Hence at least one of the two inequalities (1.33) and (1.34) is untrue. Denote the corresponding segment by $\sigma_1 : \overline{z_1^{(1)} z_2^{(1)}}$ which is one of the segments σ' and σ''. Again dividing σ_1 into two segments by its middle point and repeating the same argument, we get a segment $\sigma_2 : \overline{z_1^{(2)} z_2^{(2)}}$. Continuing in this way, we get successively a sequence of segments $\sigma_n : \overline{z_1^{(n)} z_2^{(n)}}$ $(n=0,1,2,\cdots;\sigma_0=\sigma)$ such that $\sigma_{n+1} \subset \sigma_n$ $(n=0,1,2,\cdots)$ and

$$|f(z_1^{(n)}), f(z_2^{(n)})| \geqslant (m+\varepsilon)|z_1^{(n)} - z_2^{(n)}| \, (n=0,1,2,\cdots) \tag{1.35}$$

Let z_0 be the point such that $z_0 \in \sigma_n (n=0,1,2,\cdots)$. Since

$$\lim_{z \to z_0} \frac{|f(z), f(z_0)|}{|z-z_0|} = \partial(z_0, f),$$

there is a circle Υ : $|z-z_0|<\delta$ belonging to D, such that in Υ, we have

$$|f(z), f(z_0)| = \partial(z_0, f)|z-z_0| + \eta(z)|z-z_0| \tag{1.36}$$

where $\eta(z)$ satisfies the inequality

$$|\eta(z)| < \varepsilon \tag{1.37}$$

in Υ. Evidently, when n is sufficiently large, $\sigma_n \subset \Upsilon$, consequently

$$|\eta(z_j^{(n)})| < \varepsilon \; (j=1,2)$$

and

$$|f(z_1^{(n)}), f(z_2^{(n)})| \leqslant |f(z_1^{(n)}), f(z_0)| + |f(z_0), f(z_2^{(n)})|$$
$$< \{\partial(z_0, f) + \varepsilon\}(|z_1^{(n)} - z_0| + |z_2^{(n)} - z_0|) \leqslant (m+\varepsilon)|z_1^{(n)} - z_2^{(n)}|$$

This inequality is incompatible with (1.35), and so we get a contradiction. Finally in (1.32) letting $\varepsilon \to 0$, we get (1.31).

Theorem 1.6. Let $\mathscr{F}$ be a family of meromorphic functions in a domain D. In order that the family $\mathscr{F}$ be normal in D, it is necessary and sufficient that the family

$$\mathscr{F}_* = \{\partial(z,f)\,|\,f \in \mathscr{F}\} \tag{1.38}$$

is locally uniformly bounded in D.

This theorem is due to Marty [23].

Proof. Assume that the family $\mathscr{F}_*$ is locally uniformly bounded in D. Let z_0 be a point of D. Then there are a circle Γ: $|z-z_0|<r$ belonging to D and a positive number M such that for each function $f(z) \in \mathscr{F}$, we have

$$\partial(z,f) \leqslant M \tag{1.39}$$

in Γ. By Lemma 1.7, for each function $f(z) \in \mathscr{F}$, we have

$$|f(z), f(z_0)| \leqslant M\,|z-z_0|$$

in Γ. Evidently this inequality implies that the family $\mathscr{F}$ is equicontinuous at z_0 with respect to the spherical distance. Since z_0 is arbitrary, $\mathscr{F}$ is equicontinuous in D with respect to the spherical distance, and by Theorem 1.5, $\mathscr{F}$ is normal in D.

Conversely assume that the family $\mathscr{F}$ is normal in D. Let z_0 be a point of D and suppose that we can not find a circle Γ: $|z-z_0|<r$ belonging to D and a positive number M such that for each function $f(z) \in \mathscr{F}$, we have (1.39) in Γ. Take two sequences of positive numbers r_n and $M_n (n=1,2,\cdots)$ converging respectively to 0 and $+\infty$. Then to each n corresponds a function $f_n(z) \in \mathscr{F}$ such that

$$\sup_{z \in \Gamma_n} \partial(z, f_n) > M_n, \tag{1.40}$$

where Γ_n denotes the circle $|z-z_0|<r_n$. From the sequence $f_n(z) (n=1,2,\cdots)$ we can extract a subsequence $f_{n_k}(z) (k=1,2,\cdots)$ which is a C_0-sequence in D. In particular z_0 is a C_0-point of this sequence. Let F(z) be the limit function of this sequence with respect to the spherical distance. By Theorem 1.1, two cases are possible.

1) $F(z_0) \neq \infty$. Then we can find a circle Γ_0: $|z-z_0|<r_0 (\Gamma_0 \subset D)$ and a posi-

tive integer k_0 such that the functions $f_{n_k}(z)$ $(k \geqslant k_0)$ and $F(z)$ are holomorphic in Γ_0, and as $k \to +\infty$, $f_{n_k}(z)$ converges uniformly in Γ_0 to $F(z)$. Since, for $k \geqslant k_0$, $z \in \Gamma_0$, we have

$$\partial(z, f_{n_k}) = \frac{|f'_{n_k}(z)|}{1 + |f_{n_k}(z)|^2} \leqslant |f'_{n_k}(z)|,$$

it is clear that we can find a positive number A such that when $k \geqslant k_0$, we have

$$\partial(z, f_{n_k}) \leqslant A$$

in the circle $|z - z_0| < r_0/2$. But this is incompatible with (1.40), and we get a contradiction.

2) $F(z_0) = \infty$. Then we can find a circle Γ_0: $|z - z_0| < r_0$ and a positive integer k_0 such that the functions $1/f_{n_k}(z)$ $(k \geqslant k_0)$ and $1/F(z)$ are holomorphic in Γ_0, and as $k \to +\infty$, $1/f_{n_k}(z)$ converges uniformly in Γ_0 to $1/F(z)$. As in the first case, we can find a positive number A such that when $k \geqslant k_0$, we have

$$\partial(z, \frac{1}{f_{n_k}}) \leqslant A$$

in the circle $|z - z_0| < r_0/2$. This is also incompatible with (1.40), because

$$\partial(z, f_{n_k}) = \partial(z, \frac{1}{f_{n_k}}).$$

So we conclude that the family $\mathscr{F}_*$ is locally uniformly bounded in D.

Corollary 1.5. Let $\mathscr{F}$ be a family of holomorphic functions in a domain D. If the family

$$\mathscr{F}_1 = \{f'(z) \mid f \in \mathscr{F}\}$$

is locally uniformly bounded in D, where $f'(z)$ is the derivative of $f(z)$, then the family $\mathscr{F}$ is normal in D.

This corollary follows immediately from theorem 1. 6, because for $f(z) \in \mathscr{F}$, we have

$$\partial(z,f) = \frac{|f'(z)|}{1+|f(z)|^2} \leqslant |f'(z)|.$$

2

CRITERIONS OF NORMALITY OF FAMILIES OF HOLOMORPHIC FUNCTIONS AND APPLICATIONS

2. 1. MONTEL'S THEOREM

Theorem 2. 1. Let D be a domain and a, b $(a \neq b)$ two complex numbers. Let $\mathscr{F}$ be the family of the functions $f(z)$ holomorphic in D and such that each of the equations

$$f(z) = a, \; f(z) = b$$

has no root in D. Then the family $\mathscr{F}$ is normal in D.

This theorem is due to Montel [26].

Proof. Consider first the case $a=0, b=1$. Let z_0 be a point of D and $\Gamma: |z-z_0| < \rho$ a circle belonging to D. Let $f(z)$ be a function of the family $\mathscr{F}$ and distinguish two cases:

1) $|f(z_0)| \leqslant 1$. Define $F(\zeta) = f(z_0 + \rho\zeta)$ $(|\zeta| < 1)$. By Corollary 4 in Appendix A, we must have

$$\log M(\frac{1}{2}, F) < h(1 + \overset{+}{\log} |F(0)|),$$

where h is an absolute positive constant. Hence in the cirele $\bar{\Gamma}: |z-z_0| \leqslant \rho/2$, we have $|f(z)| < e^h$.

2) $|f(z_0)| > 1$. Then we can apply the result obtained in the first case to the

function $1/f(z)$, hence in the circle $\overline{\Gamma}$, we have $|1/f(z)|<e^h$.

Consequently by Corollary 1.4, in the particular case $a=0$, $b=1$, the family $\mathscr{F}$ is normal in D. Now consider the general case. Let $f_n(z)(n=1,2,\cdots)$ be a sequence of functions of $\mathscr{F}$. Define

$$g_n(z) = \frac{f_n(z)-a}{b-a} \qquad (n = 1,2,\cdots).$$

Since $g_n(z)$ does not take the values 0 and 1, from the sequence $g_n(z)(n=1,2,\cdots)$, we can extract a subsequence $g_{n_k}(z)(k=1,2,\cdots)$ such that as $k\to+\infty$, $g_{n_k}(z)$ converges locally uniformly to a holomorphic function or to ∞ in D. Evidently this is also true for the subsequence

$$f_{n_k}(z) = a + (b-a)g_{n_k}(z) \ (k = 1,2,\cdots).$$

We are going to give some applications of Theorem 2.1. For this, we need the following lemma:

Lemma 2.1. Let $\mathscr{F}$ be a normal family of holomorphic functions in a domain D. Let σ be a bounded closed set of points belonging to D and M a positive number. Assume that for each function $f(z)\in\mathscr{F}$, we have

$$\min_{z\in\sigma}|f(z)| \leqslant M. \tag{2.1}$$

Then the family $\mathscr{F}$ is uniformly bounded on each bounded closed set E of points belonging to D.

Proof. Let E be a bounded closed set of points belonging to D. Assume that $\mathscr{F}$ is not uniformly bounded on E. Then to each positive integer n, corresponds a function $f_n(z)\in\mathscr{F}$ such that

$$\max_{z\in E}|f_n(z)| > n. \tag{2.2}$$

Consider a subsequence $f_{n_k}(z)(k=1,2,\cdots)$ of the sequence $f_n(z)(n=1,2,\cdots)$. By (2.2) and (2.1), as $k\to+\infty$, $f_{n_k}(z)$ can not converge locally uniformly to a holomorphic function or to ∞ in D. This contradicts the hypothesis that $\mathscr{F}$ is a normal family.

In the applications of Lemma 2.1, the set σ often consists of a single point.

Theorem 2.2. Let f(z) be a non-constant entire function. Then the family of entire functions

$$f_n(z) = f(2^n z) \ (n = 1,2,\cdots) \tag{2.3}$$

in not normal in the circle $|z|<2$.

Proof. Assume that the family (2.3) is normal in the circle $|z|<2$. Since $f_n(0)=f(0)(n=1,2,\cdots)$, hence by Lemma 2.1, the family (2.3) is uniformly bounded in the circle $|z|\leqslant 1$. There is then a positive number M such that for each n, we have

$$|f_n(z)| \leqslant M$$

in the circle $|z|\leqslant 1$. It follows that for each n we have

$$|f(z)| \leqslant M$$

in the circle $|z|\leqslant 2^n$. The entire function f(z) is then bounded in **C** and by Liouville's theorem, it is a constant, contrary to hypothesis.

Corollary 2.1. If f(z)is a non-constant entire function, then f(z) takes every finite value, except at most one finite value.

This is Picard's theorem on entire functions. It is an immediate consequence of Theorems 2.1 and 2.2. In fact, if f(z) does not take two finite values a and b $(a\neq b)$, then, by Theorem 2.1, the family (2.3) is normal in the circle $|z|<2$, incompatible with Theorem 2.2.

Theorem 2.3. Let f(z) be a holomorphic function in a domain $D: 0<|z|<\rho$ such that the point $z=0$ is an essential singular point of f(z). Let r be a number such that $0<2r<\rho$. Then the family of holomorphic functions in D

$$f_n(z) = f\left(\frac{z}{2^n}\right) \ (n = 1, 2\cdots) \tag{2.4}$$

is not normal in the domain $d: r/4 < |z| < 2r$.

Proof. Assume that the family (2.4) is normal in the domain d. By a theorem of Weierstrass, there is a sequence of points $\zeta_m (m = 1, 2, \cdots)$ of D such that

$$\lim_{m \to +\infty} \zeta_m = 0, \ |f(\zeta_m)| < 1 \ (m = 1, 2, \cdots).$$

We may suppose that

$$|\zeta_m| < \frac{r}{2} \ (m = 1, 2, \cdots).$$

To each point ζ_m corresponds a positive integer n_m such that

$$\frac{r}{2^{n_m+1}} \leqslant |\zeta_m| < \frac{r}{2^{n_m}}.$$

Evidently

$$\lim_{m \to +\infty} n_m = +\infty. \tag{2.5}$$

Set

$$z_m = 2^{n_m} \zeta_m, \ \frac{r}{2} \leqslant |z_m| < r, \tag{2.6}$$

then

$$f_{n_m}(z_m) = f(\zeta_m), \ |f_{n_m}(z_m)| < 1. \tag{2.7}$$

The family $f_{n_m}(z) (m = 1, 2, \cdots)$ being a subfamily of the family (2.4), is also normal in the domain d. Then from (2.6), (2.7) and Lemma 2.1, the family $f_{n_m}(z)$ ($m=1,2,\cdots$) is uniformly bounded on the circle $|z|=r$. Consequently by (2.4), there is a positive number K such that

$$|f(z)| \leqslant K \tag{2.8}$$

on the sequence of circles $|z|=r/2^{n_m}$ ($m=1,2,\cdots$). Finally by (2.5) and the principle of maximum modulus, we see that (2.8) holds for $0<|z|\leqslant r/2^{n_1}$. This is incompatible with the hypothesis that the point $z=0$ is an essential singular point of the function $f(z)$.

Corollary 2.2. Let $f(z)$ be a holomorphic function in a domain $D: 0<|z|<\rho$ such that the point $z=0$ is an essential singular point of $f(z)$. Then in D, $f(z)$ takes every finite value an infinite number of times, except at most one finite value.

This is Picard's theorem on holomorphic functions in the neighborhood of an isolated essential singular point.

Proof. Assume that there are two finite values a and b ($a\neq b$) such that $f(z)$ takes a and b only a finite number of times in the domain D. Then we can find a number ρ_0 ($0<\rho_0<\rho$) such that $f(z)$ does not take the values a and b in the domain $D_0: 0<|z|<\rho_0$. Let n_0 be a sufficiently large integer such that $r/2^{n-1}<\rho_0$ for $n\geqslant n_0$, where r is the number in Theorem 2.3. Then for $n\geqslant n_0$, the function $f_n(z)$ defined by (2.4) does not take the values a and b in the domain d defined in Theorem 2.3, hence the family $\mathscr{F}_0$ $f_n(z)$ ($n\geqslant n_0$) is normal in d, by Theorem 2.1. Since the family (2.4) differs from $\mathscr{F}_0$ only by a finite number of additional functions $f_n(z)$ ($1\leqslant n<n_0$), so the family (2.4) is also normal in d, incompatible with Theorem 2.3.

Theorem 2.4. Given a domain D, two bounded closed sets σ, E of points belonging to D and a positive number μ, we can find a positive number $A(D,\sigma,E,\mu)$ depending only on D,σ,E and μ having the following property: If $f(z)$ is a holomorphic function in D satisfying the conditions:

1^0 $f(z)$ does not take the values 0 and 1 in D;

2^0 $\min_{z\in\sigma}|f(z)|\leqslant\mu$;

then we have

$$\max_{z \in E} |f(z)| \leqslant A(D, \sigma, E, \mu). \tag{2.9}$$

This is Schottky's theorem in general form.

Proof. Let $\mathscr{F}$ be the family of the functions $f(z)$ holomorphic in D and satisfying the conditions 1^0 and 2^0. By Theorem 2.1, $\mathscr{F}$ is normal in D and then by Lemma 2.1, $\mathscr{F}$ is uniformly bounded on E, hence there is a positive number A such that (2.9) holds for each function $f(z)$ of $\mathscr{F}$. This number A has then the required property.

Given a number $r(0<r<1)$ and a positive number μ, we can apply Theorem 2.4 to $D=(|z|<1)$, $\sigma=(0)$, $E=(|z|\leqslant r)$ and μ, hence we have the following corallary.

Corollary 2.3. Let $f(z)$ be a holomorphic function in the circle $\Delta: |z|<1$, satisfying the following conditions:

1^0 $f(z)$ does not take the values 0 and 1 in Δ.

2^0 $|f(0)|\leqslant\mu$, μ being a positive number.

Then for $0<r<1$ we have

$$M(r,f) \leqslant A(r,\mu), \tag{2.10}$$

where $A(r,\mu)$ is a positive number depending only on r and μ. In particular, for $\mu=|f(0)|$, we have

$$M(r,f) \leqslant A(r,|f(0)|). \tag{2.11}$$

This is Schottky's theorem.

Theorem 2.5. Given a domain D, two bounded closed sets σ, E of points belonging to D, a positive number μ and a positive integer p, we can find a positive number $B(D,\sigma,E,\mu,p)$ depending only on D, σ, E, μ and p having the following property: If $f(z)$ is a holomorphic function in D satisfying the conditions 1^0 and 2^0 in Theorem 2.4, then we have

$$\max_{z\in E} |f^{(p)}(z)| \leqslant B(D,\sigma,E,\mu,p). \tag{2.12}$$

Proof. Let $\mathscr{F}$ be the family of functions defined in the proof of Theorem 2.4. We know that $\mathscr{F}$ is normal in D and is uniformly bounded on each bounded closed set of points belonging to D. Consider the family

$$\mathscr{F}_p = \{f^{(p)}(z) \mid f(z) \in \mathscr{F}\}.$$

We are going to show that the family $\mathscr{F}_p$ is also uniformly bounded on each bounded closed set of points belonging to D. By the finite covering theorem, it is sufficient to show that $\mathscr{F}_p$ is locally uniformly bounded in D. For this, consider a circle $\overline{\Gamma}$: $|z-z_0|\leqslant 2r$ belonging to D. Then there is a positive number M such that for each function f(z) of $\mathscr{F}$, we have

$$|f(z)| \leqslant M$$

in $\overline{\Gamma}$. By Cauchy's inequality, in the circle $|z-z_0|<r$ we have

$$|f^{(p)}(z)| \leqslant p!\frac{M}{r^p}.$$

This proves our assertion. There is then a positive number B such that (2.12) holds for each $f(z)\in\mathscr{F}$. This number B has the required property.

Corollary 2.4. Let f(z) be a holomorphic function in a circle $|z|<R$, satisfying the following conditions:

1^0 f(z) does not take the values 0 and 1 in the circle $|z|<R$.

2^0 $|f(0)|\leqslant\mu$, μ being a positive number.

Then we have

$$R|f'(0)| \leqslant B(\mu), \tag{2.13}$$

where $B(\mu)$ is a positive number depending only on μ. In particular, for $\mu =$

$|f(0)|$, we have

$$R|f'(0)| \leqslant B(|f(0)|). \tag{2.14}$$

This is Landau's theorem.

Proof. Consider first the case $R=1$. In this case, it is sufficient to apply Theorem 2.5 to the particular case where $D=(|z|<1)$, $\sigma=E=(0)$, μ and $p=1$. For the general case, we apply the result obtained in the case $R=1$ to the function $f(R\zeta)(|\zeta|<1)$.

Now consider a transcendental entire function $f(z)$. Since the point ∞ is an essential singular point of $f(z)$, we can prove, as in the proof of Theorem 2.3, that the family of entire functions

$$f_n(z) = f(2^n z) \quad (n = 1, 2, \cdots) \tag{2.15}$$

is not normal in the domain ω: $1/2<|z|<4$. Consequently by Theorem 1.4, there is a point $z_0 \in \omega$ such that the family (2.15) is not normal at z_0, Consider a circle Γ : $|z-z_0|<\delta$ belonging to ω. By Theorem 2.1, for each finite value a, there are an infinite number of positive integers n such that $f_n(z)$ takes the value a in Γ, except at most one finite value a. Evidently in the circle Γ, the function $f_n(z)$ takes the same values as those taken by the function $f(z)$ in the circle

$$c_n: |z-z_n| < r_n, \quad z_n = 2^n z_0, \quad r_n = 2^n \delta. \tag{2.16}$$

Consider two circles c_n and $c_{n+p}(p\geqslant 1)$. In order that c_n and c_{n+p} have no common point, it is sufficient that

$$r_n + r_{n+p} \leqslant |a_{n+p} - a_n|.$$

Then we see that if $\delta \leqslant |z_0|/3$, the sequence of circles $c_n(n=1,2,\cdots)$ are mutually disjoint, and hence in the open set $\Omega=\bigcup_{n=1}^{\infty} c_n$, $f(z)$ takes every finite value a an infinite number of times, except at most one finite value a. Consider the ray L: $z=z_0 t(0\leqslant t<+\infty)$. The centers of the circles $c_n(n=1,2,\cdots)$ all lie on L and Ω

belongs to an angle A : $|\arg z-\theta_0|<\varepsilon$ $(z_0=|z_0|e^{i\theta_0})$ with L as bisector, where ε may be as small as we like, provided that δ is sufficiently small. A fortiori, in A, the function f(z) takes every finite value a an infinite number of times, except at most one finite value a. L is called a Julia direction of the function f(z), because it is first discovered by Julia [16]. Hence we have the following theorem:

Theorem 2. 6. If f(z) is a transcendental entire function, then it has at least one Julia direction.

More generally consider a curve $\Gamma: z=\varphi(t)$ $(0\leqslant t<+\infty)$, where $\varphi(t)$ is a continuous complex valued function for $0\leqslant t<+\infty$, such that when t increases from 0 to $+\infty$, $|\varphi(t)|$ increases from 0 to $+\infty$. We say simply that the curve Γ satisfies the conditim (C). Let $z_0(z_0\neq 0)$ be a point. Then the curve $\Gamma_*: z=z_0\varphi(t)$ $(0\leqslant t<+\infty)$ also satisfies the condition (C), and if for a value $t_0, \varphi(t_0)=1$, then the curve Γ_* passes by the point z_0. In any case, the curve $z=\varphi_1(t)$, $\varphi_1(t)=\varphi(t)/\varphi(t_0)$ has that property.

Let $t_n(n=1,2,\cdots)$ be defined by

$$|\varphi(t_n)| = 2^n \quad (n = 1,2,\cdots) \tag{2. 17}$$

and let f(z) be a transcendental entire function. As in the above we see that the family of entire functions

$$g_n(z) = f\{\varphi(t_n)z\} \ (n = 1,2,\cdots) \tag{2. 18}$$

is not normal in the domain $\omega: 1/2<|z|<4$. Let $z_0\in\omega$ be such that the family (2. 18) is not normal at z_0 and consider the sequence of cireles

$$\Gamma_n: |z-z_0\varphi(t_n)| < \delta|\varphi(t_n)| \ (n = 1,2,\cdots). \tag{2. 19}$$

Assume $\delta\leqslant|z_0|/3$, then the circles $\Gamma_n(n=1,2,\cdots)$ are mutually disjoint, and we see as in the above, that in the open set $\bigcup_{n=1}^{\infty}\Gamma_n$, the function f(z) takes every finite value a an infinite number of times, except at most one finite value a. A fortiori, in the domain

$$\bigcup_{0<t<+\infty} (|z-z_0\varphi(t)| < \delta |\varphi(t)|)$$

this is also true. Set $\psi(t)=z_0\varphi(t)$, $\delta=|z_0|\eta$ $(0<\eta\leqslant 1/3)$. Then in the domain

$$\bigcup_{0<t<+\infty} (|z-\psi(t)| < \eta |\psi(t)|)$$

the function f(z) takes every finite value a an infinite number of times, except at most one finite value a, no matter how small in the positive number η. We say that the curve $z=\psi(t)$ is a Julia curve of f(z). Hence we have the following theorem:

Theorem 2.7. Let f(z) be a transcendental entire function. Then given any curve $z=\varphi(t)$ satisfying the condition (C), we can find a point z_0 of the domain $1/2<|z|<4$, such that the curve $z=z_0\varphi(t)$ is a Julia curve of f(z).

Now we give an extension of Theorem 2.1.

Theorem 2.8. Let D be domain and M, δ two positive numbers. Let $\mathscr{F}$ be the family of the functions f(z) satisfying the following conditions:

1^0 f(z) is holomorphic in D.

2^0 There are two finite values a(f), b(f) such that

$$|a(f)| < M,\ |b(f)| < M,\ |a(f)-b(f)| < \delta$$

and that each of the equations

$$f(z) = a(f),\ f(z) = b(f)$$

has no root in D.

Then the family $\mathscr{F}$ is nonmal in D.

Proof. Let $f_n(z)(n=1,2,\cdots)$ be a sequence of functions of the family $\mathscr{F}$. Set

$$a_n = a(f_n),\ b_n = b(f_n)\ (n = 1,2,\cdots)$$

and consider the sequence of functions

$$g_n(z) = \frac{f_n(z)-a_n}{b_n-a_n} \quad (n = 1,2,\cdots).$$

Since $g_n(z)$ does not take the values 0 and 1 in D, by Theorem 2. 1, we can extract from the sequence $g_n(z)(n=1,2,\cdots)$ a subsequence $g_{n_k}(z)(k=1,2,\cdots)$ such that as $k\to+\infty$, $g_{n_k}(z)$ converges locally uniformly to a holomorphic function or to the constant ∞ in D. In the second case, evidently as $k\to+\infty$, $f_{n_k}(z)$ converges locally uniformly to the constant ∞ in D. In the first case, the sequence $g_{n_k}(z)(k=1,2,\cdots)$ is locally uniformly bounded in D. This is also true for the sequence $f_{n_k}(z)(k=1,2,\cdots)$. By Corollary 1. 3, we can extract from the sequence $f_{n_k}(z)(k=1,2,\cdots)$ a subsequence $f_{m_j}(z)(j=1,2,\cdots)$ such that as $j\to+\infty$, $f_{m_j}(z)$ converges locally uniformly to a holomorphic function in D. This proves Theorem 2. 8.

By means of Theorem 2. 8, we can give some complements to Corollary 2. 1 and Theorems 2. 6 and 2. 7. First for the sake of convenience, we give a definition. Consider a non-constant entire function f(z) and two positive numbers $R>1$ and $\eta<1$. A circle $\Gamma: |z-z_0|<\rho$ is called a (R,η)-filling circle of f(z), if there do not exist two finite values a and b such that

$$|a| < R, |b| < R, |a-b| > \eta$$

and that f(z) does not take the values a and b in Γ. Then evidently in the circle Γ, f(z) takes every value w of the cirele $|w|<R$ except at most values which belong to a circle $|w-w_0|\leqslant\eta$ with $|w_0|<R$.

Consider a non-constant entire function f(z). By Theorem 2. 2, the family (2. 3) of entire functions is not normal in the circle $|z|<2$. Let $R>1$ and $\eta<1$ be two positive numbers. By Theorem 2. 8, the circle $|z|<2$ is a (R,η)-filling circle of the function $f_n(z)$, for at least one positive integer n. Consequently the circle $|z|<2^{n+1}$ is a (R,η)-filling circle of the function f(z).

Next consider a transcendental entire function f(z). We know that the family (2. 15) is not normal in the circle $\Upsilon: |z-z_0|<\delta$, no matter how small is the positive number δ, where z_0 is a point of the domain $1/2<|z|<4$. Let $R>1$ and $\eta<1$ be two positive numbers. Again by Theorem 2. 8, the circle Υ is a (R,η)-

filling circle of the function $f_n(z)$, for an infinite number of positive integers n. Consequently the circle c_n defined by (2. 16) is a (R,η)-filling circle of the function f(z), for an infinite number of positive integers n. In the same way we see that the circle Γ_n defined by (2. 19) is a (R,η)-filling circle of the function f(z), for an infinite number of positive integers n.

As an application of Theorem 2. 8, we give a proof of the following theorem which can also be deduced from Theorem 2 in Appendix A.

Theorem 2. 9. There exist two numbers $\delta>0$, $\lambda>0$ having the following property: If $w=f(z)$ is a holomorphic function in the circle $|z|<1$, such that

$$f(0)=0,\ M(\frac{1}{2},f)\geqslant 1, \tag{2. 20}$$

then the values taken by the function f(z) in the circle $|z|<1$ cover a domain $\rho_1<|w|<\rho_2$ with

$$\rho_1>\delta,\ \rho_2-\rho_1>\lambda. \tag{2. 21}$$

Proof. Suppose that the positive numbers δ, λ do not exist. Then for each positive integer n, there is a corresponding holomorphic function $f_n(z)$ in the circle $|z|<1$ such that

$$f_n(0)=0,\ M(\frac{1}{2},f_n)\geqslant 1 \tag{2. 22}$$

and that its values do not cover a domain $\rho_1<|w|<\rho_2$ with

$$\rho_1>\frac{1}{n},\ \rho_2-\rho_1>\frac{1}{n}.$$

À fortiori, the values of the function $f_n(z)$ do not cover either of the domains D_1: $2<|w|<4$ and D_2: $5<|w|<7$. There are then two points $w_n\in D_1$ and $w_n'\in D_2$ such that the function $f_n(z)$ do not take the values w_n and w_n' for $|z|<1$. By Theorem 2. 8, the family $f_n(z)(n=1,2,\cdots)$ is normal in the circle $|z|<1$. In view of

(2. 22), from this sequence we can extract a subsequence $f_{n_k}(z)$ $(k=1,2,\cdots)$ such that as $k\to+\infty$, $f_{n_k}(z)$ converges locally uniformly to a holomorphic function $f(z)$ in the circle $|z|<1$. Evidently

$$f(0)=0,\ M(\frac{1}{2},f)\geqslant 1.$$

Let $|z|=r$ $(0<r<1)$ be a circle on which $f(z)$ has no zero and let $m=\min_{|z|=r}|f(z)|$, $m>0$. By Rouché's theorem, in the circle $|z|<r$ the values of $f(z)$ cover the circle $|w|<m$. Then by the uniform convergence of $f_{n_k}(z)$ to $f(z)$ on the circle $|z|=r$ and the identity

$$f_{n_k}(z)-w=\{f(z)-w\}+\{f_{n_k}(z)-f(z)\},$$

we see again by Rouché's theorem, that when $k\geqslant k_0$, in the circle $|z|<r$, the values of the function $f_{n_k}(z)$ cover the circle $|w|<m/2$. Take an integer k such that

$$k\geqslant k_0,\ \frac{4}{n_k}<\frac{m}{2}.$$

Then, à fortiori, the values of the function $f_{n_k}(z)$ cover the domain $\rho_1<|w|<\rho_2$ with

$$\rho_1=\frac{2}{n_k},\ \rho_2=\frac{4}{n_k}.$$

This contradicts the above stated property of the function $f_n(z)$.

Corollary 2. 5. If $w=f(z)$ is a holomorphic function in the circle $|z|<1$ satisfying the condition (2. 20), then its values cover a circle $|w|=\rho$ whose radius ρ is greater than an absolute positive constant δ.

This is a theorem of Bohr[4]. It is an immediate consequence of Theorem 2. 9.

Corollary 2.6. If $w=f(z)$ is a holomorphic function in the circle $|z|<1$ such that $f(0)=0$, $|f'(0)|\geqslant 1$, then its values cover a circle $|w-w_0|<h$ whose radius h is greater than an absolute positive constant μ.

This is a theorem of Bloch in a form given by Landau[20]. For the proof, it is sufficient to note that the condition $|f'(0)|\geqslant 1$ and Cauchy's inequality imply that

$$\frac{1}{2}\leqslant\frac{1}{2}|f'(0)|\leqslant M(\frac{1}{2},f),$$

and to apply Theorem 2.9 to the function $2f(z)$.

2.2. MIRANDA'S THEOREM

Theorem 2.10. Let D be a domain, a and $b\neq 0$ two complex numbers, and $\nu\geqslant 1$ and integer. Let $\mathscr{F}$ be the family of the functions $f(z)$ holomorphic in D and such that each of the equations

$$f(z)=a,\ f^{(\nu)}(z)=b \tag{2.23}$$

has no root in D. Then the family $\mathscr{F}$ is normal in D.

This theorem is due to Miranda[25].

Proof. Consider first the case $a=0, b=1$. Let z_0 be a point of D and $\Gamma: |z-z_0|<\rho$ a circle belonging to D. Let $f(z)$ be a function of the family $\mathscr{F}$ and define $F(\zeta)=\frac{1}{\rho^\nu}f(z_0+\rho\zeta)\ (|\zeta|<1)$. By Corollaries 9 and 11 in Appendix A, we must have the two inequalities

$$\log M(\frac{1}{2},F)<h_\nu(1+\overset{+}{\log}|F(0)|), \tag{2.24}$$

$$\log M(\frac{1}{2},\frac{1}{F})<h_\nu(1+\overset{+}{\log}\left|\frac{1}{F(0)}\right|), \tag{2.25}$$

where h_ν is a positive number depending only on ν. Consequently if $|F(0)|=|f(z_0)|/\rho^\nu\leqslant 1$, then by (2.24), in the circle $\tilde{\Gamma}: |z-z_0|\leqslant\rho/2$, we have

$$|f(z)| < e^{A}\cdot\rho^{\nu} \leqslant e^{A}\cdot max(\rho^{\nu}, \frac{1}{\rho^{\nu}}).$$

On the other hand, if $|F(0)| = |f(z_0)|/\rho^{\nu} > 1$, then by (2. 25), in the circle $\bar{\Gamma}$, we have

$$\left|\frac{1}{f(z)}\right| < \frac{e^{A}\cdot}{\rho^{\nu}} \leqslant e^{A}\cdot max(\rho^{\nu}, \frac{1}{\rho^{\nu}}).$$

Hence by Corollary 1. 4, the family $\mathscr{F}$ is normal in D, in the case $a=0, b=1$. For the general case, it is sufficient to apply the result just obtained to the family

$$\mathscr{F}_1 = \{\frac{f(z)-a}{b} | f(z) \in \mathscr{F}\}.$$

Now we are going to give some applications of Theorem 2. 10, which are nearly parallel to those of Theorem 2. 1.

Theorem 2. 11. Let f(z) be an entire function which is not a polynomial of degree at most equal to ν, where $\nu \geqslant 1$ is an integer. Then the family of entire functions

$$f_n(z) = \frac{f(2^n z)}{2^{n\nu}} \quad (n = 1, 2, \cdots) \tag{2. 26}$$

is not normal in the circle $|z| < 2$.

Proof. Assume that the family (2. 26) is normal in the circle $|z| < 2$. Since $f_n(0) = f(0)/2^{n\nu} (n = 1, 2, \cdots)$, by Lemma 2. 1, the family (2. 26) is uniformly bounded on the circle $|z| = 1$. Hence there is a positive number K such that

$$M(r_n, f) \leqslant K r_n^{\nu}, \; r_n = 2^n \qquad (n = 1, 2, \cdots),$$

which imply, by Cauchy's inequalities, f(z) is a polynomial of degree at most equal to v, contrary to hypothesis.

Corollary 2.7. Let f(z) be an entire function which is not a polynomial of degree at most equal to v, where $v \geqslant 1$ is an integer. Then for any two finite values a and $b \neq 0$, one at least of the two equations

$$f(z) = a, \quad f^{(v)}(z) = b \tag{2.27}$$

has root in **C**.

Proof. Assume that each of the equations (2.27) has no root. Then by Theorem 2.10, the family

$$g_n(z) = \frac{f(2^n z)-a}{2^{nv}} = f_n(z) - \frac{a}{2^{nv}} \quad (n = 1, 2, \cdots)$$

is normal in the circle $|z| < 2$, where $f_n(z)$ is defined by (2.26). Then the family (2.26) is also normal in the circle $|z| < 2$, contrary to Theorem 2.11.

Theorem 2.12. Let f(z) be a transcendental entire function and $v \geqslant 1$ an integer. Then the family (2.26) is not normal in the domain $\omega: 1/2 < |z| < 4$.

Proof. Since the point ∞ is an essential singular point of the function f(z), there is a sequence of points $\zeta_m (m = 1, 2, \cdots)$ such that

$$\lim_{m \to +\infty} \zeta_m = \infty, \quad |f(\zeta_m)| < 1 \qquad (m = 1, 2, \cdots).$$

We may suppose that

$$|\zeta_m| > 2 \ (m = 1, 2, \cdots).$$

To each m corresponds n_m such that

$$2^{n_m} < |\zeta_m| \leqslant 2^{n_m + 1}.$$

Evidently $\lim_{m\to+\infty} n_m=+\infty$. Set $z_m=\zeta_m/2^{n_m}$. Then

$$1<|z_m|\leqslant 2,\ |f_{n_m}(z_m)|<\frac{1}{2^\nu}.$$

Consequently, by Lemma 2.1, if the family (2.26) is normal in the domain ω, then the sequence $f_{n_m}(z)(m=1,2,\cdots)$ is uniformly bounded on the circle $|z|=1$. Hence there is a positive number K such that

$$M(r_m,f)\leqslant Kr_m^\nu,\ r_m=2^{n_m}\qquad (m=1,2,\cdots),$$

which imply that the function $f(z)$ is a polynomial of degree at most equal to ν, contrary to the hypothesis that $f(z)$ is a transcendental entire function.

Theorem 2.13. Let $f(z)$ be a holomorphic function in a domain $R_0<|z|<+\infty$, such that the point ∞ is an essential singular point of $f(z)$. Then the family (2.26) is not normal in the domain $\Omega: R/2<|z|<4R$, where R is a number such that $R>2R_0$.

Proof. We can write

$$f(z)=F(z)+G(z),$$

where $F(z)$ is a transcendental entire function and $G(z)$ a holomorphic function in the domain $R_0<|z|<+\infty$, such that

$$\lim_{z\to\infty} G(z)=0.$$

Let $f_n(z)$ be defined by (2.26). Then

$$f_n(z)=F_n(z)+G_n(z),\ F_n(z)=\frac{F(2^nz)}{2^{n\nu}},\ G_n(z)=\frac{G(2^nz)}{2^{n\nu}}.$$

As in the proof of Theorem 1.12, we can show that the family $F_n(z)(n=1,2,$

$\cdots$) is not normal in the domain Ω. Evidently this must be also true for the family $f_n(z)(n=1,2,\cdots)$.

Corollary 2. 8. Let $f(z)$ be a holomorphic function in a domain $D: R_0<|z|<+\infty$, such that the point ∞ is an essential singular point of $f(z)$. Let $\nu\geqslant 1$ be an integer. Then for any two finite values a and $b\neq 0$, one at least of the two equations

$$f(z)=a,\ f^{(\nu)}(z)=b \tag{2.28}$$

has an inifinite number of roots in D.

Proof. Assume that each of the equations (2. 28) has only a finite number of roots in D. Let $\rho>R_0$ be a number such that each of the equations (2. 28) has no root for $\rho<|z|<+\infty$. Let n_0 be a positive integer such that for $n\geqslant n_0$, we have $2^{n-1}R>\rho$, where R is the number occuring in Theorem 2,13. Then by Theorem 2. 10, the family

$$g_n(z)=\frac{f(2^n z)-a}{2^{n\nu}}=f_n(z)-\frac{a}{2^{n\nu}}\ (n\geqslant n_0),$$

where $f_n(z)$ is defined by (2. 26), is normal in the domain Ω defined in Theorem 2. 13. This implies that the family $f_n(z)(n\geqslant n_0)$ and hence the family (2. 26) is normal in Ω, contrary to Theorem 2. 13.

Theorem 2. 14. Given a domain D, two bounded closed sets σ, E of points belonging to D, a positive number m and an integer $\nu\geqslant 1$, we can find a positive number $P(D,\sigma,E,m,\nu)$ depending only on D,σ,E,m and ν having the following property: If $f(z)$ is a holomorphic function in D satisfying the conditions:

1^0 each of the equations $f(z)=0$, $f^{(\nu)}(z)=1$ has no root in D;

$2^0 \min_{z\in\sigma}|f(z)|\leqslant m$;

then we have

$$\max_{z\in E}|f(z)|\leqslant P(D,\sigma,E,m,\nu). \tag{2.29}$$

This is a generalization of Schottky's theorem. It can be proved in basing upon Theorem 2.10 and Lemma 2.1.

Corollary 2.9. Let $f(z)$ be a holomorphic function in the circle $\Delta: |z|<1$ and $\nu \geqslant 1$ an integer, satisfying the following conditions:

1^0 Each of the equations $f(z)=0$, $f^{(\nu)}(z)=1$ has no root in D.

2^0 $|f(0)| \leqslant m$.

Then for $0<r<1$ we have

$$M(r,f) \leqslant P(r,m,\nu), \tag{2.30}$$

where $P(r,m,\nu)$ is a positive number depending only on r, m and ν. In particular, for $m=|f(0)|$, we have

$$M(r,f) \leqslant P(r,|f(0)|,\nu). \tag{2.31}$$

Theorem 2.15. Given a domain D, two bounded closed sets σ, E of points belonging to D, a positive number m and two positive integers ν and p, we can find a positive number $Q(D,\sigma,E,m,\nu,p)$ depending only on D, σ, E, m, ν and p having the following property: If $f(z)$ is a holomorphic function in D satisfying the conditions 1^0 and 2^0 in Theorem 2.14, then we have

$$\max_{z\in E} |f^{(p)}(z)| \leqslant Q(D,\sigma,E,m,\nu,p). \tag{2.32}$$

This is a generalization of Landau's theorem. It can be proved in basing upon Theorem 2.10 and Lemma 2.1, by a method similar to that used in the proof of Theorem 2.5.

Corollary 2.10. Let $f(z)$ be a holomorphic function in a circle $|z|<R$ ($R \geqslant 1$) and $\nu \geqslant 1$ an integer, satisfying the following conditions:

1^0 Each of the equations $f(z)=0, f^{(\nu)}(z)=1$ has no root in the circle $|z|<R$.

2^0 $|f(0)| \leqslant m$, m being a positive number.

Then we have

$$R\,|f^{(\nu+1)}(0)| \leqslant Q(m,\nu), \tag{2.33}$$

where $Q(m,\nu)$ is a positive number depending only on m and ν. In particular, for $m=|f(0)|$, we have

$$R\,|f^{(\nu+1)}(0)| \leqslant Q(|f(0)|,\nu). \tag{2.34}$$

Proof. In the case $R=1$, (2.33) follows from (2.32), in taking $D=(|z|<1)$, $\sigma=E=(0)$, $m,\nu,p=\nu+1$. In the general case, we apply the result obtained in the particular case $R=1$ to the function $f(R\zeta)/R^{\nu}(|\zeta|<1)$.

Now consider a transcendental entire function $f(z)$ and an integer $\nu\geqslant 1$. By Theorem 2.12, the family (2.26) is not normal in the domain $\omega: 1/2<|z|<4$. Accordingly there is a point $z_0\in\omega$ such that the family (2.26) is not normal at z_0. Consider a circle $\Upsilon: |z-z_0|<\delta$ with $\delta\leqslant|z_0|/3$, and the sequence of mutually disjoint circles $c_n(n=1,2,\cdots)$ defined by (2.16). Assume that there are two finite values a and $b\neq 0$ such that each of the equations

$$f(z)=a,\ f^{(\nu)}(z)=b \tag{2.35}$$

has no root in the circle c_n, when n is sufficiently large, say $n\geqslant n_0$. Then for $n\geqslant n_0$, the function

$$g_n(z)=\frac{f(2^n z)-a}{2^{n\nu}}=f_n(z)-\frac{a}{2^{n\nu}}$$

is such that each of the equations

$$g_n(z)=0,\ g_n^{(\nu)}(z)=b$$

has no root in the circle Υ. Hence by Theorem 2.10, the family $g_n(z)(n\geqslant n_0)$ is normal in Υ. Then we see that the family (2.26) would be normal in Υ. Consequently in the open set $\bigcup_{n=1}^{\infty}c_n$, one at least of the two equations (2.35) has an infinite number of roots. Consider the ray $L: z=z_0 t\,(0\leqslant t<+\infty)$ and an angle A:

$|\arg z-\theta_0| < \varepsilon (z_0 = |z_0|e^{i\theta_0})$ with L as bisector. As in the proof of Theorem 2.6, we see that for any two finite values a and $b \neq 0$, one at least of the two equations (2. 35) has an infinite number of roots in A, and this is true no matter how small is the positive number ε. This result is an extension of Theorem 2. 6.

Similarly we can obtain an extension of Theorem 2. 7 as follows: Consider a curve $z=\varphi(t)$ satisfying the condition (C) and let $t_n (n=1,2,\cdots)$ be defined by (2. 17). Let f(z) be a transcendental entire function and $\nu \geqslant 1$ an integer. By the same method used in the proof of Theorem 2. 12, it can be shown that the family

$$f_n(z) = \frac{f\{\varphi(t_n)z\}}{\{\varphi(t_n)\}^\nu} \quad (n = 1,2,\cdots) \tag{2.36}$$

is not normal in the domain $\omega: 1/2 < |z| < 4$. Let $z_0 \in \omega$ be such that the family (2. 36) is not normal at z_0. Consider a circle $\gamma: |z-z_0| < \delta$ with $\delta \leqslant |z_0|/3$, and the sequence of mutually disjoint circles Γ_n defined by (2. 19). As in the above, we see that in the open set $\bigcup_{n=1}^{\infty} \Gamma_n$, one at least of the two equations (2. 35) has an infinite number of roots. So we have the following result: For any two finite values a and $b \neq 0$, one at least of the two equations (2. 35) has an infinite number of roots in the domain

$$\bigcup_{0<t<+\infty} (|z-\psi(t)| < \eta|\psi(t)|), \psi(t) = z_0\varphi(t), \tag{2.37}$$

no matter how small is the positive number η.

Next we give an extension of Theorem 2. 10.

Theorem 2. 16. Let D be a domain, M, $\delta (0<\delta<M)$ two positive numbers and $\nu \geqslant 1$ an integer. Let $\mathscr{F}$ be the family of the functions f(z) satisfying the following conditions:

1^0 f(z) is holomorphic in D.

2^0 There are two finite values a(f) and $b(f) \neq 0$ such that

$$|a(f)| < M, \delta < |b(f)| < M \tag{2.38}$$

and that each of the equations

$$f(z) = a(f),\ f^{(\nu)}(z) = b(f) \tag{2.39}$$

has no root in D.

Then the family $\mathscr{F}$ is normal in D.

Proof. Let $f_n(z)(n=1,2,\cdots)$ be a sequence of functions of the family $\mathscr{F}$. Set

$$a_n = a(f_n),\ b_n = b(f_n)\ (n = 1,2,\cdots)$$

and consider the sequence of functions

$$g_n(z) = \frac{f_n(z)-a_n}{b_n}\ (n = 1,2,\cdots).$$

Since each of the equations

$$g_n(z) = 0,\ g_n^{(\nu)}(z) = 1$$

has no root in D, by Theorem 2.10, we can extract from the sequence $g_n(z)(n=1,2,\cdots)$ a subsequence $g_{n_k}(z)(k=1,2,\cdots)$ such that as $k \to +\infty$, $g_{n_k}(z)$ converges locally uniformly to a holomorphic function or the constant ∞ in D. Then we complete the proof as in the proof of Theorem 2.8.

In order to give some applications of Theorem 2.16, it is convenient to give the following definition: Consider a non-constant entire function $f(z)$, two positive numbers $R, \eta(0<\eta<R)$ and an integer $\nu \geqslant 1$. A circle $\Gamma: |z-z_0|<\rho$ is called a $(R,\eta;\nu)$-filling circle of $f(z)$, if there do not exist two finite values a and $b \neq 0$ such that

$$|a| < R,\ \eta < |b| < R$$

and that each of the equations

$$f(z) = a,\ f^{(\nu)}(z) = b$$

has no root in Γ. Then evidently, in the circle Γ, either $f(z)$ takes every value w of the circle $|w|<R$, or $f^{(\nu)}(z)$ takes every value w of the domain $\eta<|w|<R$.

Now let $f(z)$ be an entire function which is not a polynomial of degree at most equal to ν, where $\nu\geqslant 1$ is an integer. Then, by Theorem 2.11, the family (2.26) is not normal in the circle $|z|<2$. Consider the sequence of circles

$$C_n: |z| < 2^{n+1} \qquad (n = 1,2,\cdots). \tag{2.40}$$

Let $R,\eta (0<\eta<R)$ be two positive numbers. Assume that for $n\geqslant 1$, C_n is not a $(R,\eta;\nu)$-filling circle of $f(z)$, and hence there are two finite values a_n and $b_n\neq 0$ such that

$$|a_n| < R, \ \eta < |b_n| < R$$

and that each of the equations

$$f(z) = a_n, \ f^{(\nu)}(z) = b_n$$

has no root in C_n. Then by Theorem 2.16, the family

$$g_n(z) = \frac{f(2^n z)-a_n}{2^{n\nu}} = f_n(z)-\frac{a_n}{2^{n\nu}} \ (n \geqslant 1)$$

is normal in the circle $|z|<2$, and hence the family (2.26) would be normal in the circle $|z|<2$. Consequently C_n is a $(R,\eta;\nu)$-filling circle of the function $f(z)$, for at least one positive integer n.

Next let $f(z)$ be a transcentental entire function and $\nu\geqslant 1$ an integer. By Theorem 2,12, the family (2.26) is not normal in the domain $\omega: 1/2<|z|<4$. Let $z_0\in\omega$ be such that the family (2.26) is not normal st z_0. Consider a circle Υ : $|z-z_0|<\delta(\delta\leqslant|z_0|/3)$ and the sequence of circles $c_n (n=1,2,\cdots)$ defined by (2.16). In the same way as above, we see that c_n is a $(R,\eta;\nu)$-filling circle of $f(z)$, for an infinite number of positive integers n. Similarly since the family (2.36) is not normal in the domain ω, we see, in the same way, that the circle Γ_n defined by (2.19) is a $(R,\eta;\nu)$-filling circle of the function $f(z)$, for an infinite number of positive integers n, where $z_0\in\omega$ is a point such that the family (2.36) is not normal at z_0 and δ is a positive number such that $\delta \leqslant |z_0|/3$.

Finally as an application of Theorem 2. 16, we prove the following theorem:

Theorem 2. 17. Given an integer $\nu \geqslant 1$, we can find a number $\lambda_\nu > 0$ depending only on ν having the following property: If $f(z)$ is a holomorphic function in the circle $|z|<1$ such that

$$f(0) = 0, \qquad M(\frac{1}{2}, f) \geqslant 1, \tag{2.41}$$

then there are two positive numbers $R(f), \eta(f)$ such that $R(f)-\eta(f) > \lambda_\nu$ and that the circle $|z|<1$ is a $(R(f), \eta(f); \nu)$-filling circle of $f(z)$.

Proof. Assume that such a number λ_ν does not exist. Then to each positive integer n corresponds a holomorphic function $f_n(z)$ in the circle $|z|<1$ satisfying the condition (2. 41), such that there do not exist two positive numbers R, η such that $R-\eta > 1/n$ and that the circle $|z|<1$ is a $(R, \eta; \nu)$-filling circle of $f_n(z)$, à fortiori, the circle $|z|<1$ is not a $(3, 1; \nu)$-filling circle of $f_n(z)$. By Theorem 2. 16, the family $f_n(z)(n=1,2,\cdots)$ is normal in the circle $|z|<1$. Consequently from the sequence $f_n(z)(n=1,2,\cdots)$ we can extract a subsequence $f_{n_k}(z)$ $(k=1,2,\cdots)$ such that as $k \to +\infty$, $f_{n_k}(z)$ converges locally uniformly to a holomorphic function $f(z)$ in the circle $|z|<1$, satisfying the condition (2. 41). Then, as in the proof of Theorem 2. 9, we can find a circle $|z|=r(0<r<1)$, a number $m>0$ and a positive integer k_0 such that when $k \geqslant k_0$, in the circle $|z|<r$, the values of the function $f_{n_k}(z)$ cover the circle $|w|<m/2$. Take an integer k such that

$$k \geqslant k_0, \ \frac{4}{n_k} < \frac{m}{2}.$$

Then the circle $|z|<1$ is a $(4/n_k, 2/n_k; \nu)$-filling circle of the function $f_{n_k}(z)$. This contradicts the above stated property of the function $f_n(z)$.

In the definition of $(R, \eta; \nu)$-filling circle, we have assumed that the function $f(z)$ is a non-constant entire function. However it is sufficient to assume that $f(z)$ is a non-constant holomorphic function in that circle.

2.3. BLOCH'S THEOREM

For the sake of convenience we first give a definition.

Definition 2.1. Let $w=f(z)$ be a holomorphic function in a domain D, and Δ a domain in the w-plane. We say that Δ is a simply covered image domain of $f(z)$ for D, if there is a domain $d \subset D$ such that $f(z)$ is univalent in d and maps d onto Δ. This is equivalent to that the inverse function of $w=f(z)$ has a branch $\varphi(w)$ which is holomorphic in Δ such that $\varphi(w) \in D$ and $f\{\varphi(w)\}=w$ for $w \in \Delta$.

Theorem 2.18. Let D be a domain and $A>0$ a number. Let $\mathscr{F}$ be the family of the functions $f(z)$ satisfying the following conditions:

1^0 $f(z)$ is holomorphic in D.

2^0 There do not exist two numbers R_1, R_2 such that

$$R_1 > A, \; R_2 - R_1 > A$$

and that for each number $0 \leqslant \omega < 2\pi$, the domain

$$R_1 < |w| < R_2, \; \omega < \arg w < \omega + 2\pi$$

is a simply covered image domain of $f(z)$ for D.

Then the family $\mathscr{F}$ is normal in D.

Proof. Let z_0 be a point of D and $\Gamma: |z-z_0|<\rho$ a circle belonging to D. Let $S: f_n(z)(n=1,2,\cdots)$ be a sequence of functions of the family $\mathscr{F}$. By Theorem 1.4, it is sufficient to prove that from the sequence S we can extract a subsequence $f_{n_k}(z)(k=1,2,\cdots)$ such that as $k \to +\infty$, $f_{n_k}(z)$ converges uniformly to a holomorphic function or to the constant ∞ in the circle $\Gamma': |z-z_0|<\rho/4$. Distinguish two cases:

1) The sequence $f_n(z_0)(n=1,2,\cdots)$ is bounded, namely $|f_n(z_0)| \leqslant L(n=1,2,\cdots)$. By means of the auxiliary function $F_n(\zeta)=f_n(z_0+\rho\zeta)(|\zeta|<1)$, we see, by Corollary 2 in Appendix A, that in the circle $|z-z_0| \leqslant \rho/2$, we have

$$|f_n(z)| < 4e^3(A + |f_n(z_0)|)\Phi(\frac{1}{2}) \leqslant 4e^3(A+L)\Phi(\frac{1}{2}),$$

where

$$\Phi(r) = exp\left(\frac{a}{-log\, r}\right).$$

Hence the sequence S is uniformly bounded in the circle $|z-z_0| \leqslant \rho/2$, and by Corollary 1.3, we can extract from S a subsequence $f_{n_k}(z)(k=1,2,\cdots)$ such that as $k \to +\infty$, $f_{n_k}(z)$ converges uniformly to a holomorphic function in the circle Γ'.

2) The sequence $f_n(z_0)(n=1,2,\cdots)$ is unbounded. Then there is a subsequence $f_{n_k}(z_0)(k=1,2,\cdots)$ such that as $k \to +\infty$ $f_{n_k}(z_0)$ converges to ∞. Consider a point $z_1 \in \Gamma'$. Then the circle $|z-z_1| < 3\rho/4$ belongs to the circle Γ. Again by Corollary 2 in Appendix A, in the circle $|z-z_1| \leqslant \rho/4$, we have

$$|f_{n_k}(z)| < 4e^3(A + |f_{n_k}(z_1)|)\Phi(\frac{1}{3})$$

and in particular,

$$|f_{n_k}(z_0)| < 4e^3(A + |f_{n_k}(z_1)|)\Phi(\frac{1}{3}).$$

This inequality shows that as $k \to +\infty$, $|f_{n_k}(z)|$ converges uniformly to $+\infty$ in Γ'.

Corollary 2.11. Let D be a domain and $\alpha > 0$ a number. Let $\mathscr{F}$ be the family of the functions f(z) satisfying the following conditions:

1^0 f(z) is holomorphic in D.

2^0 There does not exist a circle $\Gamma: |w-w_0| < \rho$ such that $\rho > \alpha$ and that Γ is a simply covered image domain of f(z) for D.

Then the family $\mathscr{F}$ is normal in D.

This theorem is due to Bloch [2]. It is a simple consequence of Theorem 2.18 in which we take $A = 2\alpha$.

For functions which do not take the value zero, we have the following theorems:

Theorem 2.19. Let D be a domain and $K>1$ a number. Let $\mathscr{F}$ be the family of the functions $f(z)$ satisfying the following conditions:

1° $f(z)$ is holomorphic and does not take the value zero in D.

2° There is a number $\omega=\omega(f)$ such that $0\leqslant\omega<2\pi$ and that the domain

$$\frac{1}{K}<|w|<K, \quad \omega<\arg w<\omega+2\pi \tag{2.42}$$

is not a simply covered image domain of $f(z)$ for D.
Then the family $\mathscr{F}$ is normal in D.

Proof. Let z_0 be a point of D and $|z-z_0|<\rho$ a circle belonging to D. Let $f(z)$ be a function of the family $\mathscr{F}$ and define $F(\zeta)=f(z_0+\rho\zeta)$ ($|\zeta|<1$). By Corollaries 5 and 7 in Appendix A, we must have the two inequalities

$$\log M(\frac{1}{2},F)<h(1+\log K+\overset{+}{\log}|F(0)|), \tag{2.43}$$

$$\log M(\frac{1}{2},\frac{1}{F})<h(1+\log K+\overset{+}{\log}\left|\frac{1}{F(0)}\right|), \tag{2.44}$$

where h is an absolute positive constant. Consequently, if $|F(0)|=|f(z_0)|\leqslant1$, then by (2.43), in the circle $\bar{\Gamma}: |z-z_0|\leqslant\rho/2$, we have

$$|f(z)|<e^{h(1+\log K)}.$$

On the other hand, if $|F(0)|=|f(z_0)|>1$, then by (2.44), in $\bar{\Gamma}$, we have

$$\left|\frac{1}{f(z)}\right|<e^{h(1+\log K)}.$$

Hence by Corollary 1.4, the family $\mathscr{F}$ is normal in D.

Theorem 2.20. Let D be a domain, $\nu\geqslant1$ an integer and $K>1$ a number. Let $\mathscr{F}$ be the family of the functions $f(z)$ satisfying the following conditions:

1° $f(z)$ is holomorphic and does not take the value zero in D.

2° There is a number $\omega=\omega(f)$ such that $0\leqslant\omega<2\pi$ and that the domain (2.

42) is not a simply covered image domain of $f^{(v)}(z)$ for D.
Then the family $\mathscr{F}$ is normal in D.

This theorem is at the same time an extension of Bloch's theorem, and an extension of Miranda's theorem as well.

Proof. Let z_0 be a point of D and $|z-z_0|<\rho$ a circle belonging to D. Let f (z) be a function of the family $\mathscr{F}$ and define $F(\zeta)=f(z_0+\rho\zeta)/\rho^{v}(|\zeta|<1)$. By Corollaries 10 and 12 in Appendix A, we must have the two inequalities

$$\log M(\frac{1}{2},F) < h_v(1 + \log K + \overset{+}{\log} |F(0)|),$$

$$\log M(\frac{1}{2},\frac{1}{F}) < h_v(1 + \log K + \overset{+}{\log} |\frac{1}{F(0)}|),$$

where h_v is a positive number depending only on v. Then the proof is completed as in the proof of Theorem 2.10.

Now we give some applications of Theorem 2.18. Consider first a non-constant entire function f(z). By Theorem 2.2, the family (2.3) is not normal in the circle $|z|<2$. Let $A>0$ be a number. By Theorem 2.18, there is at least one positive integer n such that the function $f_n(z)=f(2^n z)$ satisfies the following condition: There exist two numbers R_1, R_2 such that

$$R_1 > A, \; R_2 - R_1 > A \tag{2.45}$$

and that for each number $0\leqslant\omega<2\pi$, the domain

$$R_1 < |w| < R_2, \; \omega < \arg w < \omega + 2\pi \tag{2.46}$$

is a simply covered image domain of $f_n(z)$ for the circle $|z|<2$. Evidently the domain (2.46) is a simply covered image domain of the function f(z) for the circle $|z|<2^{n+1}$.

Next consider a transcendental entire function f(z), and the sequence of circles $c_n(n=1,2,\cdots)$ defined by (2.16), where $z_0(\frac{1}{2}<|z_0|<4)$ is a point at which the family (2.15) is not normal and δ is a positive number such that $\delta\leqslant|z_0|/3$. As above we see that for each positive integer n_0, we can find an integer $n\geqslant n_0$ such that there exist two numbers R_1, R_2 satisfying (2.45) and having the property that for each number $0 \leqslant \omega < 2\pi$, the domain (2.46) is a simply covered image domain of f(z) for the circle c_n. This is also true for the sequence of circles $\Gamma_n (n=1,2,\cdots)$ defined by (2.19).

As applications of Theorem 2.18, we can also prove two theorems which may be considered as extensions of Schottky's theorem and Landau's theorem. However since the method of proof is the same as that already used several times, we state them without proof.

Theorem 2.21. Given a domain D, two bounded closed sets σ,E of points belonging to D and two positive numbers μ, A, we can find a positive number H (D,σ,E,μ,A) depending only on D,σ,E,μ and A having the following property: If f(z) is a holomorphic function in D satisfying the following conditins:

1^0 f(z) satisfies the condition 2^0 in Theorem 2.18;

$2^0 \min_{z\in\sigma} |f(z)| \leqslant \mu$;

then we have

$$\max_{z\in E} |f(z)| \leqslant H(D,\sigma,E,\mu,A). \tag{2.47}$$

Theorem 2.22. Given a domain D, two bounded closed sets σ, E of points belonging to D, two positive numbers μ,A and a positive integer p, we can find a positive number $K(D,\sigma,E,\mu,A,p)$ depending only on D,σ,E,μ,A and p having the following property: If f(z) is a holomorphic function in D satisfying the conditions 1^0 and 2^0 in Theorem 2.21, then we have

$$\max_{z\in E} |f^{(p)}(z)| \leqslant K(D,\sigma,E,\mu,A,p). \tag{2.48}$$

Similar applications of Theorem 2.19 and Theorem 2.20 can also be obtained.

3

CRITERIONS OF NORMALITY OF FAMILIES OF MEROMORPHIC FUNCTIONS AND APPLICATIONS

3.1. MONTEL'S THEOREM

Theorem 3.1. Let D be a domain and a, b, c three distinct values in $\hat{\mathbb{C}}$. Let $\mathscr{F}$ be the family of the functions f(z) meromorphic in D and such that each of the equations

$$f(z) = a, \quad f(z) = b, \quad f(z) = c \tag{3.1}$$

has no root in D. Then the family $\mathscr{F}$ is normal in D.

This theorem due to Montel [26] is the generalization of Theorem 2.1 to the case of families of meromorphic functions.

Proof. If one of the values a, b, c is infinite, for instance $c = \infty$, then a, b ($a \neq b$) are finite and the functions f(z) of the family $\mathscr{F}$ are holomorphic in D and such that each of the first two equations in (3.1) has no root in D, hence $\mathscr{F}$ is normal in D, by Theorem 2.1. Now assume that the values a, b, c are finite. Consider the family of functions

$$\mathscr{F}_1 = \{\frac{1}{f(z)-a} \mid f(z) \in \mathscr{F}\}.$$

The functions of $\mathscr{F}_1$ are holomorphic and do not take the values 1/(b-a) and 1/(c-a) in D. By Theorem 2.1, $\mathscr{F}_1$ is normal in D and hence is equicontinuous in D with respect to the spherical distance, by Theorem 1.5. Let z_0 be a point of D and ε a positive number. Then there is a circle γ : $|z-z_0|<\delta$ belonging to D, such that for each function $f(z)\in\mathscr{F}$, the inequality

$$\left|\frac{1}{f(z)-a},\frac{1}{f(z_0)-a}\right|<\varepsilon$$

holds in γ. This is also true for the inequality

$$|f(z)-a,f(z_0)-a|<\varepsilon.$$

Then by Lemma 1.1, we have

$$|f(z),f(z_0)|\leqslant\frac{2}{|a,\infty|^2}|f(z)-a,f(z_0)-a|<\frac{2}{|a,\infty|^2}\varepsilon$$

in γ. Hence the family $\mathscr{F}$ is equicontinuous in D with respect to the spherical distance, and by Theorem 1.5, $\mathscr{F}$ is normal in D.

In what follows, we give some applications of Theorem 3.1, which are nearly parallel to those of Theorem 2.1.

Theorem 3.2. Let f(z) be a non-constant meromorphic function in **C**. Then the family of meromorphic functions in **C**

$$f_n(z)=f(2^n z)\ (n=1,2,\cdots) \tag{3.2}$$

is not normal in the circle $|z|<2$.

Proof. Assume that the family (3.2) is normal in the circle $|z|<2$. Then by Corollary 1.4, we can find a circle Γ: $|z|<r\,(0<r<2)$ and a positive number M such that each function $f_n(z)$ of the family (3.2) satisfies in Γ one of the inequalities

$$|f_n(z)|<M,\ \left|\frac{1}{f_n(z)}\right|<M. \tag{3.3}$$

So at least one of these two inequalities holds in Γ, for an infinite number of positive integers n. Assume for instance, that this is true for the first inequality in (3. 3). Then the inequality

$$|f(z)| < M$$

holds in the circle $|z|<2^n r$ for arbitrarily large n. It follows that the function f(z) is an entire function and hence a constant, contrary to the hypothesis. We also arrive at a contradiction, if we assume that the second inequality in (3. 3) holds in Γ, for an infinite number of positive integers n.

Corollary 3. 1. If f(z) is a non-constant meromorphic function in **C**, then f(z) takes every value $a \in \hat{\mathbf{C}}$, except at most two values $a \in \hat{\mathbf{C}}$.

This is Picard's theorem on meromorphic functions in **C**. Corollary 3. 1 follows immediately from Theorems 3. 1 and 3. 2. In fact, if the function f(z) does not take three distinct values a, b, c belonging to $\hat{\mathbf{C}}$, then by Theorem 3. 1, the family (3. 2) is normal in the circle $|z|<2$, incompatible with Theorem 3. 2.

We can also deduce Corollary 3. 1 from Corollary 2. 1. In fact, assume that the function f(z) does not take three values a, b, c. If one of the values a, b, c is infinite, then f(z) would be an entire function which does not take two finite values and hence, by Corollary 2. 1, is a constant. If the values a, b, c are finite, then the function 1/(f(z)-a) is an entire function which does not take the values 1/(b-a) and 1/(c-a). Again, by Corollary 2. 1, f(z) would be a constant.

Definition 3.1. Let f(z) be a meromorphic function in a domain $\mathbf{D}: 0<|z|<\rho$. We say that the point z=0 is an essential singular point of f(z), if one of the following conditions is satisfied:

1^0 There is a number ρ_1 $(0<\rho_1<\rho)$ such that f(z) is holomorphic in the domain $D_1: 0<|z|<\rho_1$ and the point z=0 is an essential singular point of f(z).

2^0 The point z=0 is a point of accumulation of the set of the poles of f(z) in the domain D.

Corollary 3.2. Let f(z) be a meromorphic function in a domain $\mathbf{D}: 0<|z|<\rho$ such that the point z=0 is an essential singular point of f(z). Then in **D**, f(z) takes every value $a \in \hat{\mathbf{C}}$ an infinite number of times, except at most two values $a \in \hat{\mathbf{C}}$.

This is Picard's theorem on meromorphic functions in the neighborhood of an essential singular point.

Proof. Distinguish two cases. If the condition 1^0 in Definition 3.1 is satisfied, then by Corollary 2.2, in D, f(z) takes every value an infinite number of times, except at most a finite value and ∞. Now assume that the condition 2^0 in Definition 3.1 is satisfied and that in D, f(z) takes three values a, b, c only a finite number of times. Evidently the values a, b, c are necessarily all finite. Let $D_1: 0<|z|<\rho_1 (0<\rho_1<\rho)$ be a domain in which f(z) does not take the values a, b, c. The function $g(z)=1/\{f(z)-a\}$ is holomorphic in D_1 and does not take the values $1/(b-a)$ and $1/(c-a)$. By Corollary 2.2, the point $z=0$ is not an essential singular point of g(z). Since the point $z=0$ is a point of accumulation of the set of the zeros of g(z), the point $z=0$ is a removable singular point of g(z) which is then identically equal to zero in D_1. This is obviously impossible.

Lemma 3.1. Let $\mathscr{F}$ be a normal family of meromorphic functions in a domain D. Let σ be a bounded closed set of points belonging to D, a $(a \in \hat{\mathbf{C}})$ a value and $\delta (0<\delta\leqslant 1)$ a number. Assume that the following conditions are satisfied:

1^0 Each function $f(z) \in \mathscr{F}$ does not take the value a in D.

2^0 For each function $f(z) \in \mathscr{F}$, we have $\max_{z\in\sigma} |f(z), a| \geqslant \delta$. Then for each bounded closed set E of points of D, there is a number $\alpha>0$ such that for each function $f(z) \in \mathscr{F}$, we have

$$\min_{z\in E} |f(z), a| \geqslant \alpha. \tag{3.4}$$

Proof. Distinguish two cases.

1) $a=\infty$. Then the functions of the family $\mathscr{F}$ are holomorphic in D and for each function $f(z) \in \mathscr{F}$, we have

$$\min_{z\in\sigma} |f(z)| < \min_{z\in\sigma} (1+|f(z)|^2)^{\frac{1}{2}} \leqslant \frac{1}{\delta}.$$

Hence by Lemma 2. 1, for each bounded closed set E of points of D, there is a number $K>0$ such that for each function $f(z)\in\mathscr{F}$, we have

$$\max_{z\in E}|f(z)|\leqslant K$$

which implies that

$$\min_{z\in E}|f(z),\infty|\geqslant\frac{1}{(1+K^2)^{\frac{1}{2}}}.$$

2) $a\neq\infty$. Let E be a bounded closed set of points of D, and assume that there does not exist a number $\alpha>0$ having the required property for E. Then to each positive integer n corresponds a function $f_n(z)\in\mathscr{F}$ such that

$$\min_{z\in E}|f_n(z),a|<\frac{1}{n}. \tag{3. 5}$$

From the sequence $f_n(z)(n=1,2,\cdots)$ we can extract a subsequence $f_{n_k}(z)(k=1,2,\cdots)$ which is a C_0-sequence in D. By Theorem 1. 2, as $k\to+\infty$, $f_{n_k}(z)$ converges locally uniformly to a limit function F(z) in D with respect to the spherical distance. Next by the finite covering theorem, we have

$$\lim_{k\to+\infty}|f_{n_k}(z),F(z)|=0$$

uniformly on E. On the other hand by Theorem 1. 3, F(z) is a meromorphic function in D or the constant ∞. In either case, by the triangular inequality (1. 6) and Lemma 1. 4, we see easily that the function $|F(z),a|$ is continuous in D. Now given a number $\varepsilon>0$, let k_0 be a positive integer such that when $k\geqslant k_0$, we have

$$|f_{n_k}(z),F(z)|<\varepsilon \tag{3. 6}$$

on E. Then from (3. 5), (3. 6) and the inequality

$$|F(z),a|\leqslant|F(z),f_{n_k}(z)|+|f_{n_k}(z),a|,$$

we see that, when $k \geqslant k_0$, we have

$$\min_{z \in E} |F(z), a| < \varepsilon + \frac{1}{n_k},$$

hence

$$\min_{z \in E} |F(z), a| = 0$$

and there is a point $z_0 \in E$ such that $F(z_0) = a$. Accordingly $F(z)$ is a meromorphic function in D. On the other hand, by the condition 2^0 in Lemma 3.1 and the inequality

$$|f_{n_k}(z), a| \leqslant |f_{n_k}(z), F(z)| + |F(z), a|,$$

we find

$$\max_{z \in \sigma} |F(z), a| > \frac{\delta}{2}$$

which shows that $F(z)$ is not identically equal to a. Since $F(z_0) = a \neq \infty$, by Theorem 1.1, we can find a circle Υ : $|z - z_0| < \rho$ belonging to D and a positive integer k_* such that the functions $f_{n_k}(z)(k \geqslant k_*)$ and $F(z)$ are holomorphic in Υ, and

$$\lim_{\substack{k \to +\infty \\ k \geqslant k_*}} |f_{n_k}(z) - F(z)| = 0$$

uniformly in Υ. Then by a known theorem, $f_{n_k}(z) - a$ has zero in Υ, when k is sufficiently large, incompatible with the condition 1^0 in Lemma 3.1.

Theorem 3.3. Given a domain D, two bounded closed sets σ, E of points belonging to D, three distinct values $a_j (a_j \in \hat{\mathbf{C}}) (j = 1, 2, 3)$ and a number $\delta (0 < \delta \leqslant 1)$, we can find a number $\alpha(D, \sigma, E, a_1, a_2, a_3, \delta) > 0$ depending only on $D, \sigma, E, a_j (j = 1, 2, 3)$ and δ having the following property: If $f(z)$ is a meromorphic function in D satisfying the conditions:

1^0 f(z) does not take the values $a_j(j=1,2,3)$ in D;

$2^0 \max_{z\in E} |f(z), a_1| \geqslant \delta$;

then we have

$$\min_{z\in E} |f(z), a_1| \geqslant \alpha(D, \sigma, E, a_1, a_2, a_3, \delta). \tag{3.7}$$

This is a general form of Schottky's theorem in the case of meromorphic functions.

Proof. Let $\mathscr{F}$ be the family of functions meromorphic in D and satisfying the conditions 1^0 and 2^0 in Theorem 3.3. By Theorem 3.1, the family $\mathscr{F}$ is normal in D. Next by Lemma 3.1, there is a number $\alpha>0$ such that for each function $f(z)\in\mathscr{F}$, we have

$$\min_{z\in E} |f(z), a| \geqslant \alpha.$$

This number α has then the required property.

Theorem 3.4. Given a domain D, a bounded closed set E of points belonging to D and three distinct values $a_j(a_j\in\hat{\mathbf{C}})(j=1,2,3)$, we can find a positive number $\beta(D, E, a_1, a_2, a_3)$ depending only on D, E and $a_j(j=1,2,3)$ having the following property: If a function f(z) is meromorphic in D and does not take the values $a_j(j=1,2,3)$ in D, then

$$\max_{z\in E} \partial(z, f) \leqslant \beta(D, E, a_1, a_2, a_3). \tag{3.8}$$

This is a general form of Landau's theorem in the case of meromorphic functions.

Proof. Let $\mathscr{F}$ be the family of meromorphic functions in D, which do not take the values $a_j(j=1,2,3)$ in D. By Theorem 3.1, the family $\mathscr{F}$ is normal in D. Next by Theorem 1.6, the family $\mathscr{F}_*$ defined by (1.38) is locally uniformly bounded in D. Then by the finite covering theorem, $\mathscr{F}_*$ is uniformly bounded on E. Consequently there is a positive number β such that for each function $f(z) \in \mathscr{F}$, we have

$$\max_{z \in E} \partial(z, f) \leqslant \beta.$$

This number β has the required property.

Now consider a transcendental meromorphic function $f(z)$ in **C** and the family of meromorphic functions in **C**

$$f_n(z) = f(2^n z) \quad (n = 1, 2, \cdots). \tag{3.9}$$

Unlike the case of transcendental entire functions, it is not always true that the family (3.9) is not normal in the domain $\omega: 1/2 < |z| < 4$. A well known example is the function [26]

$$f(z) = \frac{\varphi(z)}{\psi(z)}, \quad \varphi(z) = \prod_{n=0}^{\infty} (1 - \frac{z}{2^n}), \quad \psi(z) = \prod_{n=0}^{\infty} (1 + \frac{z}{2^n}).$$

We have

$$f_1(z) = f(2z) = \frac{1-2z}{1+2z} f(z),$$

$$f_2(z) = f_1(2z) = \frac{(1-2z)(1-2^2 z)}{(1+2z)(1+2^2 z)} f(z),$$

and in general

$$f_n(z) = R_n(z) f(z), \quad R_n(z) = \frac{(1-2z)(1-2^2 z)\cdots(1-2^n z)}{(1+2z)(1+2^2 z)\cdots(1+2^n z)}. \tag{3.10}$$

Now assume $z \neq 0$ and write

$$(-1)^n R_n(z) = \frac{P_n(z)}{Q_n(z)}, \quad P_n(z) = \prod_{j=1}^{n} (1 - \frac{1}{2^j z}), \quad Q_n(z) = \prod_{j=1}^{n} (1 + \frac{1}{2^j z}).$$

Consider a point $z_0 \in \omega$ and distinguish two cases:

1)$f(z_0)\neq\infty$. Then we can find a circle $\overline{\Gamma}: |z-z_0|\leqslant\rho$ belonging to ω and a positive number M such that $|f(z)|<M$ in $\overline{\Gamma}$. Take a number $1/2<h<4$ such that $\overline{\Gamma}$ belongs to the domain $h<|z|<4$. Set

$$A_n = \prod_{j=1}^{n}(1-\frac{1}{2^j h}),\quad B_n = \prod_{j=1}^{n}(1+\frac{1}{2^j h})$$
$$A = \prod_{j=1}^{\infty}(1-\frac{1}{2^j h}) > 0,\quad B = \prod_{j=1}^{\infty}(1+\frac{1}{2^j h}).$$

Since for $|z|\geqslant h$ we have

$$A_n \leqslant |P_n(z)| \leqslant B_n,\quad A_n \leqslant |Q_n(z)| \leqslant B_n$$
$$A_n > A,\quad B_n < B,$$

we have

$$\frac{A}{B} < |R_n(z)| < \frac{B}{A}.$$

It follows then from (3.10), that

$$|f_n(z)| < \frac{B}{A}M \tag{3.11}$$

in $\overline{\Gamma}$.

2) $f(z_0)=\infty$. Then from the identity

$$\frac{1}{f_n(z)} = \frac{1}{R_n(z)}\frac{1}{f(z)},$$

we get the inequality

$$\left|\frac{1}{f_n(z)}\right| < \frac{B}{A}M'$$

in a circle $\overline{\Gamma}'$: $|z-z_0|\leqslant\rho'$ belonging to ω. Consequently by Corollary 1.4, the family (3. 9) is normal in ω.

Definition 3. 2. Let $f(z)$ be a transcendental meromorphic function in **C**. Let $\Gamma: z=z(t)(0\leqslant t<+\infty)$ be a curve, where $z(t)$ is a continuous complex valued function such taht

$$\lim_{t\to+\infty} z(t) = \infty. \tag{3. 12}$$

Let $a\in\hat{\mathbf{C}}$ be a value. If

$$\lim_{t\to+\infty} f\{z(t)\} = a, \tag{3. 13}$$

then a is called an asymptotic value of $f(z)$ and Γ a corresponding path of determination.

Theorem 3. 5. Let $f(z)$ be a transcendental meromorphic function in **C**. If $f(z)$ has an asymptotic value a, then the family (3. 9) is not normal in the domain $\omega: 1/2<|z|<4$.

This theorem is due to Julia [16].

Proof. Without loss of generality, we may assume $a\neq\infty$. In fact, if $a=\infty$, it is sufficient to consider the function $1/f(z)$ and the family $1/f_n(z)=1/f(2^n z)(n=1,2,\cdots)$.

Distinguish two cases:

1) $f(z)$ has only a finite number of poles. In this case we can find a positive number R such that $f(z)$ is holomorphic for $R<|z|<+\infty$ and has an essential singularity at the point $z=\infty$. Then we see, as the proof of Theorem 2. 3, that the family (3. 9) is not normal in the domain ω.

2) $f(z)$ has an infinite number of poles. In this case we can find a subsequence $f_{n_k}(z)(k=1,2,\cdots)$ of the sequence (3. 9), such that for each k, the function $f_{n_k}(z)$ has at least one pole in the region $E: 1\leqslant|z|\leqslant 2$. Let $\Gamma: z=z(t)(0\leqslant t<+\infty)$ be the curve defined in Definition 3. 2. We may assume that

$$|z(0)| < 2^{n_1}.$$

Now assume that the family (3.9) is normal in the domain ω. Then from the sequence $f_{n_k}(z)\,(k=1,2,\cdots)$ we can extract a subsequence $f_{m_h}(z)\,(h=1,2,\cdots)$ which is a C_0-sequence in ω, whose limit function $F(z)$, with respect to the spherical distance, is a meromorphic function in ω or the constant ∞. Consider a circle $|z|=r\,(1\leqslant r\leqslant 2)$. For each h, we can find t_h such that

$$|z(t_h)| = 2^{m_h} r.$$

Evidently $t_h \to +\infty$, as $h \to +\infty$. Set $z_h = z(t_h)/2^{m_h}$, then

$$|z_h| = r, \quad f_{m_h}(z_h) = f\{z(t_h)\}.$$

Hence, by (3.13), we have

$$\lim_{h\to+\infty} f_{m_h}(z_h) = a.$$

On the other hand, by Theorem 1.2 and the finite covering theorem, as $h\to+\infty$, $f_{m_h}(z)$ converges uniformly to $F(z)$ in E with respect to the spherical distance. Then from the inequality

$$|F(z_h), a| \leqslant |F(z_h), f_{m_h}(z_h)| + |f_{m_h}(z_h), a|,$$

we see that $\min\limits_{|z|=r}|F(z),a|=0$ and hence $F(z)$ must be a meromorphic function in ω and takes the value a on the circle $|z|=r$. Since $r\,(1\leqslant r\leqslant 2)$is arbitrary, hence $F(z)$ is equal identically to the constant a. But this is incompatible with the fact that for each h, the function $f_{m_h}(z)$ has at least one pole in E, so we get a contradiction.

Theorem 3.6. Let f(z) be a transcendental meromorphic function in **C**. In order that the family (3.9) be not normal in the domain $\omega: 1/2<|z|<4$, it is necessary and sufficient that the function $|z|\partial(z,f)$ is unbounded in **C**, where $\partial(z,f)$ is the spherical derivative of f(z).

Proof. First of all, from (3.9) we have

$$\partial(z,f_n) = 2^n\partial(2^nz,f). \tag{3.14}$$

Assume that the family (3.9) is normal in the domain ω. Then by Theorem 1.6 and the finite covering theorem, the sequence $\partial(z,f_n)(n=1,2,\cdots)$ is uniformly bounded in the region $E: 1\leqslant|z|\leqslant 2$, namely there is a positive number $\mathbf{M}$ such that for each $n\geqslant 1$, we have

$$\partial(z,f_n) \leqslant M$$

in E. Hence from (3.14),

$$2^n|z|\partial(2^nz,f) \leqslant |z|M \leqslant 2M$$

in E. This is equivalent to the inequality

$$|z|\partial(z,f) \leqslant 2M$$

for $2^n\leqslant|z|\leqslant 2^{n+1}$. Since $n\geqslant 1$ is arbitrary, we conclude that the function $|z|\partial(z, f)$ is bounded in $\mathbf{C}$.

Conversely if there is a positive number $\mathbf{M}'$ such that

$$|z|\partial(z,f) \leqslant M'$$

in $\mathbf{C}$, then for $z\in\omega$, we have

$$\partial(z,f_n) = 2^n\partial(2^nz,f) \leqslant M'/|z| < 2M'.$$

Consequently the sequence $\partial(z,f_n)(n=1,2,\cdots)$ is uniformly bounded in ω, and by Theorem 1.6, the family (3.9) is normal in ω.

Now consider a transcendental meromorphic function $f(z)$ in $\mathbf{C}$ such that the family (3.9) is not normal in the domain $\omega: 1/2<|z|<4$. By Theorem 1.4, there is a point $z_0\in\omega$ such taht the family (3.9) is not normal at z_0. Let $\Gamma: |z-z_0|<\delta$ be a circle belonging to ω. By Theorem 3.1, for each value $a\in\hat{\mathbf{C}}$, there are an infinite number of positive integers n such that $f_n(z)$ takes the value a in Γ, except at most two values $a\in\hat{\mathbf{C}}$. Consequently, as in the case of transcendental entire functions, $c_n(n=1,2,\cdots)$being the sequence of circles (2.16) with $\delta\leqslant|z_0|/3$, then in the open set $\Omega=\bigcup_{n=1}^{\infty}c_n$, f(z) takes every value $a\in\hat{\mathbf{C}}$ an infinite number of times, except at most two values $a\in\hat{\mathbf{C}}$. Next we see that, no matter how small is the positive number ε, in the angle $A: |\arg z-\theta_0|<\varepsilon(z=|z_0|e^{i\theta_0})$ with the ray $L: z=z_0t\,(0\leqslant t<+\infty)$ as bisector, f(z) takes every value $a\in\hat{\mathbf{C}}$ an infinite number of times, except, at most two values $a\in\hat{\mathbf{C}}$. The ray L is also called a Julia direction of f(z). Hence in view of Theorem 3.6, we have the following theorem:

Theorem 3.7. If f(z) is a transcendental meromorphic function in **C** such that the function $|z|\partial(z,f)$ is unbounded in **C**, then f(z) has at least one Julia direction.

In the case of transcendental entire functions, besides Julia direction, we have also proved the existence of Julia curve (Theorem 2.7). Now consider a transcendental meromorphic function f(z)in **C**, and the family (2.18) of meromorphic functions in **C**. It is easy to see that Theorem 3.5 and Theorem 3.6 are also true for the family (2.18). Assume that the family (2.18) is not normal in the domain $\omega: 1/2<|z|<4$ and let $z_0\in\omega$ be such that the family (2.18) is not normal at z_0. Then by Theorem 3.1, we see that $\Gamma_n(n=1,2,\cdots)$ being the sequence of circles defined by (2.19) with $\delta\leqslant|z_0|/3$, then in the open set $\bigcup_{n=1}^{\infty}\Gamma_n$, f(z) takes every value $a\in\hat{\mathbf{C}}$ an infinite number of times, except at most two values $a\in\hat{\mathbf{C}}$. Hence as in the case of transcendental entire functions, we have the following theorem:

Theorem 3.8. Let f(z) be a transcendental meromorphic function in **C** such that the function $|z|\partial(z,f)$ is unbounded in **C**. Then qiven any curve $z=\varphi(t)$ satisfying the condition (C), we can find a point z_0 of the domain $1/2<|z|<4$, such that the curve $z=z_0\varphi(t)$ is a Julia curve of f(z).

Now we give an extension of Theorem 3.1.

Theorem 3.9. Let D be a domain and $\delta(0<\delta<1)$ a number. Let $\mathscr{F}$ be the family of the functions f(z) satisfying the following conditions:

1^0 f(z) is meromorphic in D.

2^0 There are three values $a_j(f) \in \hat{C}$ $(j=1,2,3)$ such that

$$|a_j(f), a_{j'}(f)| > \delta \quad (1 \leqslant j, j' \leqslant 3, j \neq j')$$

and that each of the equations

$$f(z) = a_j(f) \quad (j = 1,2,3)$$

has no root in D.

Then the family $\mathscr{F}$ is normal in D.

Proof. Consider first the family $\mathscr{F}_1$ consisting of functions $f(z) \in \mathscr{F}$ such that one of the three values $a_j(f)$ $(j=1,2,3)$ is ∞, say $a_3(f)=\infty$. If $f(z) \in \mathscr{F}_1$, then $f(z)$ is holomorphic in D, and

$$|a_j(f), \infty| = \frac{1}{(1 + |a_j(f)|^2)^{\frac{1}{2}}} > \delta, \ |a_j(f)| < \frac{1}{\delta} \quad (j = 1,2),$$
$$|a_1(f)-a_2(f)| \geqslant |a_1(f), a_2(f)| > \delta.$$

Hence by Theorem 2.8, the family $\mathscr{F}_1$ is normal in D.

Next consider the family $\mathscr{F}_2$ consisting of functions $f(z) \in \mathscr{F}$ such that the three values $a_j(f)$ $(j=1,2,3)$ are finite. Evidently among these three values, there is one, say $a_1(f)$, such that

$$|a_1(f), \infty| > \frac{\delta}{2}.$$

Consider the family

$$\mathscr{F}_* = \left\{\frac{1}{f(z)-a_1(f)} \,\middle|\, f(z) \in \mathscr{F}_2\right\}. \tag{3.15}$$

Let $g(z)=1/\{f(z)-a_1(f)\}$ be a function of the family $\mathscr{F}_*$ and set

$$b_1(g) = \infty, \quad b_j(g) = \frac{1}{a_j(f)-a_1(f)} \quad (j = 2,3).$$

Then each of the equations

$$g(z) = b_j(g) \quad (j = 1,2,3)$$

has no root in D. Moreover we have

$$|b_j(g)| < \frac{1}{\delta}, |b_j(g),\infty| > \frac{1}{(1+\frac{1}{\delta^2})^{\frac{1}{2}}} > \frac{\delta}{\sqrt{2}} \quad (j = 2,3)$$

and by Lemma 1. 1,

$$|b_2(g),b_3(g)| = |a_2(f)\text{-}a_1(f),a_3(f)\text{-}a_1(f)| > \frac{1}{2}(\frac{\delta}{2})^2\delta = \frac{\delta^3}{8}.$$

Hence by the result concerning the family $\mathscr{F}_1$, the family $\mathscr{F}_*$ is normal in D. Consequently by Theorem 1. 5, the family $\mathscr{F}_*$ is equicontinuous in D with respect to the spherical distance. Now let $f(z) \in \mathscr{F}_2$ and $z_0 \in D$. By Lemma 1. 1, we have

$$\begin{aligned}\frac{1}{2}(\frac{\delta}{2})^2|f(z),f(z_0)| &\leqslant \frac{1}{2}|a_1(f),\infty|\,|f(z),f(z_0)| \\ &\leqslant |f(z)\text{-}a_1(f),f(z_0)\text{-}a_1(f)| \\ &= \left|\frac{1}{f(z)\text{-}a_1(f)},\frac{1}{f(z_0)\text{-}a_1(f)}\right|.\end{aligned}$$

Hence the family $\mathscr{F}_2$ is equicontinuous in D with respect to the spherical distance, and therefore normal in D.

Since the two families $\mathscr{F}_1$ and $\mathscr{F}_2$ are both normal in D, so the family $\mathscr{F}$ is normal in D.

For the applications of Theorem 3. 9, it is convenient to give first some definitions. Consider a point $a_0 \in \hat{C}$ and a number η $(0<\eta<1)$. Then the set of the points $a \in \hat{C}$ such that $|a,a_0|<\eta$ is called a spherical circle with center a_0 and redius η. Let f(z) be a meromorphic function in a circle $\Gamma: |z-z_0|<r$, and $\eta(0<\eta<1/2)$ a number. We say that Γ is a η-filling circle of f(z), if there do not exist three values $a_j \in \hat{C}$ $(j=1,2,3)$ such that

$$|a_j, a_{j'}| > \eta \quad (1 \leqslant j, j' \leqslant 3, j \neq j')$$

and that f(z) does not take the values $a_j (j=1,2,3)$ in Γ. Then in Γ, f(z) takes every value $a \in \hat{C}$, except at most values a which belong to two spherical circles $|a, a_0| \leqslant \eta$ and $|a, a_0'| \leqslant \eta$.

Consider a non-constant meromorphic function f(z) in **C**. By Theorem 3. 2, the family (3. 2) is not normal in the circle $|z|<2$. let $\eta (0<\eta<1/2)$ be a number. By Theorem 3. 9, there is a positive integer n such that the circle $|z|<2$ is a η-filling circle of the function $f_n(z)$. Consequently the circle $|z|<2^{n+1}$ is a η-filling circle of the function f(z).

Next consider a transcendental meromorphic function f(z) in **C** such that the function $|z|\partial(z,f)$ is unbounded in **C**. By Theorem 3. 6, the family (3. 9) is not normal in the domain $\omega: 1/2<|z|<4$. Let $z_0 \in \omega$ be such that the family (3. 9) is not normal at z_0 and Υ : $|z-z_0|<\delta$ be a circle belonging to ω. Let $\eta(0<\eta<1/2)$ be a number. Then by Theorem 3. 9, Υ is a η-filling circle of the function $f_n(z)$, for an infinite number of positive integers n. Consequently the circle c_n defined by (2. 16) is a η-filling circle of the function f(z), for an infinite number of positive integers n. Similarly we see that for an infinite number of positive integers n, the circle Γ_n defined by (2. 19) is a η-filling circle of the function f(z), where $z_0 \in \omega$ is a point at which the family (2. 18) is not normal.

Finally as an application of Theorem 3. 9, we prove the following theorem:

Theorem 3.10. Given a number $\eta(0 < \eta < 1/2)$, we can find a positive number $\mathbf{M}(\eta)$ depending only on η, such that if f(z) is a meromorphic function in a circle Γ:$|z-z_0| < \mathbf{R}$ satisfying the condition

$$R\partial(z_0, f) > M(\eta), \tag{3. 16}$$

then the circle Γ is a η-filling circle of the function f(z).

Proof. Let $\mathscr{F}$ be the family of the functions f(z) such that f(z) is meromorphic in the circle $|z|<1$ and that the circle $|z|<1$ is not a η-filling circle of f (z). By Theorem 3. 9, the family $\mathscr{F}$ is normal in the circle $|z|<1$. Then by

Theorem 1. 6, we can find a positive number $M(\eta)$ such that for each function $f(z) \in \mathscr{F}$, we have

$$\partial(o, f) \leqslant M(\eta).$$

Consequently if a function $f(z)$ is meromorphic in the circle $|z|<1$ and is such that

$$\partial(o, f) > M(\eta), \tag{3.17}$$

then the circle $|z|<1$ is a η-filling circle of $f(z)$.

Next let $f(z)$ be a meromorphic function in a circle $\Gamma: |z-z_0|<R$ satisfying the condition (3. 16). Then the function $g(\zeta)=f(z_0+R\zeta)(|\zeta|<1)$ is meromorphic in the circle $|\zeta|<1$ and we have

$$\partial(o, g) = R\partial(z_0, f).$$

Hence $\partial(o, g)>M(\eta)$ and the circle $|\zeta|<1$ is a η-filling circle of $g(\zeta)$. Consequently the circle Γ is a η-filling circle of the function $f(z)$.

3. 2. ZALCMAN'S THEOREM

Up to now, we have proved some theorems on the normality of families of holomorphic functions or meromorphic functions. And as applications of these theorems, we have also proved some theorems on entire functions or meromorphic functions in **C**. It is interesting to note that there is a certain correspondence between these two groups of theorems. For instance, to Corollary 1. 3 corresponds Liouville's theorem that a non-constant entire function can not be bounded, to Theorem 2. 1 corresponds Picard's theorem that a non-constant entire function can not omit to take two finite values, and to Theorem 3. 1 corresponds Picard's theorem that a non-constant meromorphic function in **C** can not omit to take three values. Such a correspondence had led Bloch to the idea that, in general, if a property P can not be possessed by non-constant entire functions or non-constant meromorphic functions in **C**, a family of holomorphic functions or meromorphic functions in a domain D, which possess in common the property P in D, is normal in D. This "principle" has been useful for the discovery of new criterions of normality. In connection with it, Zalcman [39] proved rigorously the following theorem:

Theorem 3.11. Let $\mathscr{F}$ be a family of meromorphic functions in the circle $\Delta: |z|<1$. In order that $\mathscr{F}$ be not normal in Δ, it is necessary and sufficient that there exist

1^0 a number $0<r<1$,

2^0 a sequence of points $z_n(n=1,2,\cdots)$ such that $|z_n|<r(n=1,2,\cdots)$,

3^0 a sequence of positive numbers $\varepsilon_n(n=1,2,\cdots)$ such that $\lim\limits_{n\to+\infty}\varepsilon_n=0$,

4^0 a sequence of functions $f_n(z)\in\mathscr{F}(n=1,2,\cdots)$, such that the sequence of functions

$$g_n(\zeta)=f_n(z_n+\varepsilon_n\zeta)\quad(n=1,2,\cdots)\tag{3.18}$$

is locally uniformly convergent to a non-constant meromorphic function $g(\zeta)$ in $\mathbf{C}$, with respect to the spherical distance. More precisely, to each point $\zeta_0\in\mathbf{C}$ correspond a circle $\Upsilon: |\zeta-\zeta_0|<\rho$ and a positive integer n_0 such that for $n\geqslant n_0$, $g_n(\zeta)$ is meromorphic in Υ and as $n\to+\infty$, $g_n(\zeta)$ converges uniformly to $g(\zeta)$ in Υ, with respect to the spherical distance.

Proof. Assume that the family $\mathscr{F}$ is not normal in the circle Δ. Then by Theorem 1.6, there is a number $r_0(0<r_0<1)$ such that the family $\mathscr{F}_*=\{\partial(z,f)\,|\,f(z)\in\mathscr{F}\}$ is not uniformly bounded in the circle $\Gamma_0: |z|\leqslant r_0$. Hence to each integer $m\geqslant1$ corresponds a function $F_m(z)\in\mathscr{F}$ such that

$$\max_{z\in\Gamma_0}\partial(z,F_m)>m.$$

Take a number $r(r_0<r<1)$ and set

$$A_m=\max_{|z|\leqslant r}\left(1-\frac{|z|^2}{r^2}\right)\partial(z,F_m)=\left(1-\frac{|z_m'|^2}{r^2}\right)\partial(z_m',F_m),\tag{3.19}$$

where $|z_m'|\leqslant r$. Since

$$A_m \geqslant \max_{|z|\leqslant r_0}(1-\frac{|z|^2}{r^2})\partial(z,F_m) \geqslant (1-\frac{r_0^2}{r^2})m,$$

we have

$$|z_m'| < r, \lim_{m\to+\infty} A_m = +\infty.$$

Set

$$\lambda_m = \frac{1}{A_m}(1-\frac{|z_m'|^2}{r^2}) = \frac{1}{\partial(z_m',F_m)}. \tag{3.20}$$

We have

$$\frac{\lambda_m}{r-|z_m'|} = \frac{r+|z_m'|}{A_m r^2} < \frac{2}{A_m r}, \quad \lim_{m\to+\infty}\frac{\lambda_m}{r-|z_m'|} = 0.$$

Consider the sequence of functions

$$\varphi_m(\zeta) = F_m(z_m' + \lambda_m\zeta) \quad (m = 1,2,\cdots). \tag{3.21}$$

For each $m\geqslant 1$, the function $\varphi_m(\zeta)$ is meromorphic in the circle $|\zeta|<\rho_m$, where

$$\rho_m = \frac{r-|z_m'|}{\lambda_m}, \quad \lim_{m\to+\infty}\rho_m = +\infty$$

and we have

$$\partial(o,\varphi_m) = \lambda_m\partial(z_m',F_m) = 1. \tag{3.22}$$

Now take a sequence of numbers $R_n(n=1,2,\cdots)$ such that

$$R_1 > 0, R_n < R_{n+1} \quad (n = 1,2,\cdots), \quad \lim_{n\to+\infty} R_n = +\infty.$$

Consider first the circle $|\zeta|<R_1$. Take a positive integer m_0 such that for $m\geqslant$

m_0, we have

$$\rho_m > R_1, \quad \frac{\lambda_m}{r-|z'_m|}R_1 < \frac{1}{2}.$$

Then for $m \geqslant m_0$, the function $\varphi_m(\zeta)$ is meromorphic in the circle $|\zeta|<R_1$ and since in this circle, $|z'_m+\lambda_m\zeta|<r$, we have, by (3. 19) and (3. 20),

$$\begin{aligned}\partial(\zeta,\varphi_m) &= \lambda_m\partial(z'_m+\lambda_m\zeta,F_m) \leqslant \frac{\lambda_m A_m}{1-\dfrac{|z'_m+\lambda_m\zeta|^2}{r^2}} \\ &\leqslant \frac{r+|z'_m|}{r+|z'_m|+\lambda_m R_1}\,\frac{r-|z'_m|}{r-|z'_m|-\lambda_m R_1} < 2.\end{aligned}$$

Hence by Theorem 1. 6, the family $\varphi_m(\zeta)(m \geqslant m_0)$ is normal in the circle $|\zeta|<R_1$. So from the sequence $\varphi_m(\zeta)(m \geqslant m_0)$ we can extract a subsequence $S_1:\varphi_{m_k^1}(\zeta)$ $(k=1,2,\cdots)$which is a C_0-sequence in the circle $|\zeta|<R_1$. Let $\psi_1(\zeta)$ be the limit function of S_1 with respect to the spherical distance. $\psi_1(\zeta)$ is a meromorphic function in the circle $|\zeta|<R_1$ or the constant ∞. But by (3. 22) and Theorem 1. 1, we see that $\partial(o,\psi_1)=1$. It follows that $\psi_1(\zeta)$ is a non-constant meromorphic function in the circle $|\zeta|<R_1$. Similarly from the sequence S_1, we can extract a subsequence $S_2:\varphi_{m_k^2}(\zeta)(k=1,2,\cdots)$ which is a C_0-sequence in the circle $|\zeta|<R_2$. Let $\psi_2(\zeta)$ be the limit function of S_2 with respect to the spherical distance. $\psi_2(\zeta)$ is a non-constant meromorphic function in the circle $|\zeta|<R_2$. Evidently $\psi_1(\zeta)=\psi_2(\zeta)$ in the circle $|\zeta|<R_1$. Continuing this process, we get successively a sequence of sequences $S_n:\varphi_{m_k^n}(\zeta)(k=1,2,\cdots)(n=1,2,\cdots)$ such that, in general, S_{n+1} is a subsequence of S_n, that S_n is a C_0-sequence in the circle $|\zeta|<R_n$ whose limit function $\psi_n(z)$ with respect to the spherical distance is a non-constant meromorphic function in the circle $|\zeta|<R_n$ and that $\psi_n(\zeta)=\psi_{n+1}(\zeta)$ in the circle $|\zeta|<R_n$. By means of the sequence of functions $\psi_n(\zeta)$ $(n=1,2,\cdots)$, we can then define a non-constant meromorphic function $g(\zeta)$ in **C** such that $g(\zeta)=\psi_n(\zeta)$ in the circle $|\zeta|<R_n$, for each $n \geqslant 1$. Consider the diagonal sequence $\varphi_{m_n^n}(\zeta)(n=1,2,\cdots)$. Then for each $n \geqslant 1$, the functions $\varphi_{m_p^p}(\zeta)(p \geqslant n)$ are meromorphic in the circle $|\zeta|<R_n$ and as $p \to +\infty$, $\varphi_{m_p^p}(\zeta)$ converges locally uniformly to $g(\zeta)$ in the circle $|\zeta|<R_n$ with respect to the spherical distance. Finally set

$$z_n = z'_{m_n^*}, \quad \varepsilon_n = \lambda_{m_n^*}, \quad f_n(z) = F_{m_n^*}(z).$$

Then the number r, the sequence of points $z_n(n=1,2,\cdots)$, the sequence of positive numbers $\varepsilon_n(n=1,2,\cdots)$ and the sequence of functions $f_n(z)(n=1,2,\cdots)$ satisfy the required condition in Theorem 3. 11, by (3. 21).

We have proved that the condition is necessary. Conversely assume that there exist $r, z_n, \varepsilon_n\, f_n(z)(n=1,2,\cdots)$ satisfying the required condition in Theorem 3. 11. Suppose that the family $\mathscr{F}$ is normal in Δ. Then by Theorem 1. 6 and the finite covering theorem, we can find a positive number M such that for each function $f(z)\in\mathscr{F}$, we have

$$\partial(z, f) \leqslant M \tag{3. 23}$$

in the circle $|z|\leqslant(1+r)/2$, where r is the number in Theorem 3. 11. Consider a point $\zeta_0\in\mathbf{C}$. Then we can find a circle $\Upsilon: |\zeta-\zeta_0|<\rho$ and an integer $n_0\geqslant1$ such that for $n\geqslant n_0$ the function $g_n(\zeta)$ defined by (3. 18) is meromorphic in Υ and as $n\to+\infty$, $g_n(\zeta)$ converges uniformly to $g(\zeta)$ in Υ, with respect to the spherical distance. By Theorem 1. 1, we have

$$\lim_{n\to+\infty}\partial(\zeta_0, g_n) = \partial(\zeta_0, g). \tag{3. 24}$$

On the other hand, from (3. 18), we have

$$\partial(\zeta_0, g_n) = \varepsilon_n\partial(z_n + \varepsilon_n\zeta_0, f_n).$$

When n is sufficiently large, we have $|z_n+\varepsilon_n\zeta_0|<(1+r)/2$, and by (3. 23)

$$\partial(\zeta_0, g_n) \leqslant \varepsilon_n M. \tag{3. 25}$$

(3. 24 and (3. 25) yield $\partial(\zeta_0, g)=0$. Since ζ_0 is arbitrary, this is incompatible with the condition that $g(\zeta)$ is non-constant.

Definition 3. 3. Let f(z) be a meromorphic function in a domain D and $a\in\hat{\mathbf{C}}$ a value. Let $m\geqslant2$ be an integer or $m=+\infty$. We say that a is an exceptional value of weight $1-\frac{1}{m}$ of $f(z)$ in D, if in D the equation $f(z)=a$ has no root whose order of multiplicity is less than m. In particular, if $m=+\infty$, an exceptional value a of weight 1 of $f(z)$ in D is such that the equation $f(z)=a$ has no root in D.

As an application of Theorem 3.11, we give a proof of the following theorem due to Bloch [3], which generalizes those obtained by Carathéodory [5,6] and Montel [26]:

Theorem 3.12. Let D be a domain. Let $a_j \in \hat{\mathbf{C}}$ $(j=1,2,\cdots,q; 3\leqslant q\leqslant 5)$ be q distinct values and $m_j \geqslant 2 (j=1,2,\cdots,q)$ be such that for each $1\leqslant j\leqslant q$, m_j is an integer or $+\infty$, and that

$$\sum_{j=1}^{q}\left(1-\frac{1}{m_j}\right) > 2. \tag{3.26}$$

Let $\mathscr{F}$ be the family of the functions $f(z)$ satisfying the following conditions:

1^0 $f(z)$ is meromorphic in D.

2^0 For each $1\leqslant j\leqslant q$, the value a_j is an exceptional value of weight $1-\frac{1}{m_j}$ of $f(z)$ in D.

Then the family $\mathscr{F}$ is normal in D.

Proof. Consider first the case that D is the circle $\Delta: |z|<1$. Assume that the family $\mathscr{F}$ is not normal in Δ. Then by Theorem 3.11, there exist r, z_n, ε_n and $f_n \in \mathscr{F} (n=1,2,\cdots)$ satisfying the condition in Theorem 3.11. Consider the sequence of functions $g_n(\zeta)$ $(n=1,2,\cdots)$ and the non-constant meromorphic function $g(\zeta)$ in $\mathbf{C}$. Consider a value a_j. We are going to show that a_j is an exceptional value of weight $1-\frac{1}{m_j}$ of $g(\zeta)$ in $\mathbf{C}$. In fact, distinguish two cases:

1) $a_j \neq \infty$. Assume that the equation $g(\zeta)=a_j$ has a root ζ_0 whose order of multiplicity is less than m_j. Then by Theorem 1.1, we can find a circle $\Gamma_0: |\zeta-\zeta_0| < \rho_0 (0<\rho_0\leqslant\rho)$ and a positive integer $n_0'(n_0'\geqslant n_0)$ such that the functions $g_n(\zeta)$ $(n\geqslant n_0')$ and $g(\zeta)$ are holomorphic in Γ_0, and as $n\to+\infty$, $g_n(\zeta)$ converges uniformly to $g(\zeta)$ in Γ_0. Since $g(\zeta)$ is non-constant, we can find a circle $\overline{\Gamma}_0': |\zeta-\zeta_0| \leqslant\rho_0' (0<\rho_0'<\rho_0)$ in which the equation $g(\zeta)=a_j$ has only the root ζ_0. Then by the

identity

$$g_n(\zeta)-a_j = \{g_n(\zeta)-g(\zeta)\} + \{g(\zeta)-a_j\}$$

and Rouché's theorem, we see that, when n is sufficiently large, the number of the roots (with due count of multiplicity) of the equation $g_n(\zeta)=a_j$ in the circle $\gamma'_0: |\zeta-\zeta_0|<\rho'_0$ is μ, where μ is the order of multiplicity of the root ζ_0 of the equation $g(\zeta)=a_j$. Since $\mu<m_j$, hence in the circle γ'_0 the equation $g_n(\zeta)=a_j$ has a root ζ_* whose order of multiplicity $\mu_*<m_j$. By (3.18), the point $z_*=z_n+\varepsilon_n\zeta_*$ is a root of order of multiplicity μ_* in Δ of the equation $f_n(z)=a_j$. This is incompatible with the condition 2^0 in Theorem 3.12.

2) $a_j=\infty$. Assume that the equation $g(\zeta)=\infty$ has a root ζ_0 whose order of multiplicity is less than m_j. Then by Theorem 1.1, we can find a circle $\gamma_0: |\zeta-\zeta_0|<\rho_0 (0<\rho_0\leqslant\rho)$ and a positive integer $n'_0 (n'_0\geqslant n_0)$ such that the functions $1/g_n(\zeta) (n\geqslant n'_0)$ and $1/g(\zeta)$ are holomorphic in γ_0, and as $n\to+\infty$, $1/g_n(\zeta)$ converges uniformly to $1/g(\zeta)$ in γ_0. As in the first case, we see that, when n is sufficiently large, the equation $1/g_n(\zeta)=0$ has a root ζ_* whose order of multiplicity $\mu_*<m_j$, in a circle $\gamma'_0: |\zeta-\zeta_0|<\rho'_0 (0<\rho'_0<\rho_0)$. Since ζ_* is a root of order of multiplicity μ_* of the equation $g_n(\zeta)=\infty$, the point $z_*=z_n+\varepsilon_n\zeta_*$ is a root of order of multiplicity μ_* in Δ of the equation $f_n(z)=\infty$, by (3.18). Again this is incompatible with the condition 2^0 in Theorem 3.12.

Thus we have shown that for each $1\leqslant j\leqslant q$, the value a_j is an exceptional value of weight $1-\frac{1}{m_j}$ of $g(\zeta)$ in **C**. This contradicts Theorem 1 in Appendix B.

Next consider the case that D is a circle $\Gamma: |z-z_0|<r$. This case can be reduced to the case of the unit circle $\Delta: |w|<1$ by introducing the family

$$\mathscr{F}_1 = \{f(z_0+rw) | f(z)\in\mathscr{F}\}.$$

We see that the family $\mathscr{F}$ is normal in Γ.

Finally consider the general case of a domain D. Let z_0 be a point of D and $\Gamma: |z-z_0|<r$ a circle belonging to D. Evidently the functions f(z) of the family $\mathscr{F}$ satisfy à fortiori the conditions 1^0 and 2^0 in Theorem 3.12, in Γ. Hence $\mathscr{F}$ is normal in Γ. Since z_0 is arbitrary, $\mathscr{F}$ is normal in D.

As Theorem 3.1, Theorem 3.12 also has similar applications which we omit.

In the case $q=3$, the condition (3.26) may be written in the form

$$\sum_{j=1}^{3} \frac{1}{m_j} < 1.$$

If furthermore $m_j=\infty (j=1,2,3)$, we get Theorem 3.1.

Theorem 3.12 can be further generalized. For this we need the following lemma:

Lemma 3.2. Let $a_j^{(n)} \in \hat{C} (n=1,2,\cdots)(j=1,2,\cdots,q)$ be q sequences of points. Then we can find an increasing sequence of positive integers $n_p (p=1,2,\cdots)$ such that for each $1 \leqslant j \leqslant q$, the sequence $a_j^{(n_p)} (p=1,2,\cdots)$ satisfies one of the following two conditions:

1^0 $a_j^{(n_p)} \in C (p=1,2,\cdots)$, $\lim_{p \to +\infty} a_j^{n_p} = A_j (A_j \in \hat{C})$;

2^0 $a_j^{(n_p)} = \infty (p=1,2,\cdots)$, we write $\lim_{p \to +\infty} a_j^{n_p} = A_j (A_j = \infty)$.

In particular, if there is a number $\delta (0<\delta<1)$ such that for each $n \geqslant 1$, the q points $a_j^{(n)} (j=1,2,\cdots,q)$ satisfy the condition

$$|a_j^{(n)}, a_{j'}^{(n)}| > \delta \quad (1 \leqslant j, j' \leqslant q, j \neq j'),$$

then

$$|A_j, A_{j'}| \geqslant \delta \quad (1 \leqslant j, j' \leqslant q, j \neq j'). \tag{3.27}$$

Proof. Consider first the case $q=1$, and a sequence of points $b_n \in \hat{C} (n=1, 2,\cdots)$. Distinguish two cases:

1) $b_n \in C$ for an infinite number of positive integers n. In this case, we first find an increasing sequence of positive integers $m_k (k=1,2,\cdots)$ such that $b_{m_k} \in C$ $(k=1,2,\cdots)$ and then find an increasing sequence of positive integers $k_p (p=1, 2,\cdots)$ such that

$$\lim_{p\to+\infty} b_{n_p} = B, B\in\hat{C},\ n_p = m_{k_p} (p=1,2,\cdots).$$

2) $b_n = \infty (n \geqslant n_0)$. In this case, we take $n_p = n_0 + p$ $(p=1,2,\cdots)$.
Thus the existence of $n_p (p=1,2,\cdots)$ is proved for the case $q=1$. Now assume that this is true for an $q\geqslant 1$, and let us prove that it is also true for $q+1$. In fact, consider $q+1$ sequences of points $a_j^{(n)} \in \hat{C} (n=1,2,\cdots)(j=1,2,\cdots,q+1)$. By assumption, for the first q sequences $a_j^{(n)} (n=1,2,\cdots)(j=1,2,\cdots,q)$ we can find $n_p (p=1,2,\cdots)$ having the property in Lemma 3. 2. Consider the sequence $b_p = a_{q+1}^{(n_p)} (p=1,2,\cdots)$. By the case just treated, we can find an increasing sequence of positive integers $p_s (s=1,2,\cdots)$ such that the sequence $b_{p_s} (s=1,2,\cdots)$ satisfies one of the conditions 1^0 and 2^0 in Lemma 3. 2. Consequently the increasing sequence of positive integers $m_s = n_{p_s} (s=1,2,\cdots)$ has the required property for the $q+1$ sequences $a_j^{(n)} (n=1,2,\cdots)(j=1,2,\cdots,q+1)$.

The last part of Lemma 3. 2 follows from the inequality

$$\delta < |a_j^{(n_p)}, a_{j'}^{(n_p)}| \leqslant |a_j^{(n_p)}, A_j| + |A_j, A_{j'}| + |A_{j'}, a_{j'}^{(n_p)}|$$

and

$$\lim_{p\to+\infty} |a_j^{(n_p)}, A_j| = 0, \quad \lim_{p\to+\infty} |a_{j'}^{(n_p)}, A_{j'}| = 0.$$

Theorem 3. 13. Let D be a domain. Let $m_j \geqslant 2 (j=1,2,\cdots,q; 3\leqslant q\leqslant 5)$ be such that for each $1\leqslant j\leqslant q$, m_j is an integer or $+\infty$, and that the condition (3. 26) is satisfied. Let $\delta (0<\delta<1/q)$ be a number. Let e the family of the functions $f(z)$ satisfying the following conditions:

1^0 $f(z)$ is meromorphic in D;

2^0 There are q values $a_j(f) \in \hat{C} (j=1,2,\cdots,q)$ such that

$$|a_j(f), a_{j'}(f)| > \delta \quad (1\leqslant j, j' \leqslant q, j\neq j')$$

and that for each $1\leqslant j\leqslant q$, the value $a_j(f)$ is an exceptional value of weight $1-\frac{1}{m_j}$ of $f(z)$ in D.

Then the family $\mathscr{F}$ is normal in D.

Proof. In view of the proof of Theorem 3.12, it is sufficient to prove Theorem 3.13 in the case that D is the circle $\Delta: |z|<1$. Assume that the family $\mathscr{F}$ is not normal in Δ. Then by Theorem 3.11, there exist r, z_n, ε_n and $f_n(z)\in\mathscr{F}(n=1,2,\cdots)$ satisfying the condition in Theorem 3.11. Consider the sequence of functions $g_n(\zeta)(n=1,2,\cdots)$ defined by (3.18) and the non-constant meromorphic function $g(\zeta)$ in **C** set

$$a_j^{(n)} = a_j(f_n) \quad (n = 1,2,\cdots) \quad (j = 1,2,\cdots,q).$$

By Lemma 3.2, we can find an increasing sequence of positive integers $n_p(p=1,2,\cdots)$ having the property in Lemma 3.2. The corrsponding limits $A_j\in\hat{\mathbf{C}}$ $(j=1,2,\cdots,q)$ are distinct, because by the condition 2^0 in Theorem 3.13, we have (3.27).

As in the proof of Theorem 3.12, it is sufficient to show that for each $1\leqslant j\leqslant q$, the value A_j is an exceptional value of weight $1-\frac{1}{m_j}$ of $g(\zeta)$ in **C**. Consider an integer $1\leqslant j\leqslant q$ and distinguish three cases:

1) $A_j\neq\infty$. Then $a_j^{(n_p)}\in\mathbf{C}(p=1,2,\cdots)$. Assume that the equation $g(\zeta)=A_j$ has a root ζ_0 whose order of multiplicity μ is less than m_j. Then by Theorem 1.1, we can find a circle $\Gamma_0: |\zeta-\zeta_0|<\rho_0(0<\rho_0\leqslant\rho)$ and a positive integer $n_0'(n_0'\geqslant n_0)$ such that the functions $g_n(\zeta)(n\geqslant n_0')$ and $g(\zeta)$ are holomorphic in Γ_0, and as $n\to+\infty$, $g_n(\zeta)$ converges uniformly to $g(\zeta)$ in Γ_0. Let $\overline{\Gamma}_0': |\zeta-\zeta_0|\leqslant\rho_0'(0<\rho_0'<\rho_0)$ be a circle in which the equation $g(\zeta)=A_j$ has only the root ζ_0. Then by the identity

$$g_{n_p}(\zeta)-a_j^{(n_p)} = \{g_{n_p}(\zeta)-g(\zeta)\} + \{g(\zeta)-A_j\} + \{A_j-a_j^{(n_p)}\}$$

and Rouché's theorem, we see that, when p is sufficiently large, the number of the roots (with due count of multiplicity) of the equation $g_{n_p}(\zeta)=a_j^{(n_p)}$ in the circle $\Gamma_0': |\zeta-\zeta_0|<\rho_0'$ is μ. Hence the equation $g_{n_p}(\zeta)=a_j^{(n_p)}$ has a root ζ_* whose order of multiplicity $\mu_*\leqslant\mu<m_j$. By (3.18), the point $z_*=z_{n_p}+\varepsilon_{n_p}\zeta_*$ is a root of order of multiplicity μ_* in Δ of the equation $f_{n_p}(z)=a_j^{(n_p)}$, contrary to the condition 2^0 in Theorem 3.13.

2) $A_j=\infty$ and $a_j^{(n_p)}\in\mathbf{C}(p=1,2,\cdots)$. Assume that the equation $g(\zeta)=\infty$ has a root ζ_0 whose order of multiplicity μ is less than m_j. Then by Theorem 1.1, we

can find a circle $\gamma_0: |\zeta-\zeta_0|<\rho_0(0<\rho_0\leqslant\rho)$ and a positive integer $n_0'(n_0'\geqslant n_0)$ such that the functions $1/g_n(\zeta)(n\geqslant n_0')$ and $1/g(\zeta)$ are holomophic in γ_0, and as $n\to+\infty$, $1/g_n(\zeta)$ converges uniformly to $1/g(\zeta)$ in γ_0. Then by the identity

$$\frac{1}{g_{n_p}(\zeta)}-\frac{1}{a_j^{(n_p)}}=\{\frac{1}{g_{n_p}(\zeta)}-\frac{1}{g(\zeta)}\}+\frac{1}{g(\zeta)}-\frac{1}{a_j^{(n_p)}}$$

and Rouché's theorem, we see that, when p is sufficiently large, the equation $g_{n_p}(\zeta)=a_j^{(n_p)}$ has a root ζ_* whose order of multiplicity is less than m_j, in a circle $\gamma_0': |\zeta-\zeta_0|<\rho_0'(0<\rho_0'<\rho_0)$. Again by (3.18) we get a contradiction.

3) $A_j=\infty$ and $a_j^{(n_p)}=\infty(p=1,2,\cdots)$. Assume that the equation $g(\zeta)=\infty$ has a root ζ_0 whose order of multiplicity μ is less than m_j. Then as in the second case, by the identity

$$\frac{1}{g_{n_p}(\zeta)}=\{\frac{1}{g_{n_p}(\zeta)}-\frac{1}{g(\zeta)}\}+\frac{1}{g(\zeta)}$$

and Rouché's theorem, we see that, when p is sufficiently large, the equation $1/g_{n_p}(\zeta)=0$ has a root ζ_* whose order of multiplicity μ_* is less than m_j, in a circle $\gamma_0': |\zeta-\zeta_0|<\rho_0'$. Again this leads to a contradiction by (3.18).

In particular, if we take in Theorem 3.13 $q=3$ and $m_j=\infty$ $(j=1,2,3)$, we get Theorem 3.9.

Let $m_j\geqslant 2(j=1,2,\cdots q;3\leqslant q\leqslant 5)$ satisfying the conditions in Theorem 3.13 and $\eta(0<\eta<1/q)$ a number. Let $f(z)$ be a meromorphic function in a circle $\Gamma: |z-z_0|<r$. We say that Γ is a $(m_1,m_2,\cdots,m_q;\eta)$-filling circle of $f(z)$, if there do not exist q values $a_j\in\hat{C}$ $(j=1,2,\cdots,q)$ such that

$$|a_j,a_{j'}|>\eta \quad (1\leqslant j,j'\leqslant q, j\neq j')$$

and that for each $1\leqslant j\leqslant q$, the value a_j is an exceptional value of weight $1-\frac{1}{m_j}$ of $f(z)$ in Γ. Then either for every value $a\in\hat{C}$ the equation $f(z)=a$ has a root in Γ whose order of multiplicity is less than m_1; or there is a value $\alpha_1\in\hat{C}$ which is an exceptional value of weight $1-\frac{1}{m_1}$ of $f(z)$ in Γ. Describe the spherical circle $c_1: |a,\alpha_1|\leqslant\eta$. Either for every value $a\in\hat{C}-c_1$ the equation $f(z)=a$ has a root in Γ whose order of multiplicity is less than m_2; or there is a value $\alpha_2 \in \hat{C}-c_1$ which is an exceptional value of weight $1-\frac{1}{m_2}$. Continuing this process, after at most q-1 steps, we get q-1 values $\alpha_j(j=1,2,\cdots,q-1)$ such that

$$|\alpha_j,\alpha_{j'}| > \eta \quad (1 \leqslant j,j' \leqslant q-1, j \neq j'),$$

that for each $1\leqslant j\leqslant q-1$, the value α_j is an exceptional value of weight $1-\frac{1}{m_j}$ of f (z) in Γ, and that for every value $a\in \hat{C}-\bigcup_{j=1}^{q-1}(|a,\alpha_j|\leqslant\eta)$ the equation $f(z)=a$ has a root in Γ whose order of multiplicity is less than m_q.

Interesting particular cases are: $q=5$, $m_j=2(j=1,2,3,4,5)$; $q=4, m_j=3(j=1,2,3,4)$; $q=3, m_j=4(j=1,2,3)$.

Theorem 3. 14. Given $m_j \geqslant 2 \quad (j=1,2,\cdots,q; 3\leqslant q\leqslant 5)$ satisfying the conditions in Theorem 3. 13 and a number $\eta(0<\eta<1/q)$, we can find a positive number $A(m_1,m_2,\cdots,m_q;\eta)$ depending only on $m_j(j=1,2,\cdots,q)$ and η, such that if f(z) is a meromorphic function in a circle $\Gamma: |z-z_0|<R$ satisfying the condition

$$R\partial(z_0,f) > A(m_1,m_2,\cdots,m_q;\eta), \tag{3.28}$$

then the circle Γ is a $(m_1,m_2,\cdots,m_q;\eta)$-filling circle of $f(\zeta)$.

Proof. Let $\mathscr{F}$ be the family of the functions f(z) such that f(z) is meromorphic in the circle $\Delta: |z|<1$ and that Δ is not a $(m_1,m_2,\cdots,m_q;\eta)$-filling circle of f(z). By Theorem 3. 13, the family $\mathscr{F}$ is normal in Δ. Then by Theorem 1. 6, we can find a positive number $A=A(m_1,m_2,\cdots,m_q;\eta)$ such that for each function $f(z)\in\mathscr{F}$, we have

$$\partial(o,f) \leqslant A.$$

Consequently if a function f(z) is meromorphic in Δ and is such that

$$\partial(o,f) > A,$$

then Δ is a $(m_1,m_2,\cdots,m_q;\eta)$-filling circle of $f(z)$. Then the proof of Theorem 3.14 is completed as in that of Theorem 3.10.

Consider again $m_j\geqslant 2(j=1,2,\cdots,q;3\leqslant q\leqslant 5)$ satisfying the conditions in Theorem 3.13 and a number $\eta(0<\eta<1/q)$. Let $f(z)$ be a meromorphic function in a domain D. We say that D is a $(m_1,m_2,\cdots,m_q;\eta)$-filling domain of $f(z)$, if the conditions given above in the definition of $(m_1,m_2,\cdots m_q;\eta)$-filling circle are satisfied in replacing Γ by D. On the other hand, let $S\subset\hat{C}$ be a set of points, and $m\geqslant 2$ an integer or $m=+\infty$. We say that the values of the function $f(z)$ cover S with order of multiplicity less than m, if for each value $a\in S$ the equation $f(z)=a$ has a root of order of multiplicity less than m in D. In the particular case $m=+\infty$, this means simply that the values of $f(z)$ cover S.

Let S_1, S_2 be two sets of points of $\hat{C}$. We define the spherical distance $d(S_1, S_2)$ between S_1 and S_2 by the formula

$$d(S_1,S_2)=inf\{|a_1,a_2|\,|a_1\in S_1,a_2\in S_2\}.$$

Let $S_j(j=1,2,\cdots,q)$ be q sets of points of $\hat{C}$ such that

$$d(S_j,S_{j'})>\eta \quad (1\leqslant j,j'\leqslant q,j\neq j'). \tag{3.29}$$

Evidently if D is a $(m_1,m_2,\cdots,m_q;\eta)$-filling domain of $f(z)$, then among the sets $S_j(j=1,2,\cdots,q)$ there is at least one set S_j which is covered by the values of $f(z)$ with order of multiplicity less than m_j.

Theorem 3.15. Given a domain D, two bounded closed sets $\sigma_j(j=1,2)$ of points belonging to D, a value $w_0\in\hat{C}$, a number $\delta(0<\delta\leqslant 1)$, $m_j\geqslant 2(j=1,2,\cdots,q;3\leqslant q\leqslant 5)$ satisfying the conditions in Theorem 3.13 and a number $\eta(0<\eta<1/q)$, we can find a number $\alpha>0$ depending only on D, $\sigma_j(j=1,2)$, $w_0,\delta,m_j(j=1,2,\cdots,q)$ and η having the following property: If $f(z)$ is a meromorphic function in D such that $f(z)$ does not take the value w_0 in D and that

$$\max_{z\in\sigma_1}|f(z),w_0|\geqslant\delta,\quad \min_{z\in\sigma_2}|f(z),w_0|<\alpha, \tag{3.30}$$

then D is a $(m_1,m_2,\cdots,m_q;\eta)$-filling domain of $f(z)$. Consequently, among any q sets $S_j(j=1,2,\cdots,q)$ of points of $\hat{C}$ satisfying the condition (3. 29), there is at least one set S_j which is covered by the values of f(z) with order of multiplicity less than m_j.

Proof. Let $\mathscr{F}$ be the family of meromorphic functions f(z) in D such that D is not a $(m_1,m_2,\cdots,m_q;\eta)$-filling domain of f(z), that f(z) does not take the value w_0 in D and that the first inequality in (3. 30) is satisfied. By Theorem 3. 13, the family $\mathscr{F}$ is normal in D. Then by Lemma 3. 1, there is a number $\alpha>0$ such that for each function $f(z)\in\mathscr{F}$, we have

$$\min_{z\in\sigma_2}|f(z),w_0|\geqslant\alpha.$$

This number α has the required property.

Note that the condition (3. 30) is a condition on the variation of the spherical distance $|f(z),w_0|$. In particular for holomorphic functions f(z), we can take $w_0=\infty$, then the condition (3. 30) is a condition on the variation of $|f(z)|$.

As Theorem 3. 9, Theorem 3. 13 also has similar applications which we also omit.

3. 3. GU'S THEOREM

Theorem 3. 16. Let D be a domain, a and $b\neq0$ two complex numbers, and $\nu\geqslant1$ an integer. Let $\mathscr{F}$ be the family of the functions f(z) meromorphic in D and such that each of the equations

$$f(z)=a,\quad f^{(\nu)}(z)=b$$

has no root in D. Then the family $\mathscr{F}$ is normal in D.

In the case $a=0,b=1$, this theorem was conjectured by Hayman and proved by Gu [17]. It is a remarkable extension of Theorem 2. 10 to the case of families of meromorphic functions.

Proof. Consider first the case $a=0$, $b=1$. Let z_0 be a point of D and $\Gamma:|z-z_0|<\rho$ a circle belonging to D. Let f(z) be a function of the family $\mathscr{F}$ and define $F(\zeta)=\frac{1}{\rho^\nu}f(z_0+\rho\zeta)\,(|\zeta|<1)$. By Theorem 2 in Appendix B, we can find a positive number K_ν depending only on ν such that in the circle $|\zeta|<\frac{1}{32}$ one of the two inequalities

$$|F(\zeta)|<K_\nu, \quad \left|\frac{1}{F(\zeta)}\right|<K_\nu$$

holds. Set

$$K'_\nu = K_\nu max(\rho^\nu, \frac{1}{\rho^\nu}),$$

then in the circle $|z-z_0|<\frac{1}{32}\rho$, one of the two inequalities

$$|f(z)|<K'_\nu, \quad \left|\frac{1}{f(z)}\right|<K'_\nu$$

holds. Hence by Corollary 1.4, the family $\mathscr{F}$ is normal in D.

Now consider the general case. By the particular case just treated, the family $\mathscr{F}_1$ of the functions

$$g(z)=\frac{f(z)-a}{b}, \quad f(z)\in\mathscr{F}$$

is normal in D. By Corollary 1.4, the family $\mathscr{F}_2$ of the functions

$$bg(z)=f(z)-a, \quad f(z)\in\mathscr{F}$$

is normal in D, hence, by Theorem 1.5, $\mathscr{F}_2$ is equicontinuous in D with respect to the spherical distance. Since, by Lemma 1.1, we have

$$|f(z)-a, f(z_0)-a|\geqslant\frac{1}{2}|a,\infty|^2|f(z), f(z_0)|,$$

the family $\mathscr{F}$ is also equicontinuous in D with respect to the spherical distance, and hence normal in D.

Next we give an extension of Theorem 3. 16.

Theorem 3. 17. Let D be a domain, M, $\delta(0<\delta<M)$ two positive numbers and $\nu\geqslant 1$ an integer. Let $\mathscr{F}$ be the family of the functions f(z) satisfying the following conditions:

1^0 f(z) is meromorphic in D.

2^0 There are two values a(f) and b(f) such that

$$|a(f)|<M, \quad \delta<|b(f)|<M$$

and that each of the equations

$$f(z)=a(f), \quad f^{(\nu)}(z)=b(f)$$

has no root in D.

Then the family $\mathscr{F}$ is normal in D.

This theorem can be proved by the same method used in the proof of Theorem 3. 16, by introducing first the family $\mathscr{F}_1$ of the functions

$$g(z)=\frac{f(z)-a(f)}{b(f)}, \quad f(z)\in\mathscr{F},$$

and then the family $\mathscr{F}_2$ of the functions

$$b(f)g(z)=f(z)-a(f), f(z)\in\mathscr{F},$$

and finally applying Lemma 1. 1.

As an application of Theorem 3. 17, we have the following theorem:

Theorem 3. 18. Given a domain D, a bounded closed set σ of points belonging to D, two positive numbers M, $\delta(0<\delta<M)$ and an integer $\nu\geqslant 1$, we can find a positive number $\beta(D,\sigma,M,\delta,\nu)$ depending only on D,σ,M,δ,ν having the following property: If f(z) is a meromorphic function in D such that

$$\max_{z\in\sigma} \partial(z,f) > \beta(D,\sigma,M,\delta,\nu), \tag{3.31}$$

then in D either f(z) takes every value w of the circle $|w|<M$ or $f^{(\nu)}(z)$ takes every value w of the domain $\delta<|w|<M$.

Proof. Let $\mathscr{F}$ be the family of the functions f(z) satisfying the conditions 1^0 and 2^0 in Theorem 3.17. Since $\mathscr{F}$ is normal in D, by Theorem 1.6, we can find a positive number $\beta=\beta(D,\sigma,M,\delta,\nu)$ such that for each function $f(z)\in\mathscr{F}$, we have

$$\max_{z\in\sigma} \partial(z,f) \leqslant \beta.$$

This number β evidently has the required property.

Now consider a transcendental meromorphic function f(z) in **C**. We are going to find conditions under which the family

$$f_n(z) = \frac{f(2^n z)}{2^{n\nu}} \quad (n = 1,2,\cdots) \tag{3.32}$$

of meromorphic functions in **C** is not normal in the domain $\omega: 1/2<|z|<4$. Define

$$\mu(r,f) = \min_{|z|=r} |f(z)| \quad (0<r<+\infty). \tag{3.33}$$

We say that the function f(z) satisfies the condition (C), if there is a set s of points of the interval $1/2<t<4$, which consists of an infinite number of points and has a point of accumulation t_0 $(1/2<t_0<4)$, such that for each $t\in s$, we have

$$\lim_{n\to+\infty} \frac{\mu(2^n t,f)}{(2^n t)^\nu} = 0. \tag{3.34}$$

Evidently if

$$\lim_{r\to+\infty} \frac{\mu(r,f)}{r^\nu} = 0, \tag{3.35}$$

then f(z) satisfies the condition (C). In particular if there is a continuous curve $z=z(t)(0\leqslant t<+\infty)$ with $\lim\limits_{t\to+\infty} z(t)=\infty$, such that the function $f\{z(t)\}$ is bounded when t is sufficiently large, $t\geqslant t_0$, then (3. 35) holds. À fortiori, if f(z) has a finite asymptotic value, then (3. 35) holds.

Theorem 3. 19. Let f(z) be a transcendental meromorphic function in **C**. If f(z) satisfies the condition (C), then the family (3. 32) is not normal in the domain $\omega: 1/2<|z|<4$.

Proof. Distinguish two cases:

1) f(z) has only a finite number of poles. In this case we can find a positive number R such that f(z) is holomorphic for $R<|z|<+\infty$ and has an essential singularity at the point $z=\infty$. Then by the method used in the proof of Theorem 2. 12, we see that the family (3. 32) is not normal in ω.

2) f(z) has an infinite number of poles. In this case we can find a subsequence $f_{n_k}(z)(k=1,2,\cdots)$ of the sequence (3. 32), such that for each k, the function $f_{n_k}(z)$ has at least one pole in the region $E: 1\leqslant|z|\leqslant 2$. Now assume that the family (3. 32) is normal in ω. Then from the sequence $f_{n_k}(z)(k=1,2,\cdots)$ we can extract a subsequence $f_{m_h}(z)$ $(h=1,2,\cdots)$ which is a C_0-sequence in ω. Let F (z) be its limit function with respect to the spherical distance. Consider the set s defined above and let $t\in s$. Then as $h\to+\infty$, $f_{m_h}(z)$ converges uniformly to F(z) on the circle $|z|=t$ with respect to the spherical distance. On the other hand, by (3. 32), we have

$$\min_{|z|=t}|f_{m_h}(z)| = \frac{1}{2^{m_h\nu}}\mu(2^{m_h}t,f).$$

Then from (3. 34) and the inequality

$$|F(z),0| \leqslant |F(z),f_{m_h}(z)| + |f_{m_h}(z),0|,$$

we see that we must have

$$\min_{|z|=t}|F(z),0| = 0$$

and hence on the circle $|z|=t$, $F(z)$ has a zero. It follows that $F(z)\equiv 0$, but this is incompatible with the fact that, as $h\to+\infty$, $f_{m_h}(z)$ converges uniformly to $F(z)$ in the region E with respect to the spherical distance and that for each h, the function $f_{m_h}(z)$ has at least one pole in E.

Consider again a transcendental meromorphic function $f(z)$ in **C** and the family (2. 36) of meromorphic functions in **C**. We see easily that Theorem 3. 19 is also true for the family (2. 36). We have the following theorem:

Theorem 3. 20. Let $f(z)$ be a transcendental meromorphic function in **C**. If $f(z)$ satisfies the condition (C), then the family (2. 36) is not normal in the domain $\omega: 1/2<|z|<4$.

On the basis of Theorems 3. 19 and 3. 20, we can give applications of Theorems 3. 16 and 3. 17 to transcendental meromorphic functions in **C**, similar to those of Theorems 2. 10 and 2. 16 to entire functions. In the proofs we need the following lemma:

Lemma 3. 3. Let $f_n(z)(n=1,2,\cdots)$ be a sequence of meromorphic functions in a domain D and $a_n(n=1,2,\cdots)$ be a bounded sequence of complex numbers. If the family $\mathscr{F}$: $f_n(z)(n=1,2,\cdots)$ is normal in D, then the family $\mathscr{F}_1$: $f_n(z)+a_n(n=1,2,\cdots)$ is also normal in D.

Proof. Since $\mathscr{F}$ is normal in D, by Theorem 1. 5, $\mathscr{F}$ is equicontinuous in D with respect to the spherical distance. Let

$$|a_n|\leqslant M \quad (n=1,2,\cdots).$$

By Lemma 1. 1,

$$\begin{aligned}|f_n(z),f_n(z_0)| &= |(f_n(z)+a_n)-a_n,(f_n(z_0)+a_n)-a_n|\\ &\geqslant \frac{1}{2}|a_n,\infty|^2|f_n(z)+a_n,f_n(z_0)+a_n|\\ &\geqslant \frac{1}{2}\frac{1}{1+M^2}|f_n(z)+a_n,f_n(z_0)+a_n|,\end{aligned}$$

hence the family $\mathscr{F}_1$ is also equicontinuous in D with respect to the spherical distance, and therefore normal in D.

4

CLOSED FAMILIES OF MEROMORPHIC FUNCTIONS

4. 1. CLOSED FAMILIES OF HOLOMORPHIC FUNCTIONS

Definition 4. 1. Let $\mathscr{F}$ be a family of holomorphic functions in a domain D. We say that $\mathscr{F}$ is closed in D, if from every sequence of functions $f_n(z)(n=1,2,\cdots)$ of $\mathscr{F}$, we can extract a subsequence $f_{n_k}(z)(k=1,2,\cdots)$ such that as $k\to+\infty$, $f_{n_k}(z)$ converges locally uniformly in D to a function $f(z)\in\mathscr{F}$.

Evidently a closed family $\mathscr{F}$ is a normal family.

Definition 4. 2. Let $\mathscr{F}$ be a closed family of holomorphic functions in a domain D. Let $\varphi(f)$ be a functional defined for $f\in\mathscr{F}$, such that to each $f\in\mathscr{F}$ corresponds a complex number $\varphi(f)$. We say that $\varphi(f)$ is continuous, if it satisfies the following condition: For every sequence of functions $f_n(z)(n=1,2,\cdots)$ of the family $\mathscr{F}$, such that as $n\to+\infty$, $f_n(z)$ converges locally uniformly in D to a function $F(z)\in\mathscr{F}$, we have

$$\lim_{n\to+\infty}\varphi(f_n)=\varphi(F).$$

It is clear that, according to this definition, if $\varphi(f)$ is a continuous functional defined on $\mathscr{F}$, then $|\varphi(f)|$, $\mathrm{Re}\{\varphi(f)\}$ and $\mathrm{Im}\{\varphi(f)\}$ are also continuous functionals defined on $\mathscr{F}$. On the other hand, if $\varphi(f),\varphi_j(f)(j=1,2)$ are continuous functionals defined on $\mathscr{F}$, then $a\varphi(f)$, $\varphi_1(f)+\varphi_2(f)$, $\varphi_1(f)\varphi_2(f)$ are continuous functionals defined on $\mathscr{F}$, where a is a constant. If furthermore $\varphi_2(f)\neq0$ for $f\in\mathscr{F}$, then $\varphi_1(f)/\varphi_2(f)$ is also a continuous functional defined on $\mathscr{F}$.

Theorem 4. 1. Let $\mathscr{F}$ be a closed family of holomorphic functions in a domain D and let $\varphi(f)$ be a continuous functional defined on $\mathscr{F}$. Then $\varphi(f)$ is bounded on $\mathscr{F}$.

Proof. We are going to show that there is a positive number A such that

$$|\varphi(f)| < A$$

for $f \in \mathscr{F}$. In fact, on the contrary, to each positive integer n corresponds a function $f_n(z) \in \mathscr{F}$ such that

$$|\varphi(f_n)| \geqslant n.$$

Since $\mathscr{F}$ is closed, from the sequence $f_n(z)(n=1,2,\cdots)$ we can extract a subsequence $f_{n_k}(z)(k=1,2,\cdots)$ such that as $k \to +\infty$, $f_{n_k}(z)$ converges locally uniformly in D to a function $f(z) \in \mathscr{F}$. Next since $\varphi(f)$ is continuous, we have

$$\lim_{k \to +\infty} \varphi(f_{n_k}) = \varphi(f),$$

which contradicts the inequality

$$|\varphi(f_{n_k})| \geqslant n_k.$$

Theorem 4.2. Let $\mathscr{F}$ be a closed family of holomorphic functions in a domain D and let $\varphi(f)$ be a real continuous functional defined on $\mathscr{F}$. Set

$$M = \sup_{f \in \mathscr{F}} \varphi(f), \quad m = \inf_{f \in \mathscr{F}} \varphi(f).$$

Then there exist two functions $g(z)$ and $h(z)$ of $\mathscr{F}$, such that

$$\varphi(g) = M, \quad \varphi(h) = m. \tag{4.1}$$

In other words, the functional $\varphi(f)$ has a maximum and a minimum on $\mathscr{F}$.

Proof. To each positive integer n correspond two functions $g_n(z)$ and $h_n(z)$ of $\mathscr{F}$, such that

$$M-\frac{1}{n}<\varphi(g_n)\leqslant M,\ m\leqslant\varphi(h_n)<m+\frac{1}{n}.$$

From the two sequences $g_n(z)(n=1,2,\cdots)$ and $h_n(z)(n=1,2,\cdots)$ we can extract respectively two subsequences $g_{p_k}(z)(k=1,2,\cdots)$ and $h_{q_k}(z)(k=1,2\cdots)$ such that as $k\to+\infty$, $g_{p_k}(z)$ and $h_{q_k}(z)$ converge locally uniformly in D respectively to two functions g(z) and h(z) of $\mathscr{F}$. We have

$$\lim_{k\to+\infty}\varphi(g_{p_k})=\varphi(g),\ \lim_{k\to+\infty}\varphi(h_{q_k})=\varphi(h).$$

On the other hand, the inequalities

$$M-\frac{1}{p_k}<\varphi(g_{p_k})\leqslant M,\qquad m\leqslant\varphi(h_{q_k})<m+\frac{1}{q_k}$$

imply

$$\lim_{k\to+\infty}\varphi(g_{p_k})=M,\qquad \lim_{k\to+\infty}\varphi(h_{q_k})=m.$$

Consequently the two functions g(z) and h(z) satisfy the condition (4.1).

Now we introduce another notion defined as follows:

Definition 4.3. Let $\mathscr{F}$ be a closed family of holomorphic functions in a domain D. Let $\Omega(f)$ be an operator defined for $f\in\mathscr{F}$, such that to each $f\in\mathscr{F}$ corresponds a holomorphic function $\Omega(f)$ in D. We say that $\Omega(f)$ is continuous, if it satisfies the following condition: For every sequence of functions $f_n(z)(n=1,2,\cdots)$ of the family $\mathscr{F}$, such that as $n\to+\infty$, $f_n(z)$ converges locally uniformly in D to a function $F(z)\in\mathscr{F}$, the corresponding sequence of functions $\Omega(f_n)(n=1,2,\cdots)$ is such that as $n\to+\infty$, $\Omega(f_n)$ converges locally uniformly in D to the corresponding function $\Omega(F)$.

The notion of continuous operator and that of continuous functional are in close relationship. This is shown by the following theorem:

Theorem 4.3. Let $\mathscr{F}$ be a closed family of holomorphic functions in a domain D and $\Omega(f)$ a continuous operator defined on $\mathscr{F}$. Let z_0 be a point of D and E a bounded closed set of points belonging to D. Then the following functionals defined on $\mathscr{F}$ are all continuous:

$$\varphi_0(f) = \Omega(f)(z_0)$$
$$\varphi_1(f) = \max_{z\in E} |\Omega(f)(z)|, \quad \varphi_2(f) = \min_{z\in E} |\Omega(f)(z)|$$
$$\varphi_3(f) = \max_{z\in E} Re\{\Omega(f)(z)\}, \quad \varphi_4(f) = \min_{z\in E} Re\{\Omega(f)(z)\}$$
$$\varphi_5(f) = \max_{z\in E} Im\{\Omega(f)(z)\}, \quad \varphi_6(f) = \min_{z\in E} Im\{\Omega(f)(z)\}.$$

Proof. Let $f_n(z)(n=1,2,\cdots)$ be a sequence of functions of $\mathscr{F}$, such that as $n\to+\infty$, $f_n(z)$ converges locally uniformly in D to a function $F(z)\in\mathscr{F}$. Set

$$g_n(z) = \Omega(f_n)(z)(n = 1,2,\cdots), \quad G(z) = \Omega(F)(z).$$

By hypothesis, as $n\to+\infty$, $g_n(z)$ converges locally uniformly in D to the function $G(z)$. Since

$$\varphi_0(f_n) = g_n(z_0)(n = 1,2,\cdots), \qquad \varphi_0(F) = G(z_0),$$

we have

$$\lim_{n\to+\infty} \varphi_0(f_n) = \varphi_0(F).$$

By the finite covering theorem. as $n\to+\infty$, $g_n(z)$ converges uniformly to $G(z)$ on the set E. Hence, given arbitrarily a number $\varepsilon>0$, we can find a positive integer N such that for $n\geqslant N$, the inequality

$$|g_n(z)-G(z)| < \varepsilon$$

holds on E. À fortiori, we have

$$|G(z)|-\varepsilon < |g_n(z)| < |G(z)| + \varepsilon$$

which implies

$$\varphi_1(F)-\varepsilon < \varphi_1(f_n) < \varphi_1(F) + \varepsilon,\ \varphi_2(F)-\varepsilon < \varphi_2(f_n) < \varphi_2(F) + \varepsilon$$

and hence

$$\lim_{n\to+\infty} \varphi_1(f_n) = \varphi_1(F),\ \lim_{n\to+\infty} \varphi_2(f_n) = \varphi_2(F).$$

Similarly from the inequalities

$$Re\{G(z)\}-\varepsilon < Re\{g_n(z)\} < Re\{G(z)\} + \varepsilon,$$
$$Im\{G(z)\}-\varepsilon < Im\{g_n(z)\} < Im\{G(z)\} + \varepsilon,$$

we see that

$$\lim_{n\to+\infty} \varphi_j(f_n) = \varphi_j(F)(j = 3,4,5,6).$$

The basic continuous operators defined on a closed family $\mathscr{F}$ of holomorphic functions in a domain D, are

$$\Omega_0(f) = f,\ \Omega_p(f) = f^{(p)}(p = 1,2,\cdots),$$

where $f^{(p)}$ is the p-th derivative of f. Other continuous operators can be obtained by means of the following rules in which $\mathscr{F}$ always denotes a closed family of holomorphic functions in a domain D.

1)If $\Omega(f)$, $\Omega_j(f)(j=1,2)$ are continuous operators defined on $\mathscr{F}$, then $a(z)\Omega(f)$, $\Omega_1(f)+\Omega_2(f)$ and $\Omega_1(f)\Omega_2(f)$ are continuous operators defined on $\mathscr{F}$, where $a(z)$ is a holomorphic function in D. If furthermore for each $f\in\mathscr{F}$, the corresponding function $\Omega_2(f)$ has no zero in D, then the operator $\Omega_1(f)/\Omega_2(f)$ is also continuous.

2)If $\Omega(f)$ is a continuous operator defined on $\mathscr{F}$ and $p\geqslant 1$ an integer, then the operator $\Omega(f)^{(p)}$ is continuous, where $\Omega(f)^{(p)}$ denotes the p-th derivative of the function $\Omega(f)$.

3)Let $\Omega(f)$ be a continuous operator defined on $\mathscr{F}$. Let z_0 be a point of D and $m \geqslant 1$ an integer. If for each $f \in \mathscr{F}$, the function $\Omega(f)$ has a zero of order m at z_0, then the operator

$$\frac{\Omega(f)}{(z-z_0)^m}$$

defined on $\mathscr{F}$ is continuous.

Proof. Let $f_n(z)(n=1,2,\cdots)$ be a sequence of functions of the family $\mathscr{F}$, such that as $n \to +\infty$, $f_n(z)$ converges locally uniformly in D to a function $F(z) \in \mathscr{F}$. Set

$$g_n(z) = \frac{\Omega(f_n)}{(z-z_0)^m}(n = 1,2,\cdots), \ G(z) = \frac{\Omega(F)}{(z-z_0)^m}.$$

We are going to show that the sequence $g_n(z)(n=1,2,\cdots)$ is such that as $n \to + \infty$, $g_n(z)$ converges locally uniformly in D to $G(z)$. Consider first a point z_1 of D with $z_1 \neq z_0$. Then we can find a circle $c: |z-z_1| \leqslant r$ which belongs to D-(z_0), such that as $n \to +\infty$, $\Omega(f_n)$ converges uniformly to $\Omega(F)$ in c. Evidently as $n \to +\infty$, $g_n(z)$ converges uniformly to $G(z)$ in c. Next consider the point z_0. Let c' : $|z-z_0| \leqslant r'$ be a circle belonging to D, such that as $n \to +\infty$, $\Omega(f_n)$ converges uniformly to $\Omega(F)$ in c'. Then on the circle $|z-z_0|=r'$ and hence also in c', as $n \to +\infty$, $g_n(z)$ converges uniformly to $G(z)$.

4)Assume that the domain D is simply connected and let z_0 be a point of D. Let $\Omega(f)$ be a continuous operator defined on $\mathscr{F}$, such that for each $f \in \mathscr{F}$, the function $\Omega(f)$ has no zero in D and $\Omega(f)(z_0)=1$. Then the operator

$$\Lambda(f) = log\{\Omega(f)\}, \ \Lambda(f)(z_0) = 0$$

defined on $\mathscr{F}$ is continuous.

Proof. Let $f_n(z)(n=1,2,\cdots)$ be a sequence of functions of $\mathscr{F}$, such that as $n \to +\infty$, $f_n(z)$ converges locally uniformly in D to a function $F(z) \in \mathscr{F}$. Then, by hypothesis, as $n \to +\infty$, $g_n = \Omega(f_n)$ converges locally uniformly in D to

$G=\Omega(F)$. We have

$$\Lambda(f_n)=\int_{z_0}^{z}\frac{g'_n(\zeta)}{g_n(\zeta)}d\zeta\quad(n=1,2,\cdots),\Lambda(F)=\int_{z_0}^{z}\frac{G'(\zeta)}{G(\zeta)}d\zeta.$$

Consider a point z_1 of D and let c: $|z-z_1|\leqslant r$ be a circle belonging to D, such that as $n\to+\infty$, g_n converges uniformly to G in c. Take a rectifiable curve Γ in D, joining z_0 and z_1. Then for $z\in c$ we have

$$\Lambda(f_n)=\int_{\Gamma}\frac{g'_n(\zeta)}{g_n(\zeta)}d\zeta+\int_{z_1}^{z}\frac{g'_n(\zeta)}{g_n(\zeta)}d\zeta\ (n=1,2,\cdots),\tag{4.2}$$

$$\Lambda(F)=\int_{\Gamma}\frac{G'(\zeta)}{G(\zeta)}d\zeta+\int_{z_1}^{z}\frac{G'(\zeta)}{G(\zeta)}d\zeta,\tag{4.3}$$

where in (4.2) and (4.3), the path of integration of the second integral is the line segment joining z_1 and z. Noting that as $n\to+\infty$, $g'_n(z)/g_n(z)$ converges uniformly to $G'(z)/G(z)$ on Γ and in c, we see that, given arbitrarily a number $\varepsilon>0$, we can find a positive integer N such that for $n\geqslant N$, the inequality

$$|\Lambda(f_n)-\Lambda(F)|<(L+r)\varepsilon$$

holds for $z\in c$, where L is the length of Γ. Hence as $n\to+\infty$, $\Lambda(f_n)$ converges uniformly to $\Lambda(F)$ in c. So as $n\to+\infty$, $\Lambda(f_n)$ converges locally uniformly to $\Lambda(F)$ in D.

5) Let $\Omega(f)$ be a continuous operator defined on $\mathscr{F}$ and let $\Phi(w)$ be a holomorphic function in a domain Δ. Assume that for each $f\in\mathscr{F}$, the function $\Omega(f)$ takes in D only values belonging to Δ. Then the operator $\Phi\{\Omega(f)\}$ defined on $\mathscr{F}$ is continuous.

Proof. Let $f_n(z)(n=1,2,\cdots)$ be a sequence of functions of $\mathscr{F}$, such that as $n\to+\infty$, $f_n(z)$ converges locally uniformly in D to a function $F(z)\in\mathscr{F}$. Set

$$g_n=\Omega(f_n)(n=1,2,\cdots),G=\Omega(F)$$
$$\psi_n=\Phi(g_n)(n=1,2,\cdots),\Psi=\Phi(G).$$

We are going to show that as $n\to+\infty$, ψ_n converges locally uniformly in D to Ψ. Consider a point z_0 of D and the point $w_0=G(z_0)\in\Delta$. Let $c:|z-z_0|<r$ be a circle belonging to D, such that as $n\to+\infty$, g_n converges uniformly to G in c. On the other hand, let $\gamma:|w-w_0|\leqslant\rho$ be a cirele belonging to Δ. Next let $c':|z-z_0|<r'$ $(0<r'<r)$ be a circle such that

$$|G(z)-w_0|<\rho/2$$

for $z\in c'$, and let N_1 be a positive integer such that when $n\geqslant N_1$, we have

$$|g_n(z)-G(z)|<\rho/2$$

for $z\in c$. Then for $n\geqslant N_1$ and $z\in c'$, we have

$$|g_n(z)-w_0|<\rho.$$

Now given arbitrarily a number $\varepsilon>0$, we can find a number $\delta>0$ such that for $w_1,w_2\in\gamma$, $|w_1-w_2|<\delta$, we have

$$|\Phi(w_1)-\Phi(w_2)|<\varepsilon.$$

Next let N_2 be a positive integer such that when $n\geqslant N_2$, we have

$$|g_n(z)-G(z)|<\delta$$

for $z\in c$. Set $N=\max(N_1,N_2)$. Then we see that for $n\geqslant N$, $z\in c'$, we have

$$|\psi_n(z)-\Psi(z)|=|\Phi\{g_n(z)\}-\Phi\{G(z)\}|<\varepsilon.$$

Hence as $n\to+\infty$, ψ_n converges uniformly to Ψ in c'. Since the point z_0 of D is arbitrary, as $n\to+\infty$, ψ_n converges locally uniformly in D to Ψ. Consequently the operator $\Phi\{\Omega(f)\}$ is continuous.

In particular, 5) holds when $\Phi(w)$ is an entire function, for instance $\Phi(w)=e^w$.

4.2. EXAMPLES

1)Schottky's theorem and Landau's theorem

Lemma 4.1. Let $w_0 \neq 0, 1$ be a complex number. Let $\mathscr{F}$ be the family of holomorphic functions $f(z)$ in the circle $\Delta: |z|<1$ satisfying the following conditions:

1^0 $f(z)$ does not take the values 0 and 1 in Δ.

2^0 $f(0)=w_0$.

Then the family $\mathscr{F}$ is closed in Δ.

Proof. By Theorem 2.1, the family $\mathscr{F}$ is normal in Δ. Hence from a sequence of functions $f_n(z)(n=1,2,\cdots)$ of $\mathscr{F}$, we can extract a subsequence $f_{n_k}(z)$ $(k=1,2,\cdots)$ such that as $k \to +\infty$, $f_{n_k}(z)$ converges locally uniformly in Δ either to a holomorphic function $f(z)$ in Δ or to ∞. The second case can not occur, because $f_{n_k}(0)=w_0(k=1,2,\cdots)$. So the first case is true. It remains to show that the limit function $f(z)$ satisfies the conditions 1^0 and 2^0. The condition 2^0 is evidently satisfied. For the condition 1^0, distinguish two cases:

1)$f(z)$ is constant. Then $f(z) \equiv w_0$ and condition 1^0 is satisfied.

2)$f(z)$ is non-constant. In this case, on the basis of Rouché's theorem and of the identities

$$f_{n_k}(z) = \{f_{n_k}(z)-f(z)\} + f(z), f_{n_k}(z)-1 = \{f_{n_k}(z)-f(z)\} + \{f(z)-1\},$$

we see that $f(z)$ satisfies the condition 1^0.

Now let $w_0 \neq 0, 1$ be a complex number and r $(0<r<1)$ a number. Let $\mathscr{F}$ be the family of functions defined in Lemma 4.1 corresponding to the number w_0. Consider the functional $\varphi(f)$ defined for $f \in \mathscr{F}$ by the formula

$$\varphi(f) = \max_{|z| \leq r} |f(z)|.$$

By Theorems 4.3, 4.1 and 4.2, the functional $\varphi(f)$ is bounded on $\mathscr{F}$ and if we define

$$\Phi(w_0, r) = \sup_{f \in \mathscr{F}} \varphi(f),$$

then there exists a function $f_0(z)\in\mathscr{F}$ such that

$$\varphi(f_0) = \Phi(w_0,r).$$

On the other hand, consider the functional $\psi(f)$ defined for $f\in\mathscr{F}$, by the formula

$$\psi(f) = |f'(0)|.$$

Again by Theorems 4. 3, 4. 1 and 4. 2, the functional $\psi(f)$ is bounded on $\mathscr{F}$ and if we define

$$\Psi(w_0) = \sup_{f\in\mathscr{F}}\psi(f),$$

then there exists a function $f_1(z)\in\mathscr{F}$ such that

$$\psi(f_1) = \Psi(w_o).$$

Now let $f(z)$ be a holomorphic function in Δ, which does not take the values 0 and 1 in Δ. Then we have

$$|f(z)| \leqslant \Phi\{f(0), |z|\} \tag{4. 4}$$

for $z\in\Delta$, provided that we define $\Phi(w_0, 0) = |w_0|$. Theoretically, this is the precise form of Schottky's theorem. However it is very difficult to find out the exact value of the number $\Phi(w_0,r)$ and an expression of the function $f_0(z)$. The situation is similar for other extremal values and extremal functions which we shall introduce in this chapter and afterwards.

On the other hand, we have

$$|f'(0)| \leqslant \Psi\{f(0)\}. \tag{4. 5}$$

If $g(z)$ is a holomorphic function in a circle $|z|<R$, which does not take the values 0 and 1 in this circle, then we can apply (4. 5) to the function $f(z)=g(Rz)$ $(|z|<1)$. Hence

$$R\,|g'(0)| \leqslant \Psi\{g(0)\}$$

and if $g'(0) \neq 0$,

$$R \leqslant \frac{\Psi\{g(0)\}}{|g'(0)|}.$$

This is Landau's theorem in precise form

2) Bloch constant

Lemma 4. 2. Let $\mathscr{F}$ be the family of holomorphic functions $f(z)$ in the circle $\Delta: |z|<1$ satisfying the following conditions:

1^0 $f(0)=0, f'(0)=1$.

2^0 There does not exist a circle $\Gamma: |w-w_0|<r$ with $r>2$, which is a simply covered image domain of $f(z)$ for Δ. (see Definition 2. 1)

Then the family $\mathscr{F}$ is closed in Δ.

Proof. First of all, by Corollary 2. 11, the family $\mathscr{F}$ is normal in Δ. Consequently, in view of the condition 1^0, from every sequence of functions $f_n(z)(n=1,2,\cdots)$ of the family $\mathscr{F}$, we can extract a subsequene $f_{n_k}(z)(k=1,2,\cdots)$ such that as $k\to+\infty$, $f_{n_k}(z)$ converges locally uniformly in Δ to a holomorphic function $F(z)$ in Δ. Evidently $F(z)$ satisfies the condition 1^0. We are going to show that $F(z)$ also satisfies the condition 2^0. In fact, consider a circle $\Gamma: |w-w_0|<r$ with $r>2$. Assume that there is a domain $d\subset\Delta$, such that $F(z)$ is univalent in d and maps d onto Γ. Let $z=G(w)$ be the inverse function of $w=F(z)$, which is holomorphic and univalent in Γ and maps Γ onto d. Hence d is necessarily simply connected, because Γ is so. Take two numbers ρ_1,ρ_2 such that $2<\rho_1<\rho_2<r$. The function $z=G(w)$ maps the circle $|w-w_0|=\rho_2$ onto a Jordan curve $c_2\subset d$. Since d is simply connected, the interior d_2 of c_2 belongs to d. The function $z=G(w)$ maps the circle $|w-w_0|<\rho_2$ onto d_2. Hence the point $z_0=G(w_0)\in d_2$.

Since as $k\to+\infty$, $f_{n_k}(z)$ converges uniformly on c_2 to $F(z)$, so as $k\to+\infty$, $|f_{n_k}(z)-w_0|$ converges uniformly on c_2 to $|F(z)-w_0|$. But on c_2 we have $|F(z)-w_0|=\rho_2$, hence we can find a positive integer K such that, when $k\geqslant K$, we have

$$||f_{n_k}(z)-w_0|-\rho_2| < min(\rho_2-\rho_1, r-\rho_2)$$

on c_2, hence when $k \geqslant K$, we have

$$\rho_1 < |f_{n_k}(z)-w_0| < r \tag{4.6}$$

on c_2. On the other hand, from the identity

$$f_{n_k}(z)-w_0 = \{f_{n_k}(z)-F(z)\} + \{F(z)-w_0\}$$

and Rouché's theorem, we can find a positive integer K' such that, when $k \geqslant K'$, the number of the zeros of the function $f_{n_k}(z)-w_0$ in d_2 is equal to the number of the zeros of $F(z)-w_0$ in d_2. Hence the number of the zeros of $f_{n_k}(z)-w_0$ in d_2 is equal to 1, when $k \geqslant K'$. Now let k_0 be a positive integer such that $k_0 \geqslant \max(K, K')$. Consider a point w of the circle Γ_1: $|w-w_0| < \rho_1$ and denote by $\nu(f_{n_{k_0}}-w, d_2)$ the number of the zeros of the function $f_{n_{k_0}}(z)-w$ in d_2. We know already that

$$\nu(f_{n_{k_0}}-w_0, d_2) = 1.$$

On the other hand, from (4.6) we have

$$\rho_1 < |f_{n_{k_0}}(z)-w_0| < r$$

on c_2, which shows that the function $w=f_{n_{k_0}}(z)$ maps c_2 onto a curve λ_2 outside the circle Γ_1. So for each point $w \in \Gamma_1$, we have

$$\nu(f_{n_{k_0}}-w, d_2) = 1.$$

We can then define for $w \in \Gamma_1$, a function $z = g_{k_0}(w)$ whose value $z \in d_2$ and is such that

$$f_{n_{k_0}}\{g_{k_0}(w)\} = w$$

for $w\in\Upsilon_1$. We see easily that the function $g_{k_0}(w)$ is holomorphic in Υ_1.

The function $z=g_{k_0}(w)$ maps Υ_1 onto a domain $d_1\subset d_2\subset d$ and conversely the function $w=f_{n_{k_0}}(z)$ maps d_1 onto Υ_1. Since Υ_1 is the circle $|w-w_0|<\rho_1$ with $\rho_1>2$, this contradicts the fact that $f_{n_{k_0}}(z)$ satisfies the condition 2^0 in Lemma 4. 2.

Definition 4. 4. Let $\mathscr{F}_0$ be the family of holomorphic functions $f(z)$ in Δ satsfying the condition 1^0 in Lemma 4. 2. A number $b>0$ is said to have the property (P) for a funetion $f(z)\in\mathscr{F}_0$, if there is a circle $|w-w_0|<b$ which is a simply covered image domain of $f(z)$ for Δ. A number $b>0$ is said to have the property (P) for $\mathscr{F}_0$, if b has the property (P) for each function $f(z)\in\mathscr{F}_0$.

Since the function $z\in\mathscr{F}_0$, evidently if a number $b>0$ has the property (P) for $\mathscr{F}_0$, then $b\leqslant 1$.

Theorem 4. 4. Let $\mathscr{F}_0$ be the family of holomorphic functions in Δ defined in Definition 4. 4. There exists a number $B(0<B\leqslant 1)$ satisfying the following conditions:

1^0 Each number b such that $0<b<B$ has the property (P) for $\mathscr{F}_0$.

2^0 There is a function $F(z)\in\mathscr{F}_0$ such that each number $b>B$ does not have the property (P) for $F(z)$.

This number B evidently unique is called the Bloch constant.

Proof. Denote by S the set of the numbers $b>0$ having the property (P) for $\mathscr{F}_0$. S is non-empty, by Corollary 2 in Appendix A. Let B be the least upper bound of S. Then $0<B\leqslant 1$. Evidently this number B satisfies the condition 1^0 in Theorem 4. 4. We are going to show that B also satisfies the condition 2^0 in Therorem 4. 4. Consider an integer $n\geqslant 1$. Then $B+\frac{1}{n}\notin S$, hence there is a function $f_n(z)\in\mathscr{F}_0$ such that the number $B+\frac{1}{n}$ does not have the property (P) for $f_n(z)$. Since $B+\frac{1}{n}\leqslant 2$, $f_n(z)\in\mathscr{F}$, where $\mathscr{F}$ is the family of functions defined in Lemma 4. 2. Since $\mathscr{F}$ is closed in Δ, from the sequence $f_n(z)(n=1,2,\cdots)$, we can extract a subsequence $f_{n_k}(z)(k=1,2,\cdots)$ such that as $k\to+\infty$, $f_{n_k}(z)$ converges locally uniformly to a function $F(z)\in\mathscr{F}$ in Δ. Let us show that $F(z)$ satisfies the condition 2^0 in Theorem 4. 4. In fact, consider a number $b>B$ and assume, on the contrary, that b has the property (P) for $F(z)$. So there is a domain $d \subset \Delta$ such that $F(z)$ is univalent in d and maps d onto a circle $|w-w_0|<b$. Take two numbers ρ_1, ρ_2 such that $B<\rho_1<\rho_2<b$. Then as in the proof of Lemma 4. 2, we can find two positive integers K and K′ such that if k_0 is a positive integer such that $k_0 \geqslant \max(K, K')$, then the function $f_{n_{k_0}}(z)$ maps a domain $d_1 \subset \Delta$ onto the circle $|w-w_0|<\rho_1$, and $f_{n_{k_0}}(z)$ is univalent in d_1. Take an integer k_0 such that

$$k_0 \geqslant max(K, K'), B + \frac{1}{n_{k_0}} < \rho_1.$$

Then there is a domain $d_1' \subset d_1$ such that $f_{n_{k_0}}(z)$ maps d_1' onto the circle $|w-w_0|< B+\frac{1}{n_{k_0}}$, and so we get a contradiction.

By the same method, we can also obtain similar results concerning respectively Theorem 2. 9, Corollary 2. 5 and Corollary 2. 6.

4. 3. CLOSED FAMILIES OF MEROMORPHIC FUNCTIONS

Definition 4. 5. Let $\mathscr{F}$ be a family of meromorphic functions in a domain D. We say that $\mathscr{F}$ is closed in D, if from every sequence of functions $f_n(z)(n=1,2,\cdots)$ of $\mathscr{F}$, we can extract a subsequence $f_{n_k}(z)(k=1,2,\cdots)$ such that as $k \to +\infty$, $f_{n_k}(z)$ converges locally uniformly in D to a function $f(z) \in \mathscr{F}$, with respect to the spherical distance.

Evidently of a family $\mathscr{F}$ of meromorphic functions in a domain D is closed in D, then it is normal in D, and if furthermore all the functions of $\mathscr{F}$ are nonidentically equal to zero, then the family

$$\mathscr{G} = \{g(z) = 1/f(z) | f(z) \in \mathscr{F}\}$$

is also closed in D.

Lemma 4. 3. Let $\mathscr{F}$ be a family of holomorphic functions in a domain D. In order thar $\mathscr{F}$ is closed in D according to Definition 4. 5, it is necessary and sufficient that $\mathscr{F}$ is closed in D according to Definition 4. 1.

Proof. The sufficiency of the condition follows from the inequality

$$|f_{n_j}(z), f(z)| \leqslant |f_{n_j}(z) - f(z)|.$$

The necessity of the condition follows from Corollary 1. 1.

Definition 4. 6. Let $\mathscr{F}$ be a closed family of meromorphic functions in a domain D. Let $\varphi(f)$ be a functional defined for $f \in \mathscr{F}$, such that to each $f \in \mathscr{F}$ corresponds a value $\varphi(f) \in \hat{C}$. We say that $\varphi(f)$ is continuous, if it satisfies the following condition: For every sequence of functions $f_n(z)$ $(n = 1, 2, \cdots)$ of the family $\mathscr{F}$, such that as $n \to +\infty$, $f_n(z)$ converges locally uniformly in D to a function $F(z) \in \mathscr{F}$, with respect to the spherical distance, we have

$$\lim_{n \to +\infty} |\varphi(f_n), \varphi(F)| = 0. \tag{4.7}$$

It is clear that if $\varphi(f)$ is a continuous functional defined on $\mathscr{F}$, then $\psi(f) = 1/\varphi(f)$ is also a continuous functional defined on $\mathscr{F}$. This results immediately from the equality

$$|\psi(f_n), \psi(F)| = \left|\frac{1}{\varphi(f_n)}, \frac{1}{\varphi(F)}\right| = |\varphi(f_n), \varphi(F)|.$$

Theorem 4. 5. Let $\mathscr{F}$ be a closed family of meromorphic functions in a domain D and let $\varphi(f)$ be a continuous functional defined on $\mathscr{F}$. Let $w \in \hat{C}$ and set

$$\Lambda = \sup_{f \in \mathscr{F}} |\varphi(f), w|, \quad \lambda = \inf_{f \in \mathscr{F}} |\varphi(f), w|.$$

Then there exist two functions $g(z)$ and $h(z)$ of $\mathscr{F}$, such that

$$|\varphi(g), w| = \Lambda, \quad |\varphi(h), w| = \lambda.$$

Proof. To each positive integer n corresponds two functions $g_n(z)$ and

$h_n(z)$ of $\mathscr{F}$ such that

$$\Lambda - \frac{1}{n} < |\varphi(g_n),w| \leqslant \Lambda, \ \lambda \leqslant |\varphi(h_n),w| < \lambda + \frac{1}{n}.$$

From the two sequences $g_n(z)(n=1,2,\cdots)$ and $h_n(z)(n=1,2,\cdots)$ we can extract respectively two subsequences $g_{p_k}(z)(k=1,2,\cdots)$ and $h_{q_k}(z)(k=1,2,\cdots)$ such that as $k\to+\infty$, $g_{p_k}(z)$ and $h_{q_k}(z)$ converge locally uniformly in D respectively to two functions $g(z)$ and $h(z)$ of $\mathscr{F}$, with respect to the spherical distance. We have

$$\lim_{k\to+\infty} |\varphi(g_{p_k}),\varphi(g)| = 0, \ \lim_{k\to+\infty} |\varphi(h_{q_k}),\varphi(h)| = 0.$$

On the other hand, the inequalities

$$\Lambda - \frac{1}{p_k} < |\varphi(g_{p_k}),w| \leqslant \Lambda, \ \lambda \leqslant |\varphi(h_{q_k}),w| < \lambda + \frac{1}{q_k}$$

yield

$$\lim_{k\to+\infty} |\varphi(g_{p_k}),w| = \Lambda, \ \lim_{k\to+\infty} |\varphi(h_{q_k}),w| = \lambda.$$

We have

$$|\varphi(g),w| \leqslant |\varphi(g),\varphi(g_{p_k})| + |\varphi(g_{p_k}),w|,$$
$$|\varphi(g_k),w| \leqslant |\varphi(g_{p_k}),\varphi(g)| + |\varphi(g),w|.$$

Let $k\to+\infty$, we get

$$|\varphi(g),w| \leqslant \Lambda, \ \Lambda \leqslant |\varphi(g),w|,$$

hence

$$|\varphi(g),w| = \Lambda.$$

Similarly we get

$$|\varphi(h), w| = \lambda.$$

Lemma 4. 4. Let $f_n(z)(n=1,2,\cdots)$ be a sequence of meromorphic functions in a domain D, such that as $n \to +\infty$, $f_n(z)$ converges locally uniformly in D to a meromorphic function $f(z)$ in D, with respect to the spherical distance. Then as $n \to +\infty$, $\partial(z, f_n)$ converges locally uniformly in D to $\partial(z, f)$.

Proof. Let z_0 be a point of D and distinguish two cases:

1) $f(z_0) \neq \infty$. Then by Theorem 1. 1, we can find a circle $\Gamma_0: |z-z_0| < r_0$ belonging to D and a positive integer n_0 such that the functions $f_n(z)(n \geqslant n_0)$ and $f(z)$ are holomorphic in Γ_0, and as $n \to +\infty$, $f_n(z)$ converges uniformly to $f(z)$ in Γ_0. Consequently by the formulas

$$\partial(z, f_n) = \frac{|f'_n(z)|}{1 + |f_n(z)|^2}, \quad \partial(z, f) = \frac{|f'(z)|}{1 + |f(z)|^2},$$

we see that as $n \to +\infty$, $\partial(z, f_n)$ converges uniformly to $\partial(z, f)$ in Γ_0.

2) $f(z_0) = \infty$. Then by Theorem 1. 1, we see that as $n \to +\infty$, $\partial(z, \frac{1}{f_n})$ converges uniformly to $\partial(z, \frac{1}{f})$ in a circle $\Gamma_0: |z-z_0| < r_0$. Since

$$\partial(z, f_n) = \partial(z, \frac{1}{f_n}), \partial(z, f) = \partial(z, \frac{1}{f}),$$

as $n \to +\infty$, $\partial(z, f_n)$ converges uniformly to $\partial(z, f)$ in Γ_0.

Theorem 4. 6. Let $\mathscr{F}$ be a closed family of meromorphic functions in a domain D. Let E be a bounded closed set of points of D. Then the functionals

$$\varphi_1(f) = \max_{z \in E} \partial(z, f), \quad \varphi_2(f) = \min_{z \in E} \partial(z, f)$$

defined of $\mathscr{F}$ are continuous.

Proof. Let $f_n(z)(n=1,2,\cdots)$ be a sequence of functions of $\mathscr{F}$, such that as $n\to+\infty$, $f_n(z)$ converges locally uniformly in D to a function $f(z)\in\mathscr{F}$, with respect to the spherical distance. By Lemma 4. 4 and the finite covering theorem, as $n\to+\infty$, $\partial(z,f_n)$ converges uniformly on E to $\partial(z,f)$. Consequently, given arbitrarily a number $\varepsilon>0$, we can find a positive integer N such that when $n\geqslant N$, we have

$$\partial(z,f)-\varepsilon<\partial(z,f_n)<\partial(z,f)+\varepsilon$$

on E. This yields, for $n\geqslant N$,

$$\varphi_1(f)-\varepsilon<\varphi_1(f_n)<\varphi_1(f)+\varepsilon,$$
$$\varphi_2(f)-\varepsilon<\varphi_2(f_n)<\varphi_2(f)+\varepsilon.$$

Hence

$$\lim_{n\to+\infty}\varphi_1(f_n)=\varphi_1(f),\ \lim_{n\to+\infty}\varphi_2(f_n)=\varphi_2(f).$$

Theorem 4. 7. Let $\mathscr{F}$ be a closed family of meromorphic functions in a domain D. Let $\varphi(f)$ be a continuous functional defined on $\mathscr{F}$. If $\varphi(f)\neq\infty$ for each $f\in\mathscr{F}$, then for every sequence of functions $f_n(z)(n=1,2,\cdots)$ of $\mathscr{F}$, such that as $n\to+\infty$, $f_n(z)$ converges locally uniformly in D to a function $F(z)\in\mathscr{F}$, with respect to the spherieal distance, we have

$$\lim_{n\to+\infty}\varphi(f_n)=\varphi(F). \tag{4.8}$$

Examples of such functionals $\varphi(f)$ are the functionals $\varphi_1(f)$ and $\varphi_2(f)$ defined in Theorem 4. 6.

Proof. First we show that $\varphi(f)$ is bounded. In fact, by Theorem 4. 5, there exists a function $h(z)\in\mathscr{F}$ such that

$$|\varphi(h),\infty| = \lambda,\ \lambda = \inf_{f\in\mathscr{F}} |\varphi(f),\infty|.$$

Since $\varphi(h)\neq\infty$, we have

$$\lambda = \frac{1}{(1+|\varphi(h)|^2)^{\frac{1}{2}}} > 0.$$

Hence for each $f\in\mathscr{F}$,

$$\frac{1}{(1+|\varphi(f)|^2)^{\frac{1}{2}}} \geqslant \lambda,\ |\varphi(f)| < (1+|\varphi(f)|^2)^{\frac{1}{2}} \leqslant \frac{1}{\lambda}.$$

which shows that $\varphi(f)$ is bounded.

Next let $f_n(z)(n=1,2,\cdots)$ be a sequence of functions of $\mathscr{F}$, such that as $n\to+\infty$, $f_n(z)$ converges locally uniformly in D to a function $F(z)\in\mathscr{F}$, with respect to the spherical distance. Then by hypothesis we have (4. 7) which togather with the inequality

$$\begin{aligned}|\varphi(f_n)-\varphi(F)| &= \{(1+|\varphi(f_n)|^2)(1+|\varphi(F)|^2)\}^{\frac{1}{2}}|\varphi(f_n),\varphi(F)| \\ &\leqslant (1+\frac{1}{\lambda^2})|\varphi(f_n),\varphi(F)|\end{aligned}$$

yield (4. 8).

On the basis of Theorem 4. 7, the following corollaries can be easily proved as in the case of continuous functionals defined on a closed family $\mathscr{F}$ of holomorphic functions in a domain D.

Corollary 4. 1. Let $\mathscr{F}$ be a closed family of meromorphic functions in a domain D. Let $\varphi(f),\varphi_j(f)(j=1,2)$ be continuous functionals defined on $\mathscr{F}$, such that $\varphi(f)\neq\infty$, $\varphi_j(f)\neq\infty(j=1,2)$ for each $f\in\mathscr{F}$. Then the following assertions hold:

1^0 $|\varphi(f)|$, $\mathrm{Re}\{\varphi(f)\}$ and $\mathrm{Im}\{\varphi(f)\}$ are continuous functionals defined on $\mathscr{F}$.

2^0 $a\varphi(f),\varphi_1(f)+\varphi_2(f)$, $\varphi_1(f)\varphi_2(f)$ are continuous functionals defined on $\mathscr{F}$, where a is a constant. If furthermore $\varphi_2(f)\neq 0$ for $f\in\mathscr{F}$, then $\varphi_1(f)/\varphi_2(f)$ is also a continuous functional defined on $\mathscr{F}$.

Corollary 4.2. Let $\mathscr{F}$ be a closed family of meromorphic functions in a domain D. Let $\varphi(f)$ be a continuous functional defined on $\mathscr{F}$, such that $\varphi(f)\neq\infty$ is a real number for each $f\in\mathscr{F}$. Set

$$M=\sup_{f\in\mathscr{F}}\varphi(f),\ m=\inf_{f\in\mathscr{F}}\varphi(f).$$

Then there exist two functions $g(z)$ and $h(z)$ of $\mathscr{F}$, such that

$$\varphi(g)=M,\ \varphi(h)=m.$$

Definition 4.7. Let $\mathscr{F}$ be a closed family of meromorphic functions in a domain D. Let $\Omega(f)$ be an operator defined for $f\in\mathscr{F}$, such that to each $f\in\mathscr{F}$ corresponds a meromorphic function $\Omega(f)$ in D. We say that $\Omega(f)$ is continuous, if it satisfies the following condition: For every sequence of functions $f_n(z)(n=1,2,\cdots)$ of $\mathscr{F}$, such that as $n\to+\infty$, $f_n(z)$ converges locally uniformly in D to a function $F(z)\in\mathscr{F}$, with respect to the spherical distance, the corresponding sequence of functions $\Omega(f_n)(n=1,2,\cdots)$ is such that as $n\to+\infty$, $\Omega(f_n)$ converges locally uniformly in D to the corresponding function $\Omega(F)$, with respect to the spherical distance.

Theorem 4.8. Let $\mathscr{F}$ be a closed family of meromorphic functions in a domain D and $\Omega(f)$ a continuous operator defined on $\mathscr{F}$. Let z_0 be a point of D, E a bounded closed set of points of D and w a point of $\hat{\mathbf{C}}$. Then the following functionals defined on $\mathscr{F}$ are all continuous:

$$\varphi_0(f)=\Omega(f)(z_0),$$
$$\varphi_1(f)=\max_{z\in E}|\Omega(f)(z),w|,\varphi_2(f)=\min_{z\in E}|\Omega(f)(z),w|,$$
$$\varphi_3(f)=\max_{z\in E}\partial(z,\Omega(f)),\varphi_4(f)=\min_{z\in E}\partial(z,\Omega(f)).$$

Proof. Let $f_n(z)(n=1,2,\cdots)$ be a sequence of functions of $\mathscr{F}$, such that as $n\to+\infty$, $f_n(z)$ converges locally uniformly in D to a function $F(z)\in\mathscr{F}$, with respect to the spherical distance. Then, by hypothesis, as $n\to+\infty$, $\Omega(f_n)$ converges locally uniformly on D to $\Omega(F)$, with respect to the spherical distance. Hence in particular, we have

$$\lim_{n\to+\infty}|\Omega(f_n)(z_0),\Omega(F)(z_0)|=0,$$

that is

$$\lim_{n\to+\infty}|\varphi_0(f_n),\varphi_0(F)|=0.$$

On the other hand, by the finite covering theorem, as $n\to+\infty$, $\Omega(f_n)$ converges uniformly to $\Omega(F)$ on E, with respect to the spherical distance. This togather with the inequality

$$||\Omega(f_n)(z),w|-|\Omega(F)(z),w||\leqslant|\Omega(f_n)(z),\Omega(F)(z)|$$

imply that given arbitrarily a number $\varepsilon>0$, we can find a positive integer N such that when $n\geqslant N$, we have

$$|\Omega(F)(z),w|-\varepsilon<|\Omega(f_n)(z),w|<|\Omega(F)(z),w|+\varepsilon$$

on E, which yields

$$\varphi_1(F)-\varepsilon<\varphi_1(f_n)<\varphi_1(F)+\varepsilon,\ \varphi_2(F)-\varepsilon<\varphi_2(f_n)<\varphi_2(F)+\varepsilon.$$

Consequently

$$\lim_{n\to+\infty}\varphi_1(f_n)=\varphi_1(F),\ \lim_{n\to+\infty}\varphi_2(f_n)=\varphi_2(F).$$

Finally by Lemma 4. 4, as $n\to+\infty$, $\partial\{z,\Omega(f_n)\}$ converges locally uniformly in D to $\partial\{z,\Omega(F)\}$. Then as in the above, we see that

$$\lim_{n\to+\infty}\varphi_3(f_n)=\varphi_3(F),\ \lim_{n\to+\infty}\varphi_4(f_n)=\varphi_4(F).$$

Again consider a closed family $\mathscr{F}$ of meromorphic functions in a domain D. A trivial continuous operator defined for $f \in \mathscr{F}$ is the identity operator $\Omega_0(f)=f$. If all the functions of $\mathscr{F}$ are non-identically equal to zero, then the operator $\Omega_*(f)=1/f$ is also continuous, by the formula

$$\left|\frac{1}{f_n},\frac{1}{F}\right| = |f_n, F|.$$

Unlike the case of closed families of holomorphic functions, for an integer $p \geqslant 1$, the operator $\Omega_p(f)=f^{(p)}$ is not always continuous, where $f^{(p)}$ is the p-th derivative of f. As an example, consider the sequence of functions $z^2+1/n\,(n=1,2,\cdots)$. Evidently as $n \to +\infty$, z^2+1/n converges uniformly in **C** to the function z^2. Hence as $n \to +\infty$, the function

$$f_n(z) = \frac{1}{z^2+\frac{1}{n}} = \frac{n}{nz^2+1}$$

converges uniformly in **C** to the function $f_0(z)=1/z^2$, with respect to the spherical distance. So the family **F** of meromorphic functions in $\mathbf{C}: f_n(z)\,(n=(0,1,2,\cdots)$ is closed in **C**. Next we are going to show that the operator $\Omega_1(f)=f'$ defined for $f \in \mathscr{F}$ is not continuous. In fact, we have

$$\Omega_1(f_n) = f_n'(z) = \frac{-2n^2z}{(nz^2+1)^2} \quad (n=1,2,\cdots),$$

$$\Omega_1(f_0) = f_0'(z) = \frac{-2}{z^3}.$$

At the point $z=0$, we have

$$\Omega_1(f_n)(0) = 0 \ (n=1,2,\cdots), \ \Omega_1(f_0)(0) = \infty.$$

Consequently there is no circle $\Gamma: |z| < r$ such that

$$\lim_{n \to +\infty} |\Omega_1(f_n)(z), \Omega_1(f_0)(z)| = 0$$

uniformly in γ.

Concerning the continuity of the operator $\Omega_p(f)=f^{(p)}$ we have the following theorem:

Theorem 4.9. Let $\mathscr{F}$ be a closed family of meromorphic functions in a domain D. Let $p\geqslant 1$ be an integer. In order that the operator $\Omega_p(f)=f^{(p)}$ defined for $f\in\mathscr{F}$, be continuous, it is necessary and sufficient that for each point $z_0\in D$, we can find a circle $\Gamma: |z-z_0|<r$ belonging to D having the following property: For each function $f\in\mathscr{F}$, the corresponding function $\Omega_p(f)=f^{(p)}$ either has no zero in Γ or has no pole in Γ.

Proof. Assume that the operator $\Omega_p(f)=f^{(p)}$ is continuous. Let $z_0\in D$ and suppose that we can not find a circle $\Gamma: |z-z_0|<r$ having the required property. Consider a circle $|z-z_0|<r_0$ belonging to D. Then to each integer $n\geqslant 1$ corresponds a function $f_n\in\mathscr{F}$ such that the function $f_n^{(p)}$ has a zero z_n and a pole z_n' in the circle $\Gamma_n: |z-z_0|<r_0/n$. From the sequence of functions $f_n (n=1,2,\cdots)$ we can extract a subsequence $f_{n_k} (k=1,2,\cdots)$ such that as $k\to+\infty$, f_{n_k} converges locally uniformly in D to a function $F\in\mathscr{F}$, with respect to the spherical distance. By hypothesis, as $k\to+\infty$, $f_{n_k}^{(p)}$ converges locally uniformly in D to $F^{(p)}$, with respect to the spherical distance. There is then a circle $\gamma: |z-z_0|<\rho$ belonging to D such that

$$\lim_{k\to+\infty}|f_{n_k}^{(p)}(z),F^{(p)}(z)|=0$$

uniformly in γ. À fortiori, we have

$$\lim_{k\to+\infty}|f_{n_k}^{(p)}(z_{n_k}),F^{(p)}(z_{n_k})|=\lim_{k\to+\infty}|0,F^{(p)}(z_{n_k})|=0.$$

On the other hand, we have

$$\lim_{k\to+\infty}|F^{(p)}(z_{n_k}),F^{(p)}(z_0)|=0.$$

Hence $F^{(p)}(z_0)=0$. Similarly we find that $F^{(p)}(z_0)=\infty$. This is evidently impossible.

Conversely assume that the condition in Theorem 4. 9 is satisfied. Let $f_n(z)$ $(n=1,2,\cdots)$ be a sequence of functions of $\mathscr{F}$, such that as $n\to+\infty$, $f_n(z)$ converges locally uniformly in D to a function $F(z)\in\mathscr{F}$, with respect to the spherical distance. Consider a point $z_0\in D$ and let $\Gamma: |z-z_0|<r$ be a circle having the property mentioned in Theorem 4. 9. In replacing, if necessary, r by a smaller number, we may suppose that as $n\to+\infty$, $f_n(z)$ converges uniformly to $F(z)$ in Γ, with respect to the spherical distance. Now distinguish two cases:

1) $F(z_0)\neq\infty$. Then by Theorem 1. 1, we can find a circle $\Gamma_0: |z-z_0|<r_0(0<r_0\leqslant r)$ and a positive integer n_0 such that the functions $f_n(z)(n\geqslant n_0)$ and $F(z)$ are holomorphic in Γ_0, and as $n\to+\infty$, $f_n(z)$ converges uniformly to $F(z)$ in Γ_0. It follows that as $n\to+\infty$, $f_n^{(p)}(z)$ converges uniformly to $F^{(p)}(z)$ in the circle Γ_0' $|z-z_0|<r_0/2$. À fortiori as $n\to+\infty$, $f_n^{(p)}(z)$ converges uniformly to $F^{(p)}(z)$ in Γ_0', with respect to the spherical distance.

2) $F(z_0)=\infty$. Then by Theorem 1. 1, we can find a circle $\Gamma_0: |z-z_0|<r_0(0<r_0\leqslant r)$ and a positive integer n_0 such that the functions $1/f_n(z)(n\geqslant n_0)$ and $1/F(z)$ are holomorphic in Γ_0, and as $n\to+\infty$, $1/f_n(z)$ converges uniformly to $1/F(z)$ in Γ_0. Evidently this implies that when $n\geqslant n_0'(n_0'\geqslant n_0)$, $f_n(z)$ has pole in Γ_0. Consequently when $n\geqslant n_0'$, $f_n^{(p)}(z)$ has pole in Γ_0, and hence has no zero in Γ_0. In replacing, if necessary, r_0 by a smaller number, we may suppose that in the domain $0<|z-z_0|<r_0$, $F^{(p)}(z)$ has no zero and no pole. Then by the finite covering theorem, we see that we can find an integer $n_0''\geqslant n_0'$ such that when $n\geqslant n_0''$, $f_n^{(p)}(z)$ is finite on the circle $\gamma: |z-z_0|=r_0/2$ and as $n\to+\infty$, $f_n^{(p)}(z)$ converges uniformly to $F^{(p)}(z)$ on γ. Then as $n\to+\infty$, $1/f_n^{(p)}(z)$ converges uniformly to $1/F^{(p)}(z)$ on γ. Since the functions $1/f_n^{(p)}(z)(n\geqslant n_0')$ and $1/F^{(p)}(z)$ are holomorphic in Γ_0, hence as $n\to+\infty$, $1/f_n^{(p)}(z)$ converges uniformly to $1/F^{(p)}(z)$ in the circle $\Gamma_0': |z-z_0|<r_0/2$. À fortiori, as $n\to+\infty$, $f_n^{(p)}(z)$ converges uniformly to $F^{(p)}(z)$ in Γ_0', with respect to the spherical distance.

Finally we prove the following theorem:

Theorem 4. 10. Let $\mathscr{F}$ be a closed family of meromorphic functions in a domain D. Let $\Omega(f)$ be a continuous operator defined for $f\in\mathscr{F}$. Let $a(z)$ be a holomorphic function in D and $b(z)$ a function which is holomorphic and has no zero in D. Then $\Omega(f)+a(z)$ and $b(z)\Omega(f)$ are continuous operators defined for $f\in\mathscr{F}$.

Proof. Let $f_n(z)(n=1,2,\cdots)$ be a sequence of functions of $\mathscr{F}$ such that as $n\to+\infty$, $f_n(z)$ converges locally uniformly in D to a function $F(z)\in\mathscr{F}$, with respect to the spherical distance. By hypothesis, as $n\to+\infty$, $\Omega(f_n)$ converges locally uniformly in D to $\Omega(F)$, with respect to the spherical distance. Setting $a=a(z)$, we have, by Lemma 1.1, the inequality

$$|\Omega(f_n),\Omega(F)| = |\{\Omega(f_n)+a\}-a,\{\Omega(F)+a\}-a|$$
$$\geqslant \frac{1}{2}|a,\infty|^2|\Omega(f_n)+a,\Omega(F)+a|. \tag{4.9}$$

Consider a point $z_0\in D$. Then we can find a circle Υ : $|z-z_0|<\rho$ belonging to D and a positive number M such that $|a(z)|<M$ in Υ. Then in Υ we have

$$|a,\infty|^2 > \frac{1}{1+M^2}. \tag{4.10}$$

From (4.9) and (4.10), we see that

$$\lim_{n\to+\infty}|\Omega(f_n)+a,\Omega(F)+a| = 0$$

uniformly in a circle Υ' : $|z-z_0|<\rho'$ $(0<\rho'\leqslant\rho)$. This proves that the operator $\Omega(f)+a(z)$ is continuous.

To show that the operator $b(z)\Omega(f)$ is continuous, distinguish two cases:

1) $\Omega(F)(z_0)\neq\infty$. Then by Theorem 1.1, we can find a circle Γ_0: $|z-z_0|<r_0$ and a positive integer n_0 such that the functions $\Omega(f_n)(n\geqslant n_0)$ and $\Omega(F)$ are holomorphic in Γ_0, and as $n\to+\infty$, $\Omega(f_n)$ converges uniformly to $\Omega(F)$ in Γ_0. Hence as $n\to+\infty$, $b\Omega(f_n)$ converges uniformly to $b\Omega(F)$ in Γ_0, where

$$b = b(z).$$

2) $\Omega(F)(z_0)=\infty$. Then by Theorem 1.1, we can find a circle Γ_0: $|z-z_0|<r_0$ and a positive integer n_0 such that the functions $1/\Omega(f_n)(n\geqslant n_0)$ and $1/\Omega(F)$ are holomorphic in Γ_0, and as $n\to+\infty$, $1/\Omega(f_n)$ converges uniformly to $1/\Omega(F)$ in Γ_0. Hence as $n\to+\infty$, $1/b\Omega(f_n)$ converges uniformly to $1/b\Omega(F)$ in Γ_0.

So in each case we have

$$\lim_{n\to+\infty} |b\Omega(f_n), b\Omega(F)| = 0$$

uniformly in Γ_0. This proves that the operator $b(z)\Omega(f)$ is continuous.

4.4. EXAMPLES

1) Schottky's theorem

Lemma 4.5. Let $a_j \in \hat{C}(j=1,2,3)$ be three distinct values. Let $w_0 \in \mathbf{C}$ be a value such that $w_0 \neq a_j (j=1,2,3)$. Let $\mathscr{F}$ be the family of meromorphic functions $f(z)$ in the circle $\Delta: |z|<1$ satisfying the following conditions:

1^0 $f(z)$ does not take the values $a_j (j=1,2,3)$ in Δ.

2^0 $f(0)=w_0$.

Then the family $\mathscr{F}$ is closed in Δ.

Proof. By Theorem 3.1 the family $\mathscr{F}$ is normal in Δ. Let $f_n(z)(n=1,2,\cdots)$ be a sequence of functions of $\mathscr{F}$. From this sequence we can extract a subsequence $f_{n_k}(z)(k=1,2,\cdots)$ which is a C_0-sequence in D. Let $F(z)$ be its limit function with respect to the spherical distance. By Theorem 1.2, as $k\to+\infty$, $f_{n_k}(z)$ eonverges locally uniformly to $F(z)$ in D, with respect to the spherical distance. It remains to show that $F(z) \in \mathscr{F}$. In fact, by the condition 2^0 in Lemma 4.5, we have $F(0)=w_0$. It follows that $F(z)$ is a meromorphic function in Δ. Distinguish two cases:

1) $F(z) \equiv w_0$. Then evidently $F(z) \in \mathscr{F}$.

2) $F(z)$ is non-constant. We are going to show that $F(z)$ satisfies the condition 1^0 in Lemma 4.5. In fact, consider a value a_j and assume, on the contrary, that $F(z_0)=a_j$ at a point $z_0 \in \Delta$. If $a_j \neq \infty$, then by Theorem 1.1, we can find a circle $\Gamma_0: |z-z_0|<r_0$ and a positive integer k_0 such that the functions $f_{n_k}(z)(k\geqslant k_0)$ and $F(z)$ are holomorphic in Γ_0, and as $k\to+\infty$, $f_{n_k}(z)$ converges uniformly to $F(z)$ in Γ_0. Consequently by the identity

$$f_{n_k}(z)-a_j = \{f_{n_k}(z)-F(z)\} + \{F(z)-a_j\}$$

and Rouché's theorem, $f_{n_k}(z)-a_j$ has zero in Γ_0, when k is sufficiently large, incompatible with the condition 1^0 in Lemma 4.5. If $a_j=\infty$, we see also by Theorem 1.1, that $f_{n_k}(z)$ has pole in a circle $\Gamma_0: |z-z_0|<r_0$, when k is sufficiently

large, also incompatible with the condition 1^0. Hence $F(z)\in\mathscr{F}$.

Now consider four dsitinct values $a_j(j=1,2,3)$, w_0 in $\hat{C}$ and a number $r(0\leqslant r<1)$. Let us define three positive numbers $\Phi_j(a_1,a_2,a_3,w_0,r)(j=1,2,3)$ depending only on $a_j(j=1,2,3)$, w_0 and r as follows: Distinguish two cases:

1) $w_0\neq\infty$. Let $\mathscr{F}$ be the closed family of meromorphic functions in the circle $\Delta:|z|<1$ defined in Lemma 4.5. let $\varphi_j(f)(j=1,2,3)$ be the three functionals

$$\varphi_j(f)=\min_{|z|\leqslant r}|f(z),a_j|(j=1,2,3)$$

defined for $f\in\mathscr{F}$. By Theorem 4.8, $\varphi_j(f)(j=1,2,3)$ are continuous. we define

$$\Phi_j(a_1,a_2,a_3,w_0,r)=\inf_{f\in\mathscr{F}}\varphi_j(f)(j=1,2,3).$$

By Corollary 4.2, there exist three functions $f_j(z)\in\mathscr{F}(j=1,2,3)$ such that

$$\varphi_j(f_j)=\Phi_j(a_1,a_2,a_3,w_0,r)(j=1,2,3)$$

which imply that

$$\Phi_j(a_1,a_2,a_3,w_0,r)>0\qquad(j=1,2,3).$$

For each $f\in\mathscr{F}$ we have

$$\min_{|z|\leqslant r}|f(z),a_j|\geqslant\Phi_j(a_1,a_2,a_3,w_0,r)(j=1,2,3).$$

Consequently if f(z) a meromorphic function in the circle $\Delta:|z|<1$, such that in Δ, f(z) does not take three distinct values $a_j\in\hat{C}(j=1,2,3)$ and that $f(0)\neq\infty$, then in Δ we have

$$|f(z),a_j|\geqslant\Phi_j(a_1,a_2,a_3,f(0),|z|)(j=1,2,3).\tag{4.11}$$

2) $w_0=\infty$. In this case we define

$$\Phi_j(a_1,a_2,a_3,\infty,r)=\Phi_j(\frac{1}{a_1},\frac{1}{a_2},\frac{1}{a_3},o,r)(j=1,2,3).\tag{4.12}$$

If f(z) is a meromorphic function in the circle $\Delta: |z|<1$ such that in Δ, f(z) does not take three distinct values $a_j \in \hat{\mathbf{C}}(j=1,2,3)$ and that $f(0)=\infty$, then the function $g(z)=1/f(z)$ is meromorphic and does not take the values $1/a_j(j=1,2,3)$ in Δ and $g(0)=0$. Hence by (4.11), in Δ we have

$$|g(z),\frac{1}{a_j}| \geqslant \Phi_j(\frac{1}{a_1},\frac{1}{a_2},\frac{1}{a_3},0,|z|)(j=1,2,3).$$

Since

$$|f(z),a_j| = |g(z),\frac{1}{a_j}|,$$

we see from (4.12) that (4.11) also hold in Δ.

The three inequalities (4.11) togather constitute a precise form of Schottky's theorem in the case of meromorphic functions.

2)Filling circle

First give a definition: Let f(z) be a meromorphic function in a circle $\Gamma: |z-z_0|<r$ ang $\eta(0<\eta<1/2)$ a number. We say that Γ is a η-filling circle of f(z), if there do not exist three values $a_j \in \hat{\mathbf{C}}(j=1,2,3)$ such that

$$|a_j,a_{j'}| \geqslant \eta \ (1 \leqslant j,j' \leqslant 3, j \neq j')$$

and that f(z) does not take the values $a_j(j=1,2,3)$ in Γ.

This definition is essentially the same as the definition already given in connection with Theorem 3.10. The only difference is that here we assume $|a_j,a_{j'}| \geqslant \eta$ instead of $|a_j,a_{j'}|>\eta$. We make this modification, in view of the proof of the following lemma:

Lemma 4.6. Let $\eta(0<\eta<1/2)$ be a number. Let $\mathscr{F}$ be the family of meromorphic functions f(z) in the circle $\Delta: |z|<1$ satisfying the following conditions:

1^0 Δ is not a η-filling circle of f(z).

2^0 $\partial(0,f) \geqslant 1$.

Then the family $\mathscr{F}$ is closed in Δ.

Proof. First of all, the family $\mathscr{F}$ is non-empty. In fact, the function $f(z)=z$ does not take the values $a_1=1$, $a_2=-1$, $a_3=i$ in Δ, and $|a_1,a_2|=1$, $|a_2,a_3|=|a_3,a_1|=\sqrt{2}/2$. Moreover we have $\partial(0,z)=1$. Hence $f(z)=z$ belongs to $\mathscr{F}$.

By Theorem 3. 9, the family $\mathscr{F}$ is normal in Δ. Let $f_n(z)(n=1,2,\cdots)$ be a sequence of functions of $\mathscr{F}$. From this sequence we can extract a subsequence $f_{n_k}(z)(k=1,2,\cdots)$ which is a C_0-sequence in Δ. By Theorem 1. 2, there is a function $F(z)$ defined in Δ such that as $k\to+\infty$, $f_{n_k}(z)$ converges locally uniformly to $F(z)$ in Δ, with respect to the spherical distance. Since $\partial(0,f_{n_k})\geqslant 1$, we see as in the proof of Lemma 4. 4, that $\partial(0,F)\geqslant 1$. Hence $F(z)$ is a non-constant meromorphic function in Δ. It remains to show that $F(z)$ satisfies the condition 1^0 in Lemma 4. 6. Since $f_{n_k}(z)$ satisfies the condition 1^0 in Lemma 4. 6, there are three values $a_j^{(k)}\in\hat{C}(j=1,2,3)$ such that

$$|a_j^{(k)},a_{j'}^{(k)}|\geqslant\eta\ (1\leqslant j,j'\leqslant 3,j\neq j')$$

and that $f_{n_k}(z)$ does not take the values $a_j^{(k)}(j=1,2,3)$ in Δ. So we have three sequences of values $a_j^{(k)}(k=1,2,\cdots)(j=1,2,3)$. By Lemma 3. 2, we can find an increasing sequence of positive integers $k_p(p=1,2,\cdots)$ such that for each $1\leqslant j\leqslant 3$, the sequence $a_j^{(k_p)}(p=1,2,\cdots)$ satisfies one of the following two conditions:

1) $a_j^{k_p}\in C(p=1,2,\cdots)$, $\lim\limits_{p\to+\infty}a_j^{(k_p)}=A_j(A_j\in\hat{C})$;

2) $a_j^{(k_p)}=\infty(p=1,2,\cdots)$, we write $\lim\limits_{p\to+\infty}a_j^{(k_p)}=A_j(A_j=\infty)$.

We have

$$|A_j,A_{j'}|\geqslant\eta\ (1\leqslant j,j'\leqslant 3,j\neq j').$$

To complete the proof of Lemma 4. 6, we are going to show that the function $F(z)$ does not take the values $A_j(j=1,2,3)$ in Δ. For instance, consider the value A_1. Distinguish three cases:

α) $A_1\neq\infty$, $a_1^{(k_p)}\in C(p=1,2,\cdots)$. Assume that $F(z)$ takes the value A_1 at a point $z_0\in\Delta$. Then by Theorem 1. 1, we can find a circle Γ_0: $|z-z_0|<r_0$ belonging to Δ and a positive integer k_0 such that the functions $f_{n_k}(z)(k\geqslant k_0)$ and $F(z)$ are holomorphic in Γ_0 and as $k\to+\infty$, $f_{n_k}(z)$ converges uniformly to $F(z)$ in Γ_0. Write $m_p=n_{k_p}$. Then by the identity

$$f_{m_p}(z)-a_1^{(k_p)} = \{f_{m_p}(z)-F(z)\} + \{F(z)-A_1\} + \{A_1-a_1^{(k_p)}\}$$

and Rouché's theorem, we see that, when p is sufficiently large, $f_{m_p}(z)-a_1^{(k_p)}$ has zero in Γ_0. This is incompatible with the fact that $f_{n_k}(z)$ does not take the value $a_1^{(k)}$ in Δ.

Two other cases are:

β) $A_1=\infty$, $a_1^{(k_p)}\in \mathbf{C}$ $(p=1,2,\cdots)$.

γ) $A_1=\infty$, $a_1^{(k_p)}=\infty$ $(p=1,2,\cdots)$.

In these two cases, it can be shown similarly that $F(z)$ also does not take the value A_1 in Δ, by a method already used in the proof of Theorem 3.13.

Given a number $\eta(0<\eta<1/2)$, consider again the family $\mathscr{F}$ of meromorphic functions $f(z)$ in the circle $\Delta: |z|<1$ defined in Lemma 4.6. Consider the functional

$$\varphi(f) = \partial(o,f)$$

defined for $f\in\mathscr{F}$. By Theorem 4.6, $\varphi(f)$ is continuous. Setting

$$M_0(\eta) = \sup_{f\in\mathscr{F}} \varphi(f),$$

by Corollary 4.2, there exists a function $f_0(z)\in\mathscr{F}$ such that

$$\varphi(f_0) = M_0(\eta).$$

Let $\mathscr{F}_0$ be the family of meromorphic functions $f(z)$ in Δ, such that $f(z)$ satisfies the condition 1^0 in Lemma 4.6. Then evidently for each $f\in\mathscr{F}_0$, we have

$$\partial(o,f) \leqslant M_0(\eta).$$

Consequently if $f(z)$ is a meromorphic function in Δ, such that

$$\partial(o,f) > M_0(\eta),$$

then Δ is a η-filling circle of $f(z)$. Next by the same method used in the proof of Theorem 3.10, we conclude that if $f(z)$ is a meromorphic function in a circle Γ:

$|z-z_0|<R$ satisfying the condition

$$R\partial(z_0, f) > M_0(\eta),$$

then the circle Γ is a η-filling circle of the function $f(z)$.

This result is the precise form of Theorem 3. 10. It can also be considered as an extension of Landau's theorem.

4. 5. A COVERING THEOREM

Lemma 4. 7. Let $\mathscr{F}$ be a closed family of meromorphic functions in a domain D. Let z_0 be a point of D. Assume that we can find a positive number A such that for each function $f(z) \in \mathscr{F}$ we have

$$|f(z_0)| \leqslant A.$$

Then we can find a circle $\Upsilon: |z-z_0|<r$ belonging to D and a positive number B such that for each function $f(z) \in \mathscr{F}$ we have

$$|f(z)| \leqslant B$$

in Υ.

Proof. Assume that such a circle Υ and positive number B can not be found. Consider a circle $|z-z_0|<\rho$ belonging to D. Then to each positive integer n correspond a function $f_n(z) \in \mathscr{F}$ and a point z_n of the circle $|z-z_0|<\rho/n$ such that

$$|f_n(z_n)| > n.$$

From the sequence of functions $f_n(z)(n=1,2,\cdots)$, we can extract a subsequence $f_{n_k}(z)(k=1,2,\cdots)$ such that as $k \to +\infty$, $f_{n_k}(z)$ converges locally uniformly in D to a function $F(z) \in \mathscr{F}$, with respect to the spherical distance. Since $|f_{n_k}(z_0)| \leqslant A(k=1,2,\cdots)$, we must have $F(z_0) \neq \infty$. Hence by Theorem 1. 1, we can find a circle $\Gamma_0: |z-z_0|<r_0$ and a positive integer k_0 such that the functions $f_{n_k}(z)(k \geqslant k_0)$

and $F(z)$ are holomorphic in Γ_0, and as $k\to+\infty$, $f_{n_k}(z)$ converges uniformly to $F(z)$ in Γ_0. It follows that the functions $f_{n_k}(z)(k\geqslant k_0)$ are uniformly bounded in a circle $|z-z_0|<r_0'$. But this is incompatible with the inequalities

$$|z_{n_k},z_0|<\frac{\rho}{n_k},\quad |f_{n_k}(z_{n_k})|>n_k\quad (k=1,2,\cdots).$$

Definition 4.8. Let $f(z)$ be a non-constant meromorphic function in a domain D and Δ a domain in the w-plane. We say that the values of $f(z)$ cover at least p ($p\geqslant 1$ integer) times the domain Δ, if for each point w of Δ, the number of the zeros of the function $f(z)-w$ in D is greater or equal to p (with due count of order of multiplicity of each zero).

Theorem 4.11. Let $\mathscr{F}$ be a closed family of meromorphic functions in a domain D. Let z_0 be a point of D and $p\geqslant 1$ an integer. Assume that each function $f(z)$ of $\mathscr{F}$ has a zero of order p at z_0, namely

$$f^{(j)}(z_0)=0\quad (j=0,1,\cdots,p-1),\quad f^{(p)}(z_0)\neq 0. \tag{4.13}$$

On the other hand, let Δ_t be a domain in the w-plane, depending on the variable t ($0<t<+\infty$) satisfying the following conditions:

1) To each number t ($0<t<+\infty$) corresponds a bounded domain Δ_t containing the point $w=0$.

2) If $t_1<t_2$, then $\overline{\Delta}_{t_1}\subset\Delta_{t_2}$, where $\overline{\Delta}_{t_1}$ denotes the closure of Δ_{t_1}.

3) If a point w and a number t are such that $w\in\Delta_t$, then there is a number t' such that $t'<t$ and $w\in\Delta_{t'}$. If a point w and a number t are such that $w\notin\overline{\Delta}_t$, then there is a number t' such that $t'>t$ and $w\notin\overline{\Delta}_{t'}$.

4) For each circle $\Gamma: |w|<\rho$, there is a number t such that $\Delta_t\subset\Gamma$.

Then one of the following two cases holds:

A. For each function $f(z)$ of $\mathscr{F}$, the values of $f(z)$ cover at least p times the domain $\Delta=\bigcup\limits_{0<t<+\infty}\Delta_t$.

B. There is a number t_0 ($0<t_0<+\infty$) having the following properties:

1^0 For each function $f(z)$ of $\mathscr{F}$, the values of $f(z)$ cover at least p times the domain Δ_{t_0}.

2^0 There is a function $f_0(z)$ of $\mathscr{F}$, such that for each number $t>t_0$, the values of $f_0(z)$ do not cover at least p times the domain Δ_t.

The proof consists of several steps:

i) By Lemma 4.7 and the condition (4.13), we can find a circle $|z-z_0|<r_0$ belonging to D, such that each function f(z) of $\mathscr{F}$ is holomorphic in this circle.

ii) There exists a number $r(0<r\leqslant r_0)$ such that each function f(z) of $\mathscr{F}$ has no zero in the domain $0<|z-z_0|<r$.

Proof. Assume that such a number r does not exist. Then to each positive integer n corresponds a function $f_n(z)$ of $\mathscr{F}$, such that $f_n(z)$ has at least one zero in the domain $0<|z-z_0|<r_0/n$. From the sequence $f_n(z)(n=1,2,\cdots)$ we can extract a subsequence $f_{n_k}(z)(k=1,2,\cdots)$ such that as $k\rightarrow+\infty$, $f_{n_k}(z)$ converges locally uniformly in D to a function $F(z)\in\mathscr{F}$, with respect to the spherical distance. By the condition (4.13) and Theorem 1.1, we can find a number $\rho(0<\rho<r_0)$ such that F(z) has no zero in the region $0<|z-z_0|\leqslant\rho$ and that as $k\rightarrow+\infty$, $f_{n_k}(z)$ converges uniformly to F(z) in the circle $|z-z_0|\leqslant\rho$. Hence, when k is sufficiently large, the function $f_{n_k}(z)$ has the same number of zeros as F(z) in the circle $|z-z_0|<\rho$, therefore p zeros in the circle $|z-z_0|<\rho$. But this contradicts the fact that $f_{n_k}(z)$ has at least p+1 zeros in the circle $|z-z_0|<r_0/n_k$, when $r_0/n_k<\rho$.

iii) There exists a number $\eta>0$ such that for each function f(z) of $\mathscr{F}$, the values of f(z) cover at least p times the circle $|w|<\eta$.

Proof. Let r be the number in ii). In view of the condition (4.13), it is sufficient to show that there is a number $\eta>0$ such that for each function f(z) of $\mathscr{F}$, we have

$$\min_{z\in c}|f(z)|\geqslant\eta,$$

where c is the circle $|z-z_0|=r/2$. In fact, on the contrary, to each positive integer n corresponds a function $f_n(z)$ of $\mathscr{F}$, such that

$$\min_{z\in c}|f_n(z)|<\frac{1}{n}.$$

From the sequence $f_n(z)(n=1,2,\cdots)$ we can extract a subsequence $f_{n_k}(z)(k=1,

$2,\cdots)$ such that as $k\to+\infty$, $f_{n_k}(z)$ converges locally uniformly in D to a function $F(z)$ of $\mathscr{F}$, with respect to the spherical distance. We have

$$\mu = \min_{z\in c} |F(z)| > 0.$$

Since the functions $f_{n_k}(z)(k=1,2,\cdots)$ and $F(z)$ are holomorphic in the circle $|z-z_0|<r_0$, by Theorem 1. 1, as $k\to+\infty$, $f_{n_k}(z)$ converges locally uniformly to $F(z)$ in the circle $|z-z_0|<r_0$, and hence uniformly on c. Hence, when k is sufficiently large, we have

$$\min_{z\in c} |f_{n_k}(z)| > \frac{\mu}{2}$$

which is incompatible with

$$\min_{z\in c} |f_{n_k}(z)| < \frac{1}{n_k},$$

when $1/n_k<\mu/2$.

iv) Denote by S the set of the numbers t satisfying the following condition: For each function $f(z)$ of $\mathscr{F}$, the values of $f(z)$ cover at least p times the domain Δ_t. By the condition 4) and the result iii), the set S is non-empty. Distinguish two cases:

a. S is undounded. In this case, for each function $f(z)$ of $\mathscr{F}$, the values of $f(z)$ cover at least p times the domain $\Delta=\bigcup_{0<t<+\infty}\Delta_t$. In fact, let $f(z)$ be a function of $\mathscr{F}$ and w a point of Δ. Then there is a number t such that $w\in\Delta_t$. Next let $t'\in S$ be a number such that $t<t'$. By the condition 2), $\overline{\Delta}_t\subset\Delta_{t'}$, hence $w\in\Delta_{t'}$. Consequently the number of the zeros in D of the function $f(z)-w$ is greater or equal to p.

b. S is bounded. Let t_0 be the least upper bound of S. First let us show that this number t_0 has the property 1^0 in B. In fact, consider a function $f(z)\in\mathscr{F}$ and a point $w\in\Delta_{t_0}$. By the condition 3), there is a number t' such that $t'<t_0$ and $w\in\Delta_{t'}$. Let $t''\in S$ such that $t'<t''$. Then $w\in\Delta_{t''}$ and hence the number of the zeros of the function $f(z)-w$ in D is greater or equal to p.

Secondly let us show that the number t_0 also has the property 2^0 in B. In fact, consider a positive integer n. Since $t_0+1/n \notin S$, there is a function $f_n(z) \in \mathscr{F}$ such that the values of $f_n(z)$ do not cover at least p times the domain $\Delta_{t_0+1/n}$. Consequently there is a point $w_n \in \Delta_{t_0+1/n}$ such that the number of the zeros of the function $f_n(z)-w_n$ in D is less than p. By the property 1^0, $w_n \notin \Delta_{t_0}$. From the sequence $f_n(z)(n=1,2,\cdots)$ we can extract a subsequence $f_{n_k}(z)(k=1,2,\cdots)$ such that as $k \to +\infty$, $f_{n_k}(z)$ converges locally uniformly in D to a function $f_0(z) \in \mathscr{F}$, with respect to the spherical distance. It remains to show that this function $f_0(z)$ satisfies the condition in 2^0 of B. In fact, consider the sequence of points w_{n_k} $(k=1,2,\cdots)$, which has at least one limiting point w_0.

It is impossible that $w_0 \notin \overline{\Delta}_{t_0}$. In fact, assume that $w_0 \notin \overline{\Delta}_{t_0}$. Then by the condition 3), there is a number t' such that $t' > t_0$ and $w_0 \notin \overline{\Delta}_{t'}$. Let $\gamma: |w-w_0| < \delta$ be a circle which contains no point of $\overline{\Delta}_{t'}$. Accordingly when $1/n_k < t'-t_0$, the circle γ contains no point of Δ_{t_0+1/n_k}. This is incompatible with the fact that $w_{n_k} \in \Delta_{t_0+1/n_k}$ and w_0 is a limiting point of the sequence of points $w_{n_k}(k=1,2,\cdots)$. So we have $w_0 \in \overline{\Delta}_{t_0}$.

Now consider a number $t > t_0$ and assume, on the contrary, that the values of the function $f_0(z)$ cover at least p times the domain Δ_t. Then since $w_0 \in \overline{\Delta}_{t_0} \subset \Delta_t$, the number of the zeros of the function $f_0(z)-w_0$ in D is greater or equal to p. Hence we can find a finite number of distinct points $z_j(j=1,2,\cdots,m)$ of D, such that z_j is a zero of order λ_j of $f_0(z)-w_0$ with

$$\sum_{j=1}^{m} \lambda_j \geqslant p.$$

Let $|z-z_j| \leqslant r_j(j=1,2,\cdots,m)$ be m circles belonging to D, exterior to each other and containing no other zeros of $f_0(z)-w_0$ except $z_j(j=1,2,\cdots,m)$. By Theorem 1.1, we may assume that, for each $j(1 \leqslant j \leqslant m)$, the functions $f_{n_k}(z)(k \geqslant k_j)$ and $f_0(z)$ are holomorphic in the circle $|z-z_j| < r_j$, and as $k \to +\infty$, $f_{n_k}(z)$ converges uniformly to $f_0(z)$ in the circle $|z-z_j| < r_j$. Now consider the circle $\Gamma_j: |z-z_j| = r_j/2$ $(1 \leqslant j \leqslant m)$ and set

$$\mu_j = \min_{z \in \Gamma_j} |f_0(z)-w_0| \, (j=1,2,\cdots,m), \ \mu = \min_{1 \leqslant j \leqslant m} \mu_j.$$

We have $\mu > 0$. Let $K \geqslant \max_{1 \leqslant j \leqslant m} k_j$ be an integer such that when $k \geqslant K$, we have

$$|f_{n_k}(z) - f_0(z)| < \frac{\mu}{2}$$

in each of the circles $|z-z_j|<r_j(j=1,2,\cdots,m)$. Next take an integer $k_0 \geqslant K$ such that

$$|w_{n_{k_0}} - w_0| < \frac{\mu}{2}.$$

Then by the identity

$$f_{n_{k_0}}(z) - w_{n_{k_0}} = \{f_{n_{k_0}}(z) - f_0(z)\} + \{f_0(z) - w_0\} + \{w_0 - w_{n_{k_0}}\}$$

and Rouché's theorem, we see that the function $f_{n_{k_0}}(z)-w_{n_{k_0}}$ has λ_j zeros in the circle Γ_j. Consequently the number N of the zeros of the function $f_{n_{k_0}}(z)-w_{n_{k_0}}$ in D is greater or equal to p. This is incompatible with the fact that $N<p$.

Theorem 4.11 is now completely proved.

Example of Δ_t: $\Delta_t=(|w|<t)$ $(0<t<+\infty)$.

Corollary 4.3. Let $\mathscr{F}$ be a closed family of meromorphic functions in a domain D. Let z_0 be a point of D, $w_0 \in \mathbf{C}$ a point and $p \geqslant 1$ an integer. Assume that for each function $f(z) \in \mathscr{F}$, the function $f(z)-w_0$ has a zero of order p at z_0. On the other hand, let Δ_t be a domain in the w-plane, depending on the variable $t(0<t<+\infty)$ satisfying the following conditions:

1) To each number $t(0<t<+\infty)$ corresponds a bounded domain Δ_t containing the point $w=w_0$.

Conditions 2) and 3) in Theorem 4.11.

4) For each circle γ : $|w-w_0|<\rho$, there is a number t such that $\Delta_t \subset \gamma$.

Then we have the same conclusion of Theorem 4.11.

To prove this corollary, we introduce the family of meromorphic functions in D, defined by

$$\mathscr{G} = \{g(z) = f(z) - w_0 | f(z) \in \mathscr{F}\}$$

which can be easily shown to be closed in D by means of Lemma 1.1. On the other hand, we introduce the domain $\tilde{\Delta}_t$ which is the image of the domain Δ_t under the mapping $\zeta = w - w_0$. It is also easy to see that the domain $\tilde{\Delta}_t (0 < t < +\infty)$ satisfies the conditions 1), 2), 3), 4) in Theorme 4.11. Then we apply Theorem 4.11 to the family $\mathscr{G}$ and the domain $\tilde{\Delta}_t (0 < t < +\infty)$.

Corollary 4.4. Let $\mathscr{F}$ be a closed family of meromorphic functions in a domain D. Let z_0 be a point of D and $p \geqslant 1$ an integer. Assume that each function $f(z) \in \mathscr{F}$ has a pole of order p at z_0. On the other hand, let Δ_t be a domain in the extended w-plane, depending on the varable t $(0 < t < +\infty)$ satisfying the following conditions:

1) To each number $t (0 < t < +\infty)$ corresponds an unbounded domain Δ_t containing the point $w = \infty$ and belonging to a domain $|w| > r$ $(0 < r = r_t < +\infty)$.

Conditions 2) and 3) in Theorem 4.11.

4) For each domain $\Gamma: |w| > R$, there is a number t such that $\Delta_t \subset \Gamma$.

Then we have the same conclusion of Theorem 4.11.

To prove this corollary, we introduce the family of meromorphic functions in D defined by

$$\mathscr{G} = \{g(z) = \frac{1}{f(z)} | f(z) \in \mathscr{F}\}$$

which is easily seen to be closed in D by means of the formula (1.7). Moreover each function $g(z) \in \mathscr{G}$ has a zero of order p at z_0. On the other hand, we introduce the domain $\tilde{\Delta}_t$ which is the image of the domain Δ_t under the mapping $\zeta = 1/w$. It is easy to see that the domain $\tilde{\Delta}_t (0 < t < +\infty)$ satisfies the conditons 1), 2), 3), 4) in Theorem 4.11. Then we apply Theorem 4.11 to the family $\mathscr{G}$ and the domain $\tilde{\Delta}_t (0 < t < +\infty)$.

5

QUASI-NORMAL FAMILIES OF MEROMORPHIC FUNCTIONS

5.1. SOME PRELIMINARY DEFINITIONS AND THEOREMS

Definition 5.1. Let $S: f_n(z)(n=1,2,\cdots)$ be a sequence of meromorphic functions in a domain D. A point z_0 of D is called a C_1-point of S, if there is a circle $|z-z_0|<r$ belonging to D, such that each point of the domain $0<|z-z_0|<r$ is a C_0-point of S. S is called a C_1-sequence in D, if each point of D is a C_1-point of S.

According to this definition, if a point z_0 of D is a C_0-point of S, then z_0 is also a C_1-point of S. Consequently if S is a C_0-sequence in D, then S is also a C_1-sequence in D. Now assume that S is a C_1-sequence in D, but is not a C_0-sequence in D. Then there exists at least one point z_0 of D which is not a C_0-point of S. We call such a point z_0 a nonC_0-point of S. Denote by E the set of nonC_0-points of S in D. E has no point of accumulation in D. In the domain $D_1=D-E$, S is a C_0-sequence. Hence by Theorems 1.2 and 1.3, as $n\to+\infty$, $f_n(z)$ converges locally uniformly to a limit function $F(z)$ in D, with respect to the spherical distance. $F(z)$ is a meromorphic function in D or the constant ∞.

In particular, if $f_n(z)$ $(n=1,2,\cdots)$ is a C_1-sequence of holomorphic functions in a domain D, but is not a C_0-sequence in D, then by Corollary 1.1, as $n\to+\infty$, $f_n(z)$ converges locally uniformly in the domain $D_1=D-E$ defined above, either to a holomorphic function or to the constant ∞. However the first alternative can not occur, because otherwise, by the principle of maximum modulus, the points of E would be C_0-points of the sequence $f_n(z)$ $(n=1,2,\cdots)$. So the second alternative must be true.

Definition 5.2. Let $\mathscr{F}$ be a family of meromorphic functions in a domain

D. We say that $\mathscr{F}$ is quasi-normal in D, if from every sequence of functions $f_n(z)$ $(n=1,2,\cdots)$ of $\mathscr{F}$, we can extract a subsequence $f_{n_k}(z)$ $(k=1,2,\cdots)$ which is a C_1-sequence in D. $\mathscr{F}$ is said to be quasi-normal at a point z_0 of D, if there is a circle $\Gamma: |z-z_0|<r$ belonging to D such that $\mathscr{F}$ is quasi-normal in Γ.

Evidently if $\mathscr{F}$ is quasi-normal in D, then $\mathscr{F}$ is quasi-normal at each point of D. Conversely we have the following theorem:

Theorem 5. 1. Let $\mathscr{F}$ be a family of meromorphic functions in a domain D. If $\mathscr{F}$ is quasi-normal at each point of D, then $\mathscr{F}$ is quasi-normal in D.

This theorem is proved in the same way as the proof of Theorem 1. 4. In order to get the proof of Theorem 5. 1, it is sufficient to replace in the proof of Theorem 1. 4, the word "normal" by "quasi-normal", the term "C_0-sequence" by "C_1-sequence" and the term "C_0-point" by "C_1-point".

Evidently if a family $\mathscr{F}$ of meromorphic functions in a domain D is normal in D, then $\mathscr{F}$ is also quasi-normal in D. Conversely we have the following theorem:

Theorem 5. 2. Let $\mathscr{F}$ be a quasi-normal family of meromorphic functions in a domain D and let δ $(0<\delta<1)$ be a number. Assume that for each function $f(z)$ of $\mathscr{F}$, there are two values $a_j(f)\in\hat{\mathbf{C}}$ $(j=1,2)$ such that

$$|a_1(f),a_2(f)|>\delta \tag{5. 1}$$

and that the function $f(z)$ does not take the values $a_j(f)$ $(j=1,2)$ in D. Then the family $\mathscr{F}$ is normal in D.

For the proof of this theorem, we need some lemmas.

Lemma 5. 1. Let $S: f_n(z)$ $(n=1,2,\cdots)$ be a C_1-sequence of holomorphic functions in a domain D. Assume that for each $n\geqslant 1$, the function $f_n(z)$ has no zero in D. Then S is a C_0-sequence in D.

Proof. Assume on the contrary that S is not a C_0-sequence in D. Let E be the set of nonC_0-points of S in D. We have seen in the above that as $n\to+\infty$, $f_n(z)$ converges locally uniformly in the domain $D_1=D-E$ to the constant ∞. Consider a point $z_0 \in E$. Let $0<|z-z_0|<r$ be a domain belonging to D_1. Then as $n \to +\infty$, $f_n(z)$ converges uniformly to ∞ on the circle $\Gamma: |z-z_0|=r/2$. Since $1/f_n(z)$ is holomorphic in D, it follows that, as $n \to +\infty$, $1/f_n(z)$ converges uniformly to zero in the circle $|z-z_0|<r/2$. So z_0 is a C_0-point of the sequence $1/f_n(z)$ $(n=1, 2, \cdots)$ and hence z_0 is also a C_0-point of the sequence S, incompatible with the fact that $z_0 \in E$.

Lemma 5. 2. Let $S: f_n(z)$ $(n=1,2,\cdots)$ be a sequence of meromorphic functions in a domain D. Let $a_n (n=1,2,\cdots)$ be a sequence of complex numbers such that as $n \to +\infty$, a_n tends to a finite limit a. If S is a C_0-sequence (C_1-sequence) in D, then the sequence $f_n(z)+a_n (n=1,2,\cdots)$ is a C_0-sequence (C_1-sequence) in D.

Proof. Consider first the case that S is a C_0-sequence in D. Let F(z) be the limit function of S, with respect to the spherical distance. Consider a point z_0 of D and distinguish two cases:

1) $F(z_0) \neq \infty$. Then by Theorem 1. 1, we can find a circle $\Gamma_0: |z-z_0|<r_0$ and a positive integer n_0 such that the functions $f_n(z) (n \geqslant n_0)$ and F(z) are holomorphic in Γ_0 and as $n \to +\infty$, $f_n(z)$ converges uniformly to F(z) in Γ_0. Consequently as $n \to +\infty$, $f_n(z)+a_n$ converges uniformly to F(z)+a in Γ_0, and z_0 is a C_0-point of the sequence $f_n(z)+a_n (n=1,2,\cdots)$.

2) $F(z_0)=\infty$. Then by Theorem 1. 1, we can find a circle $\Gamma_0: |z-z_0|<r_0$ and a positive integer n_0 such that the functions $1/f_n(z)$ $(n \geqslant n_0)$ and 1/F(z) are holomorphic in Γ_0 and as $n \to +\infty$, $1/f_n(z)$ converges uniformly to 1/F(z) in Γ_0. Let $A>0$ be a number such that

$$|a_n| \leqslant A \ (n=1,2\cdots), \ |a| \leqslant A.$$

Since z_0 is a zero of the function 1/F(z), we can find a circlr $\Gamma_1: |z-z_0|<r_1 (0<r_1 \leqslant r_0)$ and a positive integer $n_1 \geqslant n_0$, such that when $n \geqslant n_1$, we have

$$\left|\frac{1}{f_n(z)}\right|<\frac{1}{2A}, \quad \left|\frac{1}{F(z)}\right|<\frac{1}{2A}$$

in Γ_1. Then from the identity

$$\frac{1}{f_n(z)+a_n}=\frac{1}{f_n(z)}\frac{1}{1+\dfrac{a_n}{f_n(z)}},$$

we see that, as $n\to+\infty$, $1/\{f_n(z)+a_n\}$ converges uniformly to the function

$$\frac{1}{F(z)}\frac{1}{1+\dfrac{a}{F(z)}}$$

in Γ_1. Hence z_0 is a C_0-point of the sequence $1/\{f_n(z)+a_n\}$ $(n\geqslant n_1)$. Consequently z_0 is a C_0-point of the sequence $f_n(z)+a_n(n=1,2,\cdots)$. Since z_0 is arbitrary, the sequence $f_n(z)+a_n(n=1,2,\cdots)$ is therefore a C_0-sequence in D.

Next consider the case that S is a C_1-sequence in D. Let $z_0\in D$ be a C_0-point of S. In the above we have shown that z_0 is a C_0-point of the sequence $f_n(z)+a_n$ $(n=1,2,\cdots)$. Now consider a point $z_0\in D$. By hypothesis, z_0 is a C_1-point of S. So we can find a domain $\Delta: 0<|z-z_0|<r$ belonging to D, such that each point of Δ is a C_0-point of S. Hence each point of Δ is a C_0-point of the sequence $f_n(z)+a_n(n=1,2,\cdots)$ of which z_0 is therefore a C_1-point. Since z_0 is arbitrary, the sequence $f_n(z)+a_n(n=1,2,\cdots)$ is a C_1-sequence in D.

Lemma 5. 3. Let $S: f_n(z)$ $(n=1,2,\cdots)$ be a sequence of meromorphic functions in a domain D, such that $f_n(z)\not\equiv 0$ in $D(n=1,2,\cdots)$. If S is a C_0-sequence (C_1-sequence) in D, then the sequence $1/f_n(z)$ $(n=1,2,\cdots)$ is a C_0-sequence (C_1-sequence) in D.

For the proof of this lemma, it is sufficient to note that if $z_0\in D$ is a C_0-point of S, then z_0 is also a C_0-point of the sequence $1/f_n(z)$ $(n=1,2,\cdots)$.

Lemma 5. 4. Let $S: f_n(z)$ $(n=1,2,\cdots)$ be a C_1-sequence of meromorphic functions in a domain D. Assume that to each integer $n\geqslant 1$ correspond two values a_n and $b_n(a_n\neq b_n)$, finite or infinite, satisfying the following conditions:

1^0 $f_n(z)$ does not take the values a_n and b_n in D.

2^0 As $n\to+\infty$, a_n and $1/(b_n-a_n)$ tend respectively to finite limits a and λ. Then S is a C_0-sequence in D.

Proof. Evidently we can find a positive integer n_0 such that a_n is finite for $n \geqslant n_0$. Consider the sequence of functions

$$g_n(z) = \frac{1}{f_n(z)-a_n} - \frac{1}{b_n-a_n} \quad (n \geqslant n_0)$$

By Lemma 5. 2, the sequence $f_n(z)-a_n (n \geqslant n_0)$ is a C_1-sequence in D, then by Lemma 5. 3, the sequence $1/\{f_n(z)-a_n\} (n \geqslant n_0)$ is a C_1-sequence in D, finally by Lemma 5. 2, the sequence $g_n(z) (n \geqslant n_0)$ is a C_1-sequence in D. But $g_n(z)$ is holomorphic and does not take the value zero in D. Hence by Lemma 5. 1, the sequence $g_n(z) (n \geqslant n_0)$ is a C_0-sequence in D. Then by Lemma 5. 2 and 5. 3, we see finally that the sequence S is a C_0-sequence in D.

Now let us return to the proof of Theorem 5. 2. Let $f_n(z)$ $(n=1,2,\cdots)$ be a sequence of functions of the family $\mathscr{F}$. By hypothesis, we can extract from this sequence a subsequence $f_{n_k}(z) (k=1,2,\cdots)$ which is a C_1-sequence in D. Set

$$a_j^{(k)} = a_j(f_{n_k}) \ (k = 1,2,\cdots) \ (j = 1,2).$$

By Lemma 3. 2, we can find an increasing sequence of positive integers $k_p (p=1, 2,\cdots)$ such that each of the two sequences $a_j^{(k_p)} (p=1,2,\cdots)$ $(j=1,2)$ satisfies one of the conditions 1^0 and 2^0 in Lemma 3. 2. Let $A_j (j=1,2,)$ be respectively the limits of the two sequences $a_j^{(k_p)} (p=1,2,\cdots)(j=1,2)$. Since by (5. 1),

$$|a_1^{(k)}, a_2^{(k)}| > \delta,$$

we have

$$|A_1, A_2| \geqslant \delta.$$

Hence one at least of $A_j (j=1,2)$ is finite. Assume for instance, A_1 is finite. Then the sequence $a_1^{(k_p)} (p=1,2,\cdots)$ must satisfy the condition 1^0 in Lemma 3. 2.

Consider the sequence $f_{m_p}(z) (p=1,2,\cdots; m_p=n_{k_p})$ which is also a C_1-sequence in D, and the two sequences of values $a_j^{(k_p)} (p=1,2,\cdots)(j=1,2)$. For each integer $p \geqslant 1$, the function $f_{m_p}(z)$ does not take the values $a_j^{(k_p)} (j=1,2)$ in D. Moreover as $p \to +\infty$, $a_1^{(k_p)}$ and $1/\{a_2^{(k_p)}-a_1^{(k_p)}\}$ tend respectively to finite limits

A_1 and $1/(A_2-A_1)$. Consequently by Lemma 5.4, the sequence $f_{m_p}(z)(p=1,2,\cdots)$ is a C_0-sequence in D. This completes the proof of Theorem 5.2.

Corollary 5.1. Let $\mathscr{F}$ be a quasi-normal family of meromorphic functions in a domain D. Assume that there are two values $a_j \in \hat{C}(j=1,2;a_1 \neq a_2)$ such that each function f(z) of $\mathscr{F}$ does not take the values $a_j(j=1,2)$ in D. Then the family $\mathscr{F}$ is normal in D.

Corollary 5.2. Let $\mathscr{F}$ be a quasi-normal family of holomorphic functions in a domain D. Assume that there is a finite value a such that each function f(z) of $\mathscr{F}$ does not take the value a in D. Then the family $\mathscr{F}$ is normal in D.

5.2. CRITERIONS OF QUASI-NORMALITY OF FAMILIES OF MEROMORPHIC FUNCTIONS

Definition 5.3. Let $\mathscr{F}$ be a family of meromorphic functions in a domain D and $\nu \geqslant 0$ an integer. We say that $\mathscr{F}$ is quasi-normal of order ν at most in D, if from every sequence of functions $f_n(z)$ $(n=1,2,\cdots)$ of $\mathscr{F}$, we can extract a subsequence $f_{n_k}(z)(k=1,2,\cdots)$ which is a C_1-sequence in D and of which the number of nonC_0-points does not exceed ν.

In particular when $\nu=0$, the family $\mathscr{F}$ is normal in D.

Definition 5.4. A condition K is said to be a condition of normality, if it has the following properties:

1^0 If a meromorphic function f(z) in a domain D satisfies the condition K in D, then f(z) satisfies the condition K in every subdomain of D.

2^0 If each function of a family $\mathscr{F}$ of meromorphic functions in a domain D satisfies the condition K in D, then the family $\mathscr{F}$ is normal in D.

Example 1. Given three distinct values $a_j \in \hat{C}$ $(j=1,2,3)$, let K be the condition defined as follows: We say that a meromorphic function f(z) in a domain D satisfies the condition K in D, if f(z) does not take the values $a_j(j=1,2,3)$ in D. By Theorem 3.1, this condition K is a condition of normality.

Definition 5.5. Let K be a condition of normality and $\varepsilon > 0$ a number.

Let f(z) be a meromorphic function in a domain D. A circle $\Gamma: |z-z_0|<r$ is called a (K,ε)-circle of the function f(z) in D, if it has the following properties:

1^0 $\bar{\Gamma}: |z-z_0|\leqslant r$ belongs to D.

2^0 $r<\varepsilon$.

3^0 The function f(z) does not satisfy the condition K in Γ.

A finite number of (K,ε)-circles $\Gamma_j(j=1,2,\cdots,p)$ are said to be distinct, if their closures $\bar{\Gamma}_j(j=1,2,\cdots,p)$ are mutually disjoint.

Example 2. Given a number $\eta(0<\eta<1/2)$, let $K(\eta)$ be the condition defined as follows: We say that a meromorphic function f(z) in a domain D satisfies the condition $K(\eta)$ in D, if there are three values $a_j\in\hat{C}(j=1,2,3)$ such that

$$|a_j,a_{j'}|>\eta \quad (1\leqslant j,j'\leqslant 3, j\neq j')$$

and that the funciton f(z) does not take the values $a_j(j=1,2,3)$ in D. By Theorem 3.9, the condition $K(\eta)$ is a condition of normality. A $(K(\eta),\varepsilon)$-circle Γ of a meromorphic function f(z) in a domain D is a η-filling circle of f(z) in D, according to a definition given in an application of Theorem 3.9.

Theorem 5.3. Let $\mathscr{F}$ be a family of meromorphic functions in a domain D. Let K be a condition of normality and $\varepsilon>0$ a number. Let $\nu\geqslant 0$ be an integer. If each function f(z) of $\mathscr{F}$ has at most ν distinct (K,ε)-cirlces in D, then $\mathscr{F}$ is quasi-normal of order ν at most in D.

This theorem is a general criterion of quasi-normality. It enables us to deduce from every condition of normality a criterion of quasi-normality.

Proof. Distinguish two cases:

1) $\nu=0$. In this case, by hypothesis, each function f(z) of $\mathscr{F}$ has no (K,ε)-circle in D. We have to show that the family $\mathscr{F}$ is normal in D. In fact, consider a point z_0 of D. Let $\Gamma: |z-z_0|<r$ be a circle having the properties 1^0 and 2^0 in Definition 5.5. Then each function f(z) of $\mathscr{F}$ satisfies the condition K in Γ. Consequently, by Definition 5.4, $\mathscr{F}$ is normal in Γ. Since z_0 is arbitrary, hence, by Theorem 1.4, $\mathscr{F}$ is normal in D.

2) $\nu>0$. Let $f_n(z)(n=1,2,\cdots)$ be a sequence of functions of the family $\mathscr{F}$ and distinguish two cases:

$1'$) For each point z_0 of D, there are a circle $\Gamma: |z-z_0|<r$ belonging to D and

a positive integer N such that when $n \geqslant N$, the function $f_n(z)$ satisfies the condition K in Γ. Then the family $f_n(z)(n=1,2,\cdots)$ is normal in Γ, and hence is normal in D. Consequently from this sequence we can extract a subsequence $f_{n_k}(z)$ $(k=1,2,\cdots)$ which is a C_0-sequence in D.

2′) There exists a point z_0 of D such that whatever the circle $\Gamma: |z-z_0|<r$ belonging to D and the positive integer N, we can always find an integer $n \geqslant N$ such that the function $f_n(z)$ does not satisfy the condition K in Γ. For the sake of convenience, such a point z_0 will be called an exceptional point of the sequence $f_n(z)(n=1,2,\cdots)$. We are going to show that we can extract from the sequence $f_n(z)(n=1,2,\cdots)$, a subsequence $f_{\alpha_i}(z)(i=1,2,\cdots)$ having the following property (π) with respect to the exceptional point z_0: For any circle $\Gamma: |z-z_0|<r$ belonging to D, the function $f_{\alpha_i}(z)$ does not satisfy the condition K in Γ, provided that i is sufficiently large. In fact, take a sequence of positive numbers $r_i(i=1,2,\cdots)$ tending to zero, such that the circles $\Gamma_i: |z-z_0|<r_i(i=1,2,\cdots)$ all belong to D. Then we can find an increasing sequence of positive integers $\alpha_i(i=1,2,\cdots)$ such that for each i the function $f_{\alpha_i}(z)$ does not satisfy the condition K in Γ_i. It is easy to see that the sequence $f_{\alpha_i}(z)$ $(i=1,2,\cdots)$ has the required property (π) with respect to the exceptional point z_0.

The point z_0 being an exceptional point of the sequence $f_{\alpha_i}(z)(i=1,2,\cdots)$, suppose that it has another exceptional point $z_0' \neq z_0$ in D. Then from the sequence $f_{\alpha_i}(z)(i=1,2,\cdots)$ we can extract a subsequence $f_{\beta_j}(z)(j=1,2,\cdots)$ having the property (π) with respect to the point z_0'. Evidently the sequence $f_{\beta_j}(z)(j=1,2,\cdots)$ also has the property (π) with respect to the point z_0.

The points z_0 and z_0' being exceptional points of the sequence $f_{\beta_j}(z)(j=1,2,\cdots)$, suppose that it has another exceptional point $z_0'' \neq z_0, z_0'$. Then from the sequence $f_{\beta_j}(z)(j=1,2,\cdots)$ we can extract a subsequence $f_{\gamma_k}(z)(k=1,2,\cdots)$ having the property (π) with respect to the point z_0''. The sequence $f_{\gamma_k}(z)(k=1,2,\cdots)$ also has the property (π) with respect to z_0 and z_0'.

If this operation can be repeated s times, then we get a subsequence $f_{\lambda_l}(z)(l=1,2,\cdots)$ of the sequence $f_n(z)(n=1,2,\cdots)$ and s distinct points $z_0, z_0', \cdots, z_0^{(s-1)}$ of D, such that the sequence $f_{\lambda_l}(z)(l=1,2,\cdots)$ has the property (π) with respect to each of the points $z_0, z_0', \cdots, z_0^{(s-1)}$. We are going to show that $s \leqslant \nu$, where ν is the integer in Theorem 5.3. In fact, ε being the positive number in Theorem 5.3, let us take a positive number $r<\varepsilon$ such that the circles $\overline{\Gamma}_j: |z-z_0^{(j)}| \leqslant r(j=0,1,\cdots,s-1; z^{(0)}=z_0)$ belong to D and are mutually disjoint. Then we can get a positive integer l such that in each of the circles $\Gamma_j: |z-z_0^{(j)}|<r(j=0,1,\cdots,s-1)$, the function $f_{\lambda_l}(z)$ does not satisfy the condition K. So by hypothesis, we must have $s\leqslant\nu$.

Thus for a certain positive integer $s\leqslant\nu$, the subsequence $f_{\lambda_l}(z)(l=1,2,\cdots)$ obtained in repeating the operation s times, has no other exceptional point distinct from the s points $z_0, z_0', \cdots, z_0^{(s-1)}$. Then in the domain $D_1=D-(z_0,z_0',\cdots,z_0^{(s-1)})$ the sequence $f_{\lambda_l}(z)(l=1,2,\cdots)$ satisfies the condition of the first case. Hence we can extract from this sequence, a subsequence $f_{n_k}(z)(k=1,2,\cdots)$ which is a C_0-sequence in D_1. Consequently the sequence $f_{n_k}(z)(k=1,2,\cdots)$ is a C_1-sequence in D, of which only the $s\leqslant\nu$ points $z_0,z_0',\cdots,z_0^{(s-1)}$ may be nonC_0-points.

Now let us deduce some criterions of quasi-normality from Theorem 5. 3.

Corollary 5. 3. Let D be a domain, $0<\delta<1$ a number and $p_j\geqslant0(j=1,2,3)$ three integers. Let $\mathscr{F}$ be the family of the functions f(z) satisfying the following conditions:

1^0 f(z) is meromorphic in D.

2^0 There are three values $a_j(f)\in\hat{\mathbf{C}}$ $(j=1,2,3)$ such that

$$|a_j(f),a_{j'}(f)|>\delta \quad (1\leqslant j,j'\leqslant 3, j\neq j')$$

and that for each $1\leqslant j\leqslant 3$, the equation

$$f(z)=a_j(f)$$

has at most p_j distinct roots in D.

Then the family $\mathscr{F}$ is quasi-normal of order $\sum_{j=1}^{3}p_j$ at most in D.

This corollary is a generalization of Theorem 3. 9.

Proof. If $p_j=0(j=1,2,3)$, then each of the equations

$$f(z)=a_j(f) \quad (j=1,2,3) \tag{5. 2}$$

has no root in D. Hence by Theorem 3. 9, $\mathscr{F}$ is normal in D. Consider the case $\sum_{j=1}^{3}p_j>0$. Let K(δ) be the condition of normality defined in Example 2, and let ε

>0 be a number. Consider a function f(z) of the family $\mathscr{F}$. By Theorem 5. 3, it is sufficient to show that f(z) has at most $\sum_{j=1}^{3} p_j$ distinct $(K(\delta),\varepsilon)$-circles in D. In fact, assume, on the contrary, that f(z) has $q=\sum_{j=1}^{3} p_j+1$ distinct $(K(\delta),\varepsilon)$-circles $\Gamma_i(i=1,2,\cdots,q)$ in D. Consider the three corresponding values $a_j(f)(j=1,2,3)$. Then for each $1\leqslant i\leqslant q$, one at least of the equations (5. 2) has a root in Γ_i. This implies that the total number of the distinct roots of the equations (5. 2) is at least equal to q, incompatible with the condition 2^0 in Corollary 5. 3.

Corollary 5. 4. Let D be a domain, $a_j\in\hat{\mathbf{C}}(j=1,2,3)$ three distinct values and $p_j\geqslant 0(j=1,2,3)$ three integers. Let $\mathscr{F}$ be the family of the functions f(z) meromorphic in D and such that for each $1\leqslant j\leqslant 3$, the equation

$$f(z)=a_j$$

has at most p_j distinct roots in D. Then the family $\mathscr{F}$ is quasi-normal of order $\sum_{j=1}^{3} p_j$ at most in D.

This corollary is a generalization of Theorem 3. 1.

Corollary 5. 5. Let D be a domain. Let $m_j\geqslant 2(j=1,2,\cdots,q;3\leqslant q\leqslant 5)$ be such that for each $1\leqslant j\leqslant q$, m_j is an integer or $+\infty$ and that

$$\sum_{j=1}^{q}(1-\frac{1}{m_j})>2. \tag{5. 3}$$

Let $0<\delta<1/q$ be a number and $p_j\geqslant 0$ $(j=1,2,\cdots q)$ q integers. Let $\mathscr{F}$ be the family of the functions f(z) satisfying the following conditons:

1^0 f(z) is meromorphic in D.

2^0 There are q values $a_j(f)\in\hat{\mathbf{C}}(j=1,2,\cdots,q)$ such that

$$|a_j(f),a_{j'}(f)|>\delta \quad (1\leqslant j,j'\leqslant q,j\neq j')$$

and that for each $1\leqslant j\leqslant q$, the equation

$$f(z) = a_j(f) \tag{5.4}$$

has in D at most p_j distinct roots whose order of multiplicity is less than m_j. Then the family $\mathscr{F}$ is quasi-normal of order $\sum_{j=1}^{q} p_j$ at most in D.

This corollary is a generalization of Theorem 3.13.

Proof. If $p_j = 0 (j=1,2,\cdots,q)$, then for each $1 \leqslant j \leqslant q$, the equation (5.4) has no root whose order of multiplicity is less than m_j, and by Theorem 3.13, the family $\mathscr{F}$ is normal in D. Consider the case $\sum_{j=1}^{q} p_j > 0$. Let $K = K(m_1, m_2, \cdots, m_q, \delta)$ be the condition defined as follows: We say that a meromorphic function f (z) in a domain D satisfies the condition K in D, if f(z) satisfies the condition 2^0 in Theorem 3.13. By Theorem 3.13, this condition K is a condition of normality. Let $\varepsilon > 0$ be a number. Consider a function f(z) of the family $\mathscr{F}$. By Theorem 5.3, it is sufficient to show that f(z) has at most $\sum_{j=1}^{q} p_j$ distinct (K, ε)-circles in D. Assume on the contrary, that f(z) has $P = \sum_{j=1}^{q} p_j + 1$ distinct (K, ε)-circles Γ_i $(i=1,2,\cdots,P)$ in D. Consider the q corresponding values $a_j(f) (j=1,2,\cdots,q)$ satisfying the condition 2^0 in Corollary 5.5. Denote by σ_j the set of the roots of the equation (5.4), whose order of multiplicity is less than m_j. Let $E = \bigcup_{j=1}^{q} \sigma_j$. Then for each $1 \leqslant i \leqslant P$, the circle Γ_i contains a point of E. Consequently E consists of at least P distinct points, incompatible with the condition 2^0 in Corollary 5.5.

Corollary 5.6. Let D be a domain. Let $a_j \in \hat{C} (j=1,2,\cdots,q; 3 \leqslant q \leqslant 5)$ be q distinct values. Let $m_j \geqslant 2 (j=1,2,\cdots,q)$ and $p_j \geqslant 0$ $(j=1,2,\cdots,q)$ have the same meaning as in Corollary 5.5. Let $\mathscr{F}$ be the family of the functions f(z) satisfying the following conditions:

1^0 f(z) is meromorphic in D.

2^0 For each $1 \leqslant j \leqslant q$, the equation

$$f(z) = a_j$$

has in D at most p_j distinct roots whose order of multiplicity is less than m_j.

Then the family $\mathscr{F}$ is quasi-normal of order $\sum_{j=1}^{q} p_j$ at most in D.

This corollary is a generalization of Theorem 3. 12.

Corollary 5. 7. Let D be a domain. Let M, $\delta(0<\delta<M)$ be two positive numbers and $\nu\geqslant 1$ an integer. Let $p\geqslant 0$ and $q\geqslant 0$ be two integers. Let $\mathscr{F}$ be the family of the functions f(z) satisfying the following conditions:

1^0 f(z) is meromorphic in D.

2^0 There are two values a(f) and b(f) such that

$$|a(f)|<M,\quad \delta<|b(f)|<M$$

and that the equations

$$f(z)=a(f) \quad and \quad f^{(\nu)}(z)=b(f) \tag{5. 5}$$

have respectively at most p distinct roots and q distinct roots in D. Then the family $\mathscr{F}$ is quasi-normal of order p+q at most in D.

This corollary is a generalization of Theorem 3. 17.

Proof. If p=q=0, then each of the equations (5. 5) has no root in D, and by Theorem 3. 17, the family $\mathscr{F}$ is normal in D. Consider the case p+q>0. Let $K=K(M,\delta,\nu)$ be the condition defined as follows: We say that a meromorphic function f(z) in a domain D satisfies the condition K in D, if f(z) satisfies the condition 2^0 in Theorem 3. 17. By this theorem, the condition K is a condition of normality. Let $\varepsilon>0$ be a number. Consider a function f(z) of the family $\mathscr{F}$. By Theorem 5. 3, it is sufficient to show that f(z) has at most p+q distinct (K,ε)-circles in D. Assume on the contrary, that f(z) has p+q+1 distinct (K,ε)-circles $\Gamma_i(i=1,2,\cdots,p+q+1)$ in D. Consider the two corresponding values a(f) and b(f) satisfying the condition 2^0 in Corollary 5. 7. Then for each $1\leqslant i\leqslant p+q+1$, the circle Γ_i contains a root of one of the equations (5. 5), and hence the number of the distinct roots in D of the equations (5. 5) is at least equal to p+q+1, incompatible with the condition 2^0 in Corollary 5. 7.

Corollary 5. 8. Let D be a domain, a and $b\neq 0$ two complex numbers, and $\nu \geqslant 1$ an integer. Let $p \geqslant 0$ and $q \geqslant 0$ be two integers. Let $\mathscr{F}$ be the family of the functions $f(z)$ meromorphic in D and such that the equations

$$f(z) = a \quad and \quad f^{(\nu)}(z) = b$$

have respectively at most p distinct roots and q distinct roots in D. Then the family $\mathscr{F}$ is quasi-normal of order $p+q$ at most in D.

Now we introduce the notion of irreducible C_1-sequences, which plays an important role in the study of the order of a quasi-normal family of meromorphic functions.

Definition 5. 6. Let $S: f_n(z)(n=1,2,\cdots)$ be a C_1-sequence of meromorphic functions in a domain D and let z_0 a nonC_0-point of S in D. Then two cases are possible:

1^0 We can extract from the sequence S a subsequence S' of which z_0 is a C_0-point. In this case, we say that S is reducible with respect to the point z_0.

2^0 z_0 is a nonC_0-point of every subsequence of the sequence S. In this case, we say that S is irreducible with respect to the point z_0.

The sequence S is said to be an irreducible sequence in D, if in D, S has nonC_0-points with respect to each of which S is irreducible.

Lemma 5. 5. Let $S: f_n(z)(n=1,2,\cdots)$ be a C_1-sequence of meromorphic functions in a domain D. Assume that S has nonC_0-points in D. Then we can extract from S a subsequence S' which is either a C_0-sequence in D or an irreducible C_1-sequence in D.

For the proof of this lemma, we need the following lemma of which we give a proof by a method already used in the proofs of Theorems 1. 4 and 5. 1.

Lemma 5. 6. Let σ be a set consisting of an infinite number of points in a domain D. If σ has no point of accumulation in D, then σ is enumerable.

Proof. Let $z_j(j=1,2,\cdots)$ be a sequence of points of D, such that each point of D is a limiting point of this sequence $z_j(j=1,2,\cdots)$. Consider a point z_j and let R_j be the least upper bound of the set of the positive numbers r such that the circle $|z-z_j|<r$ belongs to D. Let Γ_j be the circle $|z-z_j|<R_j/2$, if $R_j<+\infty$, and the circle $|z-z_j|<1$, if $R_j=+\infty$. In this way, to each point z_j is associated a circle Γ_j with center z_j, such that its closure $\overline{\Gamma}_j\subset D$.

Now consider a point z' of D. Let $\gamma : |z-z'|<\rho$ $(0<\rho<1)$ be a circle belonging to D. By the choice of the sequence $z_j (j=1,2,\cdots)$, there is a point z_j such that $|z_j-z'|<\rho'$, $\rho'=\rho/4$. Then the circle $|z-z_j|<2\rho'$ belongs to γ. Consequently, by the definition of R_j, if $R_j<+\infty$, we have

$$2\rho' \leqslant R_j,\ \rho' \leqslant \frac{R_j}{2}$$

and the circle $|z-z_j|<\rho'$ belongs to the circle Γ_j; while if $R_j=+\infty$, then since $\rho'<1$, the circle $|z-z_j|<\rho'$ also belongs to Γ_j. But the point z' is a point of the circle $|z-z_j|<\rho'$, hence $z'\in\Gamma_j$. In particular, each point of the set σ belongs to one of the circles $\Gamma_j (j=1,2,\cdots)$. Since σ has no point of accumulation in D and $\overline{\Gamma}_j \subset D$, it follows that each circle Γ_j contains at most a finite number of points of σ. σ is therefore enumerable.

To prove Lemma 5.5, consider in general a C_1-sequence S of meromorphic functions in a domain D and a point z_0 of D. Let us define $\sigma(S,z_0)$ as follows:

1^0 If z_0 is a C_0-point of S or if z_0 is a nonC_0-point of S such that S is irreducible with respect to z_0, then we define $\sigma(S,z_0)$ to be S.

2^0 If z_0 is a nonC_0-point of S such that S is reducible with respect to z_0, then we define $\sigma(S,z_0)$ to be a subsequence S' of S, of which z_0 is a C_0-point.
Thus $\sigma(S,z_0)$ is either S or a subsequence of S; z_0 is either a C_0-point of $\sigma(S,z_0)$ or a nonC_0-point of $\sigma(S,z_0)$ which is irreducible with respect to z_0.

Now consider the sequence S in Lemma 5.5 and the set E of its nonC_0-points in D. Distinguish two cases:

1)E consists of only a finite number of points $z_j (j=1,2,\cdots,p)$. Set successively

$S_1=\sigma(S,z_1)$, $S_2=\sigma(S_1,z_2)$, $\cdots$, $S_p=\sigma(S_{p-1},z_p)$. Then it is easy to see that the sequence $S'=S_p$ has the required property in Lemma 5.5.

2)E consists of an infinite number of points. Since E has no point of accumulation in D, hence, by Lemma 5.6, E is enumerable and can be represented by a sequence of points $z_j (j=1,2,\cdots)$. Set successively

$$S_j = \sigma(S_{j-1}, z_j) \quad (j = 1, 2, \cdots; S_0 = S).$$

Let S' be the diagonal sequence deduced from $S_j(j=1,2,\cdots)$. Since for each $j \geq 1$, from the jth term onward all the following terms of S' belong to the sequence S_j, we see that S' has the required property in Lemma 5. 5.

Lemma 5. 7. Let $S: f_n(z)(n=1,2,\cdots)$ be an irreducible C_1-sequence of meromorphic functions in a domain D. Then there is a value $w_0 \in \hat{C}$ having the following properties:

1^0 For any nonC_0-point z_0 of S in D, any circle $\Gamma: |z-z_0|<r$ belonging to D and any sequence of values $w_n \in \hat{C}(n=1,2,\cdots)$ such that

$$\lim_{n \to +\infty} |w_n, W| = 0, \tag{5. 6}$$

where $W \neq w_0$, we can find a positive integer N such that when $n \geq N$, the equation $f_n(z) = w_n$ has root in Γ.

2^0 For any nonC_0-point z_0 of S in D, any circle $\Gamma: |z-z_0|<r$ belonging to D and any value $w \neq w_0$, we can find a positive integer N such that when $n \geq N$, the equation $f_n(z)=w$ has root in Γ.

For the proof of this lemma we need the following lemma:

Lemma 5. 8. Let $S: f_n(z)(n=1,2,\cdots)$ be a sequence of meromorphic functions in a circle $\Gamma: |z-z_0|<r$, such that S is a C_0-sequence in the domain $0<|z-z_0|<r$ and that z_0 is a nonC_0-point of S which is irreducible with respect to z_0. Assume that the limit function $F(z)$ of S, with respect to the spherical distance, is a meromorphic function in the domain $0<|z-z_0|<r$. Then we can find a positive integer N such that for each $n \geq N$, the function $f_n(z)$ has pole in Γ.

Proof. Let $\gamma: |z-z_0|=\rho(0<\rho<r)$ be a circle on which $F(z)$ is finite. Then there is a positive number A such that $|F(z)| \leq A$ on γ. Next let $B=A+1$ and consider the number $\mu=\mu(A,B)$ defined in Lemma 1. 2. By Theorem 1. 2 and the finite covering theorem, as $n \to +\infty$, $f_n(z)$ converges uniformly to $F(z)$ on γ, with respect to the spherical distance, hence there is a positive integer n_0 such that when $n \geq n_0$, we have

$$|f_n(z), F(z)| < \mu$$

on γ. By Lemma 1. 2, when $n \geqslant n_0$, we have

$$|f_n(z)| < B$$

on γ. Hence by the formula (1. 4), as $n \to +\infty$, $f_n(z)$ converges uniformly to $F(z)$ on γ.

Now assume that we can not find a positive integer N such that for each $n \geqslant N$, the function $f_n(z)$ has pole in the circle $\Gamma_1: |z-z_0| < \rho$. Then there is a subsequence S' : $f_{n_k}(z)(k=1,2,\cdots)$ such that for each $k \geqslant 1$, the function $f_{n_k}(z)$ has no pole in the circle Γ_1. By the principle of maximum modulus, we see that the point z_0 would be a C_0-point of S', incompatible with the condition that the sequence S is irreducible with respect to z_0.

Now let us return to the proof of Lemma 5. 7. First of all, we can find a positive integer n_0 such that $f_n(z) \not\equiv 0$ for $n \geqslant n_0$. For otherwise ther would be a subsequence $f_{n_k}(z) \equiv 0 (k=1,2,\cdots)$. This contradicts the condition that the sequence S is an irreducible C_1-sequence. Then by Lemma 5. 3, the sequence $1/f_n(z)(n \geqslant n_0)$ is also an irreducible C_1-sequence in D, whose nonC_0-points in D are those of the sequence S.

Consider the set E of the nonC_0-points in D of the sequence S. Then S is a C_0-sequence in the domain $D_1 = D\text{-}E$. Let $F(z)$ be the limit function of S, with respect to the spherical distance, defined in D_1. Distinguish two cases:

1) $F(z)$ is a non-constant meromorphic function in D_1. In this case, we are going to show that the assertion 1^0 in Lemma 5. 7 holds for any sequence of values $w_n \in \hat{C}(n=1,2,\cdots)$ satisfying the condition (5. 6), without restriction on W. In fact, let $w_n \in \hat{C}$ $(n=1,2,\cdots)$ be such a sequence and assume first that $W \neq \infty$. Then by Lemma 1. 2, we see that there is a positive integer n_0 such that $w_n \neq \infty$ for $n \geqslant n_0$ and

$$\lim_{n \to +\infty} w_n = W.$$

Consequently by Lemma 5. 2 and 5. 3, the sequence $1/\{f_n(z)-w_n\}(n \geqslant n_0', n_0' \geqslant n_0)$ is also an irreducible C_1-seuqence in D, whose set of nonC_0-points is the set E, and whose limit function in D_1, with respect to the spherical distance, is

$1/\{F(z)-W\}$. Consider a point $z_0 \in E$ and a circle $\Gamma:|z-z_0|<r$ belonging to D such that Γ contains no other point of E except z_0. Then by Lemma 5.8, we can find a positive integer $N \geqslant n_0'$, such that for $n \geqslant N$, the function $1/\{f_n(z)-w_n\}$ has pole in Γ, hence the equation $f_n(z)=w_n$ has root in Γ.

Next assume that $W=\infty$. Then we have

$$\lim_{n\to+\infty} \left| \frac{1}{w_n}, 0 \right| = 0.$$

In applying the result just obtained to the sequence of functions $1/f_n(z)(n \geqslant n_0)$ and the sequence of values $1/w_n(n \geqslant n_0)$, we see again that the equation $f_n(z)= w_n$ has root in Γ, when $n \geqslant N$.

2) $F(z)$ is a constant $w_0 \in \hat{\mathbf{C}}$. In this case, the method of the proof is the same as in case 1). In fact, consider a sequence of values $w_n \in \hat{\mathbf{C}}(n=1,2,\cdots)$ satisfying the condition (5. 6) with $W \neq w_0$. If $W \neq \infty$, then we consider again the sequence $1/\{f_n(z)-w_n\}(n \geqslant n_0')$ whose limit function in D_1, with respect to the spherical distance, is the constant $1/(w_0-W)$, and we get the same conclusion. If $W=\infty$, we again apply the result just obtained to the sequence of functions $1/f_n(z)(n \geqslant n_0)$ and the sequence of values $1/w_n(n \geqslant n_0)$.

Finally the assertion 2^0 in Lemma 5. 7 is a simple consequence of the assertion 1^0.

By means of the Lemmas 5. 5 and 5. 7, Corollary 5. 3 can be put into a precise form as follows:

Theorem 5. 4. Let D be a domain, $0<\delta<1$ a number and $p_j \geqslant 0(j=1,2,3)$ three integers such that $p_1 \leqslant p_2 \leqslant p_3$. Let $\mathscr{F}$ be the family of the functions $f(z)$ satisfying the conditions 1^0 and 2^0 in Corollary 5. 3. Then the family $\mathscr{F}$ is quasi-normal of order p_2 at most in D.

Proof. Assume $P=\sum_{j=1}^{3} p_j>0$. By Corollary 5. 3, the family $\mathscr{F}$ is quasi-normal of order P at most in D. Let S: $f_n(z)(n=1,2,\cdots)$ be a sequence of functions of $\mathscr{F}$. From S we can extract a subsequence S' which is a C_1-sequence in D and has at most P nonC_0-points in D. Assume that S' has nonC_0-points in D. Then by Lemma 5. 5, we can extract from S' a subsequence S'': $f_{n_k}(z)(k=1,2,\cdots)$ which is either a C_0-sequence in D or an irreducible C_1-sequence in D. Consider the latter case. Set

$$a_j^{(k)} = a_j(f_{n_k}) \quad (k = 1,2,\cdots) \quad (j = 1,2).$$

By Lemma 3. 2, we can find an increasing sequence of positive integers $k_\lambda(\lambda=1, 2,\cdots)$ such that each of the two sequences $a_j^{(k_\lambda)}(\lambda=1,2,\cdots)(j=1,2)$ satisfies one of the conditions 1^0 and 2^0 in Lemma 3. 2. Let $A_j(j=1,2)$ be respectively the limits of these two sequences. We have

$$\lim_{\lambda\to+\infty} |a_j^{(k_\lambda)}, A_j| = 0 \quad (j = 1,2)$$

and

$$|A_1, A_2| \geqslant \delta.$$

Consider the sequence $S''': f_{m_\lambda}(z)$ $(\lambda=1,2,\cdots; m_\lambda=n_{k_\lambda})$ which is also an irreducible C_1—points in D. Let $z_i(i=1,2,\cdots,v)$ be the nonC_0-points of S''' in D. It remains to show that $v\leqslant p_2$.

By Lemma 5. 7, there is a value $w_0\in\dot{C}$ having the property 1^0 in Lemma 5. 7, with respect to the sequence S'''. One at least of the two values $A_j(j=1,2)$ is different from w_0. for instance $A_1\neq w_0$. Consider v mutually disjoint circles $\Gamma_i: |z-z_i|<r_i(i=1,2,\cdots,v)$ belonging to D. Then for each $1\leqslant i\leqslant v$, we can find a positive integer λ_i such that when $\lambda\geqslant\lambda_i$, the equation $f_{m_\lambda}(z)=a_1^{(k_\lambda)}$ has root in Γ_i. Take a positive integer $\lambda'\geqslant\max\limits_{1\leqslant i\leqslant v}\lambda_i$. Then the equation

$$f_{m_{\lambda'}}(z) = a_1^{(k_{\lambda'})} \tag{5.7}$$

has root in each of the circles $\Gamma_i(i=1,2,\cdots,v)$. Since

$$a_1^{(k_{\lambda'})} = a_1(f_{m_{\lambda'}}),$$

by hypothesis, the equation (5. 7) has at most p_1 distinct roots. Hence $v\leqslant p_1$. This result is obtained in assuming $A_1\neq w_0$. If $A_2\neq w_0$, then we can show similarly that $v\leqslant p_2$. So we have always $v\leqslant p_2$.

Corollary 5. 9. Let D be a domain, $a_j \in \hat{C}$ $(j=1,2,3)$ three distinct values and $p_j \geqslant 0 (j=1,2,3)$ three integers such that $p_1 \leqslant p_2 \leqslant p_3$. Let $\mathscr{F}$ be the family of the functions f(z) meromorphic in D and such that for each $1 \leqslant j \leqslant 3$, the equation

$$f(z) = a_j$$

has at most p_j distinct roots in D. Then the family $\mathscr{F}$ is quasi-normal of order p_2 at most in D.

This corollary is due to Montel[26].

Corollary 5. 10. Let D be a domain, M, δ two positive numbers and $p_j \geqslant 0 (j=1,2)$ two integers such that $p_1 \leqslant p_2$. Let $\mathscr{F}$ be the family of the functions f(z) satisfying the following conditions:

1^0 f(z) is holomorphic in D.

2^0 There are two finite values $a_j(f) (j=1,2)$ such that

$$|a_j(f)| < M \quad (j = 1,2), |a_1(f) - a_2(f)| > \delta$$

and that for each $1 \leqslant j \leqslant 2$, the equation

$$f(z) = a_j(f)$$

has at most p_j distinct roots in D.

Then the family $\mathscr{F}$ is quasi-normal of order p_1 at most in D.

This corollary is a generalization of Theorem 2. 8.

Corollary 5. 10 is a simple consequence of Theorem 5. 4. It is sufficient to take in that theorem, δ the smaller of the two numbers

$$\frac{1}{(1+M^2)^{\frac{1}{2}}}, \quad \frac{\delta}{1+M^2},$$

$p_j (j=1,2,3)$ the integers $0 \leq p_1 \leq p_2$ and $a_j(f)(j=1,2,3)$ the values ∞, $a_1(f)$, $a_2(f)$.

Corollary 5.11. Let D be a domain, $a_j(j=1,2)$ two distinct finite values and $p_j \geqslant 0(j=1,2)$ two integers such that $p_1 \leqslant p_2$. Let $\mathscr{F}$ be the family of the functions f(z) holomorphic in D and such that for each $1 \leqslant j \leqslant 2$, the equation

$$f(z) = a_j$$

has at most p_j distinct roots in D. Then the family $\mathscr{F}$ is quasi-normal of order p_1 at most in D.

This corollary is also due to Montel [26].

Up to now, the criterions of quasi-normality are all deduced from corresponding conditions of normality. We are going to prove some criterions of quasi-normality of a different kind due to Chuang [9].

Consider first a meromorphic function f(z) in a domain D. A function of the form

$$M(f, f', \cdots, f^{(q)}) = a(f)^{s_0}(f')^{s_1} \cdots (f^{(q)})^{s_q},$$

where a is a constant and $s_i(i=0,1,\cdots,q)$ are non-negative integers, is called a differential monomial of f with constant coefficient. The integer

$$d(M) = \sum_{i=0}^{q} s_i$$

is called the degree of $M(f, f', \cdots, f^{(q)})$. The sum of a finite number of differential monomials of f with constant coefficient is called a differential polynomial of f with constant coefficients. In the sequel differential polynomials with constant coefficients of the following particular form play an important role:

$$P(f) = M_0(f, f', \cdots, f^{(k)}) + \sum_{k=1}^{n} M_j(f, f', \cdots, f^{(p)}), \tag{5.8}$$

where

$$M_0(f, f', \cdots, f^{(k)}) = a_0(f)^{s_0}(f')^{s_1} \cdots (f^{(k)})^{s_k}, \quad a_0 \neq 0$$

and

$$M_j(f, f', \cdots, f^{(p)}) = a_j(f)^{s_{j0}}(f')^{s_{j1}}\cdots(f^{(p)})^{s_{jp}}, \quad (j = 1, 2, \cdots, n)$$

satisfy the condition

$$d(M_j) < d(M_0) \ (j = 1, 2, \cdots, n), d = d(M_0) > 0.$$

Theorem 5.5. Let D be a domain and m a positive integer. Let P(f) be defined by (5.8). Let $\mathscr{F}$ be the family of the functions f(z) satisfying the following conditions:

1^0 f(z) is holomorphic in D.

2^0 The equation $e^{f(z)}P\{f(z)\}=1$ has at most m distinct roots in D.

Then the family $\mathscr{F}$ is quasi-normal of order k-1 at most in D.

Proof. First of all, by Corollary 14 in Appendix A, if F(z) is a holomorphic function in the circle $|z|<1$, such that the equation

$$e^{F(z)}P\{F(z)\} = 1$$

has at most m distinct roots in the circle $|z|<1$, then in the circle $|z|\leqslant 1/2$, we have

$$|F(z)| < \lambda(A + \sum_{i=0}^{k-1} \frac{|F^{(i)}(0)|}{i!}), \tag{5.9}$$

where λ is a positive constant depending only on k, p, d and where

$$A = 1 + m + |\log|a_0|| + (\sum_{j=1}^{n} \left|\frac{a_j}{a_0}\right|)^2.$$

Consider a point z_0 of the domain D and a circle $\Gamma: |z-z_0|<\delta (0<\delta<1)$ belonging to D. Let f(z) be a function of the family $\mathscr{F}$. Then the function $F(\zeta)=f(z_0+\delta\zeta)$ is holomorphic in the circle $|\zeta|<1$ and the equation

$$e^{F(\zeta)}P_1\{F(\zeta)\} = 1$$

has at most m distinct roots in the circle $|\zeta|<1$, where $P_1\{F(\zeta)\}$ is obtained in replacing in $P(f)$, f by F, a_0 by a_0/δ^{μ} and a_j by a_j/δ^{μ_j}, $\mu>0$ and $\mu_j\geqslant 0(j=1,2,\cdots,n)$ are integers. Consequently, applying the quoted corollary to the function $F(\zeta)$, we see that, in the circle $|z-z_0|\leqslant\delta/2$, we have

$$|f(z)|<\lambda(A'+\sum_{i=0}^{k-1}\frac{\delta^i|f^{(i)}(z_0)|}{i!}), \tag{5.10}$$

where

$$A'=1+m+\left|\log\frac{|a_0|}{\delta^{\mu}}\right|+(\sum_{j=1}^{n}\delta^{\mu-\mu_j}\left|\frac{a_j}{a_0}\right|)^2.$$

By (5.10) and Cauchy's inequalities, in the circle $|z-z_0|<\delta/4$, we have

$$\frac{|f^{(i)}(z)|}{i!}<(\frac{4}{\delta})^i H,$$

where

$$H=\lambda(A'+\sum_{i=0}^{k-1}\frac{|f^{(i)}(z_0)|}{i!}).$$

Hence in the circle $|z-z_0|<\delta/4$, we have

$$\begin{aligned}1+\sum_{i=0}^{k-1}\frac{|f^{(i)}(z)|}{i!}&<1+\sum_{i=0}^{k-1}(\frac{4}{\delta})^i H<(1+\sum_{i=0}^{k-1}(\frac{4}{\delta})^i)H\\&\leqslant A(1+\sum_{i=0}^{k-1}\frac{|f^{(i)}(z_0)|}{i!}),\end{aligned} \tag{5.11}$$

where

$$A=(1+\sum_{i=0}^{k-1}(\frac{4}{\delta})^i)\lambda A'.$$

Now keeping the point $z_0\in D$ fixed, let $S: f_n(z)(n=1,2,\cdots)$ be a sequence of functions of the family $\mathscr{F}$. Define

$$B(z_0,f_n)=\max_{0\leqslant i\leqslant k-1}\frac{|f_n^{(i)}(z_0)|}{i!}\quad (n=1,2,\cdots)$$

and distinguish two cases:

1) The sequence $B(z_0,f_n)$ $(n=1,2,\cdots)$ is bounded. In this case, we are going to show that the sequence S is locally uniformly bounded in D. In fact, by (5.11), in the circle $|z-z_0|<\delta/4$, we have for each $n\geqslant 1$,

$$1+\sum_{i=0}^{k-1}\frac{|f_n^{(i)}(z)|}{i!}<A(1+\sum_{i=0}^{k-1}\frac{|f_n^{(i)}(z_0)|}{i!}). \tag{5.12}$$

A fortiori the sequence S is uniformly bounded in the circle $|z-z_0|<\delta/4$. Now consider a point $z_* \in D$, $z_*\neq z_0$. Join z_0 and z_* by a polygonal line L lying in D and $0<\delta<1$ be a number such that for each point $\zeta\in L$, the circle $|z-\zeta|<\delta$ belongs to D. Then by (5.11), for any point $\zeta\in L$ and any function $f(z)\in\mathscr{F}$, we have

$$1+\sum_{i=0}^{k-1}\frac{|f^{(i)}(z)|}{i!}<A(1+\sum_{i=0}^{k-1}\frac{|f^{(i)}(\zeta)|}{i!})$$

in the circle $|z-\zeta|<\delta/4$. In the sense from z_0 to z_*, take a finite number of points $\zeta_j(j=1,2,\cdots,q)$ of L, such that

$$|\zeta_{j+1}-\zeta_j|<\frac{\delta}{4}\quad (j=1,2,\cdots,q-1),\quad \zeta_1=z_0,\ \zeta_q=z_*.$$

Then for each $n\geqslant 1$, we have

$$1+\sum_{i=0}^{k-1}\frac{|f_n^{(i)}(\zeta_{j+1})|}{i!}<A(1+\sum_{i=0}^{k-1}\frac{|f_n^{(i)}(\zeta_j)|}{i!})\ (j=1,2,\cdots,q-1).$$

Hence

$$1+\sum_{i=0}^{k-1}\frac{|f_n^{(i)}(z_*)|}{i!}<A^{q-1}(1+\sum_{i=0}^{k-1}\frac{|f_n^{(i)}(z_0)|}{i!})$$

and in the circle $|z-z_0|<\delta/4$, we have

$$1+\sum_{i=0}^{k-1}\frac{|f_n^{(i)}(z)|}{i!}<A^q\left(1+\sum_{i=0}^{k-1}\frac{|f_n^{(i)}(z_0)|}{i!}\right). \tag{5.13}$$

Consequently the sequence S is uniformly bounded in the circle $|z-z_0|<\delta/4$. Thus S is locally uniformly bounded in D and hence by Corollary 1.3 we can extract from S a subsequence S′ which is a C_0-sequence in D.

2) The sequence $B(z_0,f_n)(n=1,2,\cdots)$ is unbounded. In view of the result obtained in the first case, it is sufficient to consider the case that

$$\lim_{n\to+\infty} B(z_0,f_n)=+\infty. \tag{5.14}$$

Let n_0 be a positive integer such that $B(z_0,f_n)>1$ for $n\geqslant n_0$. Then by (5.12) and (5.13), the sequence of functions

$$\varphi_n(z)=\frac{f_n(z)}{B(z_0,f_n)}\quad(n\geqslant n_0)$$

is locally uniformly bounded in D. Consequently from the sequence $\varphi_n(z)(n\geqslant n_0)$ we can extract a subsequence

$$\varphi_{n_h}(z)=\frac{f_{n_h}(z)}{B(z_0,f_{n_h})}\quad(h=1,2,\cdots)$$

such that as $h\to+\infty$, $\varphi_{n_h}(z)$ converges locally uniformly to a holomorphic function $\varphi(z)$ in D. It is easy to see that $\varphi(z)\not\equiv 0$. In fact, evidently there is an integer $i(0\leqslant i\leqslant k-1)$ such that for an infinite number of integers h, we have

$$B(z_0,f_{n_h})=\frac{|f_{n_h}^{(i)}(z_0)|}{i!}$$

and hence

$$|\varphi_{n_h}^{(i)}(z_0)| = i!.$$

So we must have $\varphi(z) \not\equiv 0$. It is then clear that the only possible nonC_0-points of the sequence

$$f_{n_h}(z) = B(z_0, f_{n_h})\varphi_{n_h}(z) \quad (h = 1, 2, \cdots) \tag{5.15}$$

are the zeros of the function $\varphi(z)$.

Assume that $\varphi(z)$ has a zero $\zeta_0 \in D$. Then we must have $\varphi^{(k)}(z) \equiv 0$. In fact, suppose, on the contrary, that $\varphi^{(k)}(z) \not\equiv 0$. Let $\bar{\omega}$: $r_1 \leqslant |z - \zeta_0| \leqslant r_2 (0 < r_1 < r_2)$ be a region belonging to D, such that in $\bar{\omega}$ the functions $\varphi^{(i)}(z)(i = 0, 1, \cdots, k)$ have no zero. Then evidently we can find two positive numbers M and M' $(M' > M > 0)$ and a positive integer h_0 such that when $h \geqslant h_0$, we have, in $\bar{\omega}$,

$$|\varphi_{n_h}^{(i)}(z)| > M (i = 0, 1, \cdots, k), |\varphi_{n_h}^{(i)}(z)| < M' (i = 0. 1. \cdots, q), \tag{5.16}$$

where $q = \max(k, p)$.

For simplicity let us write

$$\beta_h = B(z_0, f_{n_h}), \quad \psi_h(z) = \varphi_{n_h}(z).$$

Then by (5.14) and the condition 2^0 in Theorem 5.5, the equation

$$e^{\beta_h \psi_h(z)} P\{\beta_h \psi_h(z)\} = 1 \tag{5.17}$$

has at most m distinct roots in the domain ω: $r_1 < |z - \zeta_0| < r_2$. On the other hand, by (5.14) and (5.16), we see that when $h \geqslant h_0' (h_0' \geqslant h_0)$, the equation

$$P\{\beta_h \psi_h(z)\} = 0$$

has no root in ω. Hence by Corollary 5.11, the family

$$g_h(z) = e^{\beta_h \psi_h(z)} P\{\beta_h \psi_h(z)\} (h \geqslant h_0'). \tag{5.18}$$

is normal in ω. Consider a circle Γ : $|z - \zeta_0| = r (r_1 < r < r_2)$. Since the function $\varphi(z)$

has a zero ζ_0 in the interior of Υ, we see that there are two points z' and z'' on Υ, such that

$$Re\{\varphi(z')\} < 0, Re\{\varphi(z'')\} > 0.$$

Then when $h \geqslant h_0^*(h_0^* \geqslant h_0')$ we have

$$Re\{\psi_h(z')\} < \frac{1}{2}Re\{\varphi(z')\}, \quad Re\{\psi_h(z'')\} > \frac{1}{2}Re\{\varphi(z'')\}.$$

and by (5. 18),

$$\lim_{h\to+\infty} |g_h(z')| = 0, \quad \lim_{h\to+\infty} |g_h(z'')| = +\infty.$$

Consequently the family (5. 18) can not be normal in ω. So we get a contradiction. Hence we must have $\varphi^{(k)}(z) \equiv 0$. $\varphi(z)$ is therefore identically equal to a polynomial of degree k-1 at most and the sequence $f_{n_h}(z)(h=1,2,\cdots)$ defined by (5. 15) has at most k-1 nonC_0-points in D. This completes the proof of Theorem 5. 5.

Lemma 5.9. Let f(z) be a holomorphic function in a region $\mathbf{R}/8 \leq |z-a| \leq \mathbf{R}(0 < \mathbf{R} < +\infty)$. Assume that in this region f(z) can be put in the form

$$f(z) = (\frac{z-a}{z_0-a})^m C\{1+\varphi(z)\}, |\varphi(z)| < \frac{1}{10},$$

where $m \geq 1$ is an integer, z_0 a point of the circle $|z-a| = \mathbf{R}$, $C \neq 0$ a complex number and $\varphi(z)$ a holomorphic function in the region $\mathbf{R}/8 \leq |z-a| \leq \mathbf{R}$. Then for each real number ω, there is a simply connected domain $\mathbf{D}_\omega$ belonging to the domain $\mathbf{R}/8 \leq |z-a| \leq \mathbf{R}$, such that f(z) is univalent in $\mathbf{D}_\omega$ and maps $\mathbf{D}_\omega$ onto the domain

$$\frac{1}{4}|C| < |w| < \frac{1}{2}|C|, \quad \omega < arg\ w < \omega + 2\pi. \tag{5.19}$$

As usual, to say that a function f(z) is holomorphic in the region $\mathbf{R}/8 \leq |z-a| \leq \mathbf{R}$ means that f(z) is holomorphic in a domain containing this region.

Proof. Consider a real number ω. Write

$$z_0-a = Re^{i\theta_0}, \quad C = |C|e^{i\alpha}, \quad \psi_0 = \theta_0-\frac{\alpha}{m}+\frac{\omega+\pi}{m} \tag{5.20}$$

and define

$$F(Z) = f(a+Ze^{i\psi_0}). \tag{5.21}$$

The function F(Z) is holomorphic in the region E: $R/8 \leqslant |Z| \leqslant R$ and in E we have

$$F(Z) = (\frac{Z}{R})^m e^{i(\omega+\pi)}|C|\{1+\Phi(Z)\}, |\Phi(Z)| < \frac{1}{10}. \tag{5.22}$$

Evidently it is sufficient to show that there is a simply connected domain Δ belonging to the domain $R/8 < |Z| < R$, such that the function

$$g(Z) = (\frac{Z}{R})^m \{1+\Phi(Z)\} \tag{5.23}$$

is univalent in Δ and maps Δ onto the domain

$$\frac{1}{4} < |w| < \frac{1}{2}, \quad -\pi < arg\ w < \pi. \tag{5.24}$$

In fact, since

$$F(Z) = e^{i(\omega+\pi)}|C|g(Z),$$

the function F(Z) is univalent in Δ and maps Δ onto the domain (5. 19). When Z describes Δ, the point $z=a+Ze^{i\psi_0}$ describes a domain D_ω belonging to the domain $R/8 < |z-a| < R$. It is easy to see that the domain D_ω satisfies the required conditions in Lemma 5. 9.

In (5. 23), the function

$$p(Z) = (\frac{Z}{R})^m \tag{5.25}$$

is univalent in the domain

$$E_0: \qquad \frac{R}{8} < |Z| < R, \quad -\frac{\pi}{m} < \arg Z < \frac{\pi}{m}$$

and maps E_0 onto the domain

$$H_0: \qquad \frac{1}{8^m} < |w| < 1, \quad -\pi < \arg w < \pi.$$

Consider a point w of the domain

$$H_0': \qquad \frac{1}{4} < |w| < \frac{1}{2}, \qquad -\frac{\pi}{2} < \arg w < \frac{\pi}{2}.$$

When Z is on the boundary of E_0, the point p(Z) is on the boundary of H_0. So we see that on the boundary of E_0, we have

$$|p(Z)-w| > \frac{1}{8}.$$

Then from the identity

$$g(Z)-w = \{p(Z)-w\} + p(Z)\Phi(Z)$$

and the inequality

$$|p(Z)\Phi(Z)| < \frac{1}{10},$$

we conclude, by Rouché's theorem, that the equation

$$g(Z) = w \tag{5.26}$$

has only one root $Z=h_0(w)$ in E_0. The function $Z=h_0(w)$ is holomorphic and univalent in H_0' and satisfies the identity

$$g\{h_0(w)\} = w \tag{5.27}$$

in H_0'.

If instead of the domain E_0, we start from the domain

$$E_1: \qquad \frac{R}{8} < |Z| < R, \quad \frac{1}{m}(\frac{\pi}{2}-\pi) < arg\ Z < \frac{1}{m}(\frac{\pi}{2}+\pi),$$

then it can be shown that for each point w of the domain

$$H_1': \qquad \frac{1}{4} < |w| < \frac{1}{2}, \qquad 0 < arg\, w < \pi,$$

the equation (5.26) has only one root $Z=h_1(w)$ in E_1. The function $Z=h_1(w)$ is holomorphic and univalent in H_1' and satisfies the identity

$$g\{h_1(w)\} = w$$

in H_1'. The two domains H_0' and H_1' have a common part

$$H_0' \cap H_1': \qquad \frac{1}{4} < |w| < \frac{1}{2},\ 0 < arg\ w < \frac{\pi}{2}.$$

We are going to show that in $H_0' \cap H_1'$, we have

$$h_0(w) = h_1(w). \tag{5.28}$$

In fact consider the common part of the two domains E_0 and E_1:

$$E_0 \cap E_1: \qquad \frac{R}{8} < |Z| < R, \quad \frac{1}{m}(\frac{\pi}{2}-\pi) < arg\ Z < \frac{\pi}{m}$$

and the arc

$$\sigma: \quad Z = \left(\frac{1}{3}\right)^{\frac{1}{m}} R e^{i\theta}, \quad \frac{1}{m}\frac{\pi}{4} < \theta < \frac{1}{m}\left(\frac{\pi}{4} + \frac{\pi}{10}\right)$$

belonging to $E_0 \cap E_1$. By (5. 23), if $Z \in \sigma$, we have

$$|g(Z)| < \frac{1}{3}\left(1 + \frac{1}{10}\right) < \frac{1}{2}, \quad |g(Z)| > \frac{1}{3}\left(1 - \frac{1}{10}\right) > \frac{1}{4}$$

and

$$\arg g(Z) = m\theta + \arg\{1 + \Phi(Z)\},$$

$$|\arg\{1 + \Phi(Z)\}| < \arcsin\frac{1}{10},$$

$$0 < \frac{\pi}{4} - \arcsin\frac{1}{10} < \arg g(Z) < \frac{\pi}{4} + \frac{\pi}{10} + \arcsin\frac{1}{10} < \frac{\pi}{2},$$

hence $g(Z) \in H'_0 \cap H'_1$. Consequently if we denote by s the arc described by $g(Z)$ as Z describes σ, then $s \in H'_0 \cap H'_1$ and for $w \in s$ we have (5. 28). Therefore (5. 28) holds in $H'_0 \cap H'_1$. So the function $h_1(w)$ is the analytic extension of the function $h_0(w)$ to the region

$$\frac{1}{4} < |w| < \frac{1}{2}, \quad \frac{\pi}{2} \leqslant \arg w < \pi.$$

Similarly if we start from the domain

$$E_{-1}: \frac{R}{8} < |Z| < R, \quad \frac{1}{m}\left(-\frac{\pi}{2} - \pi\right) < \arg Z < \frac{1}{m}\left(-\frac{\pi}{2} + \pi\right),$$

it can be shown that for each point w of the domain

$$H'_{-1}: \quad \frac{1}{4} < |w| < \frac{1}{2}, \quad -\pi < \arg w < 0$$

the equation (5. 26) has only one root $Z = h_{-1}(w)$ in E_{-1}. The function $Z = h_{-1}(w)$ is holomorphic and univalent in H'_{-1} and satisfies the identity

$$g(h_{-1}(w)) = w$$

in H'_{-1}. Furthermore it can also be shown that the function $h_{-1}(w)$ is the analytic extension of the function $h_0(w)$ to the region

$$\frac{1}{4} < |w| < \frac{1}{2}, \quad -\pi < arg\ w \leqslant -\frac{\pi}{2}.$$

Thus after analytic extension the function $h_0(w)$ is holomorphic and univalent in the domain (5. 24) and satisfies the identity (5. 27) in the domain (5. 24). The function $h_0(w)$ maps the domain (5. 24) onto a simply connected domain Δ belonging to the domain $R/8 < |Z| < R$. This domain Δ satisfies the required conditions.

Definition 5. 7. Let $L>0$ be a number and $k\geqslant 1$ an integer. A function $f(z)$ holomorphic in a domain D is said to satisfy the condition $\Gamma(L,k)$ in D, if there is a domain $\rho_1<|w|<\rho_2$ with $\rho_1>L$, $\rho_2-\rho_1>L$ having the following property: For any real number $\omega(0\leqslant\omega<2\pi)$, there exist k mutually disjoint simply connected domains $\Delta_j(\omega)(j=1,2,\cdots,k)$ belonging to D, such that for each $1\leqslant j\leqslant k$, the function $f(z)$ is univalent in $\Delta_j(\omega)$ and maps $\Delta_j(\omega)$ onto the domain

$$\rho_1 < |w| < \rho_2, \ \omega < arg\ w < \omega + 2\pi.$$

On the other hand, the function $f(z)$ is said to satisfy the condition $C(L,k)$ in D, if there is a circle $|w-w_0|<\rho$ with $\rho>L$ having the following property: There exist k mutually disjoint simply connected domains $\Delta_j(j=1,2,\cdots,k)$ belonging to D such that for each $1\leqslant j\leqslant k$, the function $f(z)$ is univalent in Δ_j and maps Δ_j onto the circle $|w-w_0|<\rho$.

Theorem 5. 6. Let D be a domain, $L>0$ a number and $k\geqslant 1$ an integer. Then the family $\mathscr{F}$ of the functions $f(z)$ wihich are holomorphic in D and do not satisfy the condition $\Gamma(L,k)$ in D is quasi-normal of order k-1 at most in D.

This theorem is a generalization of Theorem 2. 18.

Proof. First of all, by Corollary 13 in Appendix A, if $F(z)$ is a holomorphic function in the circle $|z|<1$, such that $F(z)$ does not satisfy the condition $\Gamma(L,k)$ in the circle $|z|<1$, then in the cricle $|z|\leqslant 1/2$, we have

$$|F(z)| < \lambda(L + \sum_{i=0}^{k-1} \frac{|F^{(i)}(0)|}{i!}),$$

where λ is a positive constant depending only on k.

Consider a point z_0 of the domain D and a circle Γ: $|z-z_0|<\delta(0<\delta<1)$ belonging to D. Let $f(z)$ be a function of the family $\mathscr{F}$. Then the function $F(\zeta)=f(z_0+\delta\zeta)$ is holomorphic and does not satisfy the condition $\Gamma(L,k)$ in the circle $|\zeta|<1$. Consequently in the circle $|z-z_0|\leqslant\delta/2$, we have

$$|f(z)| < \lambda(L + \sum_{i=0}^{k-1} \frac{|f^{(i)}(z_0)|}{i!}).$$

Next, as in the proof of Theorem 5. 5, we find that in the circle $|z-z_0|<\delta/4$, we have

$$1 + \sum_{i=0}^{k-1} \frac{|f^{(i)}(z)|}{i!} < A(1 + \sum_{i=0}^{k-1} \frac{|f^{(i)}(z_0)|}{i!}), \tag{5. 29}$$

where

$$A = (1 + \sum_{i=0}^{k-1} (\frac{4}{\delta})^i)\lambda(L + 1).$$

Now keeping the point $z_0\in D$ fixed, let $S: f_n(z)(n=1,2,\cdots)$ be a sequence of functions of the family $\mathscr{F}$. As in the proof of Theorem 5. 5, define

$$B(z_0,f_n) = \max_{0\leqslant i\leqslant k-1} \frac{|f_n^{(i)}(z_0)|}{i!} \quad (n = 1,2,\cdots)$$

and distinguish two cases:

1) The sequence $B(z_0,f_n)(n=1,2,\cdots)$ is bounded. In this case, by (5. 29), we see as in the proof of Theorem 5. 5, that the sequence S is locally uniformly bounded in D, and hence we can extract from S a subsequence S′ which is a C_0-

sequence in D.

2) The sequence $B(z_0,f_n)(n=1,2,\cdots)$ is unbounded. Then it is sufficient to consider the case that

$$\lim_{n\to+\infty} B(z_0,f_n)=+\infty. \tag{5.30}$$

Let n_0 be a positive integer such that $B(z_0,f_n)>1$ for $n\geqslant n_0$. As in the proof of Theorem 5.5, we see that the sequence of functions

$$\varphi_n(z)=\frac{f_n(z)}{B(z_0,f_n)} \quad (n\geqslant n_0)$$

is locally uniformly bounded in D. Consequently from the sequence $\varphi_n(z)(n\geqslant n_0)$ we can extract a subsequence

$$\varphi_{n_h}(z)=\frac{f_{n_h}(z)}{B(z_0,f_{n_h})} \quad (h=1,2,\cdots)$$

such that as $h\to+\infty$, $\varphi_{n_h}(z)$ converges locally uniformly to a holomorphic function $\varphi(z)$ in D. As in the proof of Theorem 5.5, we see that $\varphi(z)\not\equiv 0$ and so we need only to show that $\varphi(z)$ has at most k-1 distinct zeros in D.

Assume, on the contrary, that $\varphi(z)$ has k distinct zeros $z_j(j=1,2,\cdots,k)$ in D, of orders $m_j(j=1,2,\cdots,k)$ respectively. Let $|z-z_j|\leqslant r_j(j=1,2,\cdots,k)$ be k mutually disjoint circles belonging to D, such that for each $1\leqslant j\leqslant k$, we have

$$\varphi(z)=a_j(z-z_j)^{m_j}\{1+\psi_j(z)\},a_j\neq 0,|\psi_j(z)|<\frac{1}{100}$$

in the circle $|z-z_j|\leqslant r_j$. Then we can find a number $A>0$ and k circles Γ_j: $|z-z_j|=\rho_j(0<\rho_j<r_j)(j=1,2,\cdots,k)$ such that

$$\max_{z\in\Gamma_j}|\varphi(z)|=A \quad (j=1,2,\cdots,k).$$

Let $\zeta_j\in\Gamma_j$ be such that

$$|\varphi(\zeta_j)| = A \quad (j = 1,2,\cdots,k).$$

Then for each $1 \leqslant j \leqslant k$, we have

$$\varphi(z) = \varphi(\zeta_j)\left(\frac{z-z_j}{\zeta_j - z_j}\right)^{m_j}\{1 + \Psi_j(z)\}, |\Psi_j(z)| < \frac{1}{49} \tag{5.31}$$

in the circle $|z-z_j| \leqslant r_j$.

Consider the regions $\overline{E}_j$: $\rho_j/8 \leqslant |z-z_j| \leqslant \rho_j$ $(j=1,2,\cdots,k)$. Set

$$\mu_j = \min_{z \in \overline{E}_j} |\varphi(z)| \quad (j = 1,2,\cdots,k), \quad \mu = \min_{1 \leqslant j \leqslant k} \mu_j.$$

and let H be a positive integer such that when $h \geqslant H$, we have

$$|\varphi_{n_h}(z) - \varphi(z)| < \frac{\mu}{100}$$

in each of the circles $|z-z_j| \leqslant r_j (j=1,2,\cdots,k)$. It follows that when $h \geqslant H$, we have

$$\varphi_{n_h}(z) = \varphi(z)\{1 + \alpha_h(z)\}, \tag{5.32}$$

where $\alpha_h(z)$ satisfies the inequality

$$|\alpha_h(z)| < \frac{1}{100}$$

in each of the regions $\overline{E}_j(j=1,2,\cdots,k)$. From (5.31) and (5.32), we find that, when $h \geqslant H$, then we have for each $1 \leqslant j \leqslant k$,

$$\varphi_{n_h}(z) = \varphi(\zeta_j)\left(\frac{z-z_j}{\zeta_j - z_j}\right)^{m_j}\{1 + \beta_{hj}(z)\}, |\beta_{hj}(z)| < \frac{1}{25} \tag{5.33}$$

in the region $\overline{E}_j$. By (2.33), we can apply Lemma 5.9 to the function $\varphi_{n_h}(z)$ for the region $\overline{E}_j$. Consequently for each real number ω, there is a simply connected domain $\Delta_{hj}(\omega)$ belonging to the domain $E_j: \rho_j/8 < |z-z_j| < \rho_j$, such that $\varphi_{n_h}(z)$ is univalent in $\Delta_{hj}(\omega)$ and maps $\Delta_{hj}(\omega)$ onto the domain

$$\frac{1}{4}A < |w| < \frac{1}{2}A, \quad \omega < \arg w < \omega + 2\pi.$$

Then by the identity

$$f_{n_h}(z) = B(z_0, f_{n_h})\varphi_{n_h}(z),$$

the function $f_{n_h}(z)$ is univalent in $\Delta_{hj}(\omega)$ and maps $\Delta_{hj}(\omega)$ onto the domain

$$\frac{1}{4}AB(z_0, f_{n_h}) < |w| < \frac{1}{2}AB(z_0, f_{n_h}), \quad \omega < \arg w < \omega + 2\pi.$$

Since the domains $\Delta_{hj}(\omega)$ $(j=1,2,\cdots,k)$ belong to the domain D and are mutually disjoint, we arrive at a contradiction, in taking an integer $h \geqslant H$ such that

$$\frac{1}{4}AB(z_0, f_{n_h}) > L,$$

where L is the number in Theorem 5. 6.

Theorem 5. 6 contains Theorem 2. 18 as a particular case, because when $k = 1$, the family $\mathscr{F}$ defined in Theorem 5. 6 is normal in D.

Corollary 5. 12. Let D be a domain, $L > 0$ a number and $k \geqslant 1$ an integer. Then the family $\mathscr{F}$ of functions $f(z)$ which are holomorphic in D and do not satisfy the condition $C(L, k)$ in D is quasi-normal of order $k-1$ at most in D.

This corollary which is a generalization of Corollary 2. 11, is an immediate consequence of Theorem 5. 6, because the functions $f(z)$ of the family $\mathscr{F}$ defined in Corollary 5. 12, à fortiori, do not satisfy the condition $\Gamma(2L, k)$ in D. On the other hand, Corollary 5. 12 contains Corollary 2. 11 as a particular case.

5.3. SOME APPLICATIONS OF CRITERIONS OF QUASI-NORMALITY

Lemma 5.10. Let $\mathscr{F}$ be a family of holomorphic functions in a domain D. Let σ be a bounded closed set of points belonging to D. Let $M>0$ and $\delta>0$ be two numbers. Assume that the following conditions are satisfied:

1^0 The family $\mathscr{F}$ is quasi-normal of order ν at most in D, where $\nu\geqslant 0$ is an integer.

2^0 For each function $f(z)\in\mathscr{F}$, there are $\nu+1$ points $z_j=z_j(f)\in\sigma(j=1,2,\cdots,\nu+1)$ such that

$$|z_j-z_{j'}|\geqslant\delta \quad (1\leqslant j,j'\leqslant\nu+1, j\neq j') \tag{5.34}$$

and that

$$|f(z_j)|\leqslant M \quad (j=1,2,\cdots,\nu+1).$$

Then the family $\mathscr{F}$ is normal in D. Consequently, by Lemma 2.1, the family $\mathscr{F}$ is uniformly bounded on each bounded closed set of points belonging to D.

This lemma is a generalization of Lemma 2.1. In this lemma it is tacitly assumed that there exist $\nu+1$ points $z_j\in\sigma(j=1,2,\cdots,\nu+1)$ satisfying the condition (5.34).

Proof. Let $S: f_n(z)(n=1,2,\cdots)$ be a sequence of functions of the family $\mathscr{F}$. By hypothesis, we can extract from S a subsequence $S': f_{n_k}(z)(k=1,2,\cdots)$ of which the number of nonC_0-points does not exceed ν. Set

$$z_j^{(k)}=z_j(f_{n_k})(k=1,2,\cdots)(j=1,2,\cdots,\nu+1).$$

By Lemma 3.2, we can find an increasing sequence of positive integers $k_p(p=1,2,\cdots)$ such that for each $1\leqslant j\leqslant\nu+1$, we have

$$\lim_{p\to+\infty} z_j^{(k_p)}=\zeta_j,$$

where $\zeta_j\in\sigma \quad (j=1,2,\cdots,\nu+1)$ and

$$|\zeta_j-\zeta_{j'}|\geqslant\delta \quad (1\leqslant j,j'\leqslant\nu+1, j\neq j').$$

Consider the sequence $S'': f_{m_p}(z)\,(p=1,2,\cdots;m_p=n_{k_p})$ of which the number of nonC_0-points also does not exceed ν. We are going to show that S'' is a C_0-sequence in D. This will complete the proof of Lemma 5. 10.

In fact, assume that S'' has nonC_0-points $a_i(i=1,2,\cdots,\mu;\mu\leqslant\nu)$ in D. In the domain $D_1=D-(a_1,a_2,\cdots a_\mu)$ S'' is a C_0-sequence. By a remark made at the beginning of this chapter, as $p\to+\infty$, $f_{m_p}(z)$ converges locally uniformly to ∞ in D_1. Among the points $\zeta_j(j=1,2,\cdots,\nu+1)$ there is at least one ζ_1, for example, belonging to D_1. Then there is a circle $|z-\zeta_1|<r$ belonging to D_1, such that as $p\to+\infty$, $f_{m_p}(z)$ converges uniformly to ∞ in this circle. But this incompatible with the fact that

$$|f_{m_p}(z_1^{(k_p)})|\leqslant M,\qquad \lim_{p\to+\infty} z_1^{(k_p)}=\zeta_1.$$

Theorem 5. 7. Given a domain D, two integers $p_j\geqslant 0$ $(j=1,2,)$, two bounded closed sets σ, E of points belonging to D and $M>0,\delta>0$ two numbers, we can find a positive number $A(D,p_1,p_2,\sigma,E,M,\delta)$ depending only on $D,p_1,p_2,\sigma,E,M,\delta$ having the following property: If $f(z)$ is a holomorphic function in D satisfying the conditions:

1^0 the equations $f(z)=0$ and $f(z)=1$ have respectively at most p_1 distinct roots and p_2 distinct roots in D;

2^0 there are $q+1$ points $z_j\in\sigma$ $(j=1,2,\cdots,q+1;q=\min(p_1,p_2))$ such that

$$|z_j-z_{j'}|\geqslant\delta\quad(1\leqslant j,j'\leqslant q+1,j\neq j')$$

and that

$$|f(z_j)|\leqslant M\quad(j=1,2,\cdots,q+1);$$

then we have

$$\max_{z\in E}|f(z)|\leqslant A(D,p_1,p_2,\sigma,E,M,\delta). \tag{5. 35}$$

This theorem is a generalization of Theorem 2. 4.

Proof. Let $\mathscr{F}$ be the family of the functions $f(z)$ holomorphic in D and satisfying the conditions 1^0 and 2^0 in Theorem 5. 7. By Corollary 5. 11, $\mathscr{F}$ is quasi-normal of order q at most in D. Next by Lemma 5. 10, $\mathscr{F}$ is uniformly bounded on the set E. Consequently there is a positive number A such that (5. 35) holds for each function $f(z)\in\mathscr{F}$. This number A evidently has the required property.

Theorem 5. 8. Given a domain D, two integers $p_j\geqslant 0$ $(j=1,2)$, two bounded closed sets σ, E of points belonging to D, two number $M>0,\delta>0$ and an integer $n\geqslant 1$, we can find a positive number $B(D,p_1,p_2,\sigma,E,M,\delta,n)$ depending only on $D,p_1,p_2,\sigma,E,M,\delta,n$ having the following property: If $f(z)$ is a holomorphic function in D satisfying the conditions 1^0 and 2^0 in Theorem 5. 7, then we have

$$\max_{z\in E}|f^{(n)}(z)|\leqslant B(D,p_1,p_2,\sigma,E,M,\delta,n). \tag{5. 36}$$

This theorem is a generalization of Theorem 2. 5.

Proof. Let $\mathscr{F}$ be the family of functions defined in the proof of Theorem 5. 7. We already know by Lemma 5. 10, that $\mathscr{F}$ is uniformly bounded on each bounded closed set of points belonging to D. Then we complete the proof of Theorem 5. 8 by the method used in the proof of Theorem 2. 5.

Lemma 5. 11. Let $\mathscr{F}$ be a family of holomorphic functions in a domain D. Let σ be a bounded closed set of points belonging to D. Let $R>0$ and $M>0$ be two numbers. Let $q\geqslant 0$ be an integer. Assume that the following conditions are satisfied:

1^0 The family $\mathscr{F}$ is quasi-normal in D.

2^0 For each function $f(z)\in\mathscr{F}$, there is a value $w_0=w_0(f)$ with $|w_0|\leqslant R$, such that the equation $f(z)=w_0$ has at most q roots in D, with due count of order of multiplicity.

3^0 For each function $f(z)\in\mathscr{F}$, there is a point $z_0=z_0(f)\in\sigma$ such that $|f^{(j)}(z_0)|\leqslant M$ $(j=0,1,\cdots,q)$.

Then the family $\mathscr{F}$ is normal in D. Consequently by Lemma 2. 1, the family $\mathscr{F}$

is uniformly bounded on each bounded closed set of points belonging to D.

This lemma is also a generalization of Lemma 2. 1.

Proof. Let $S: f_n(z)(n=1,2,\cdots)$ be a sequence of functions of the family $\mathscr{F}$. By hypothesis, we can extract from S a subsequence S' $: f_{n_k}(z)(k=1,2,\cdots)$ which is a C_1-sequence in D. Set

$$z_0^{(k)} = z_0(f_{n_k})\ (k = 1,2,\cdots)$$

and let $k_p(p=1,2,\cdots)$ be an increasing sequence of positive integers such that

$$\lim_{p\to+\infty} z_0^{(k_p)} = \zeta_0, \quad \zeta_0 \in \sigma.$$

Consider the sequence S'': $f_{m_p}(z)(p=1,2,\cdots; m_p=n_{k_p})$ which is also a C_1-sequence in D. It remains to show that S'' is a C_0-sequence in D.

In fact, assume that S'' has nonC_0-points in D, of which the set we denote by E. In the domain $D_1=D-E$, S'' is a C_0-sequence. By a remark made at the beginning of this chapter, as $p\to+\infty$, $f_{m_p}(z)$ converges locally uniformly to ∞ in D_1. Set

$$\zeta_p = z_0^{(k_p)} = z_0(f_{m_p})(p = 1,2,\cdots)$$

and consider the point ζ_0. Let $0<|z-\zeta_0|\leqslant\rho$ be a region belonging to D_1. Then as $p\to+\infty$, $f_{m_p}(z)$ converges uniformly to ∞ on the circle $|z-\zeta_0|=\rho$. We have

$$f_{m_p}(z) - w_0(f_{m_p}) = \{f_{m_p}(z)-g_p(z)\} + \{g_p(z)-w_0(f_{m_p})\}, \qquad (5.37)$$

where

$$g_p(z) = \sum_{j=0}^{q} \frac{f_{m_p}{}^{(j)}(\zeta_p)}{j!}(z-\zeta_p)^j.$$

Let P be a positive integer such that when $p\geqslant P$, we have $|\zeta_p-\zeta_0|<\rho$ and $|f_{m_p}(z)|>2K+R$ on the circle $|z-\zeta_0|=\rho$, where

$$K = \sum_{j=0}^{q} \frac{M}{j!}(2\rho)^j.$$

Then by the conditions 2^0 and 3^0 in Lemma 5. 11, we have

$$|f_{m_p}(z)-g_p(z)| > |g_p(z)-w_0(f_{m_p})|$$

on the circle $|z-\zeta_0|=\rho$. Since ζ_p is a zero of the function $f_{m_p}(z)-g_p(z)$ of order at least equal to $q+1$, we see from the identity (5. 37) and Rouché's theorem, that when $p \geqslant P$, the function $f_{m_p}(z)-w_0(f_{m_p})$ has at least $q+1$ zeros in the circle $|z-\zeta_0|<\rho$, incompatible with the condition 2^0 in Lemma 5. 11.

Theorem 5. 9. Given a domain D, two integers $p_j \geqslant 0$ $(j=1,2)$, two bounded closed sets σ, E of points belonging to D and a number $M>0$, we can find a positive number $A'(D,p_1,p_2,\sigma,E,M)$ depending only on D,p_1,p_2,σ,E,M having the following property: If $f(z)$ is a holomorphic function in D satisfying the conditons:

1^0 the equations $f(z)=0$ and $f(z)=1$ have respectively at most p_1 roots and p_2 roots in D, with due count of order of multiplicity;

2^0 there is a point $z_0 \in \sigma$ such that $|f^{(j)}(z_0)| \leqslant M$ $(j=0,1,\cdots,q; q=\min(p_1, p_2))$;

then we have

$$\max_{z\in E} |f(z)| \leqslant A'(D,p_1,p_2,\sigma,E,M). \tag{5. 38}$$

This theorem is also a generalization of Theorem 2. 4.

Proof. Let $\mathscr{F}$ be the family of the functions $f(z)$ holomorphic in D and satisfying the conditions 1^0 and 2^0 in Theorem 5. 9. By Corollary 5. 11, $\mathscr{F}$ is quasi-normal of order q at most in D. Then by Lemma 5. 11, $\mathscr{F}$ is uniformly bounded on the set E. Hence there is a positive number A′ such that (5. 38) holds for each function $f(z) \in \mathscr{F}$. This number A′ has evidently the required property.

Theorem 5. 10. Given a domain D, two integers $p_j \geqslant 0$ $(j=1,2)$, two bounded closed sets σ, E of points belonging to D, a number $M>0$ and an integer $n \geqslant 1$, we can find a positive number $B'(D, p_1, p_2, \sigma, E, M, n)$ depending only on $D, p_1, p_2, \sigma, E, M, n$ having the following property: If $f(z)$ is a holomorphic function in D satisfying the conditions 1^0 and 2^0 in Theorem 5. 9, then we have

$$\max_{z \in E} |f^{(n)}(z)| \leqslant B'(D, p_1, p_2, \sigma, E, M, n).$$

This theorem is also a generalization of Theorem 2. 5.

Now we are going to consider the particular case of holomorphic functions in the unit circle.

Let $p_j \geqslant 0 (j=1,2)$ be two integers and let $0<r_0<1$, $\mu>0$, $\delta>0$ be three numbers. Let $\mathscr{F}$ be the family of the functions $f(z)$ holomorphic in the circle Δ: $|z|<1$ and satisfying the following conditions:

1^0 The equations $f(z)=0$ and $f(z)=1$ have respectively at most p_1 distinct roots and p_2 distinct roots in Δ.

2^0 There are $q+1$ points $z_j = z_j(f)$ $(j=1,2,\cdots,q+1; q=\min(p_1,p_2))$ of the circle $|z| \leqslant r_0$, such that

$$|z_j - z_{j'}| \geqslant \delta \ (1 \leqslant j, j' \leqslant q+1, j \neq j')$$

and that

$$|f(z_j)| \leqslant \mu \ (j = 1, 2, \cdots, q+1).$$

We know already that the family $\mathscr{F}$ is normal in Δ. More precisely $\mathscr{F}$ is uniformly bounded on each bounded closed set of points belonging to Δ.

Consequently from a sequence of functions $f_n(z)$ $(n=1,2,\cdots)$ of the family $\mathscr{F}$, we can extract a subsequence $f_{n_k}(z)$ $(k=1,2,\cdots)$ such that as $k \to +\infty$, $f_{n_k}(z)$ converges locally uniformly to a holomorphic function $F(z)$ in Δ. Two cases are possible.

1) $F(z)$ is non-constant. In this case, $F(z) \in \mathscr{F}$. In fact, first of all, by Rouché's theorem, we see as usual that $F(z)$ satisfies the condition 1^0. Next set

$$z_j^{(k)} = z_j(f_{n_k}) \quad (k = 1, 2, \cdots) \quad (j = 1, 2, \cdots, q+1).$$

By Lemma 3.2, we can find an increasing sequence of positive integers $k_p (p=1, 2,\cdots)$ such that for each $1\leqslant j\leqslant q+1$, we have

$$\lim_{p\to+\infty} z_j^{(k_p)} = \zeta_j \tag{5.39}$$

where ζ_j belongs to the circle $|z|\leqslant r_0$ and

$$|\zeta_j-\zeta_{j'}| \geqslant \delta \quad (1\leqslant j,j' \leqslant q+1, j\neq j').$$

Consider the sequence $f_{m_p}(z)(p=1,2,\cdots;m_p=n_{k_p})$. As $p\to+\infty$, $f_{m_p}(z)$ also converges locally uniformly to $F(z)$ in Δ. On the other hand, we have

$$|f_{m_p}\{z_j^{(k_p)}\}| \leqslant \mu.$$

Then from (5.39) we see that

$$|F(\zeta_j)| \leqslant \mu \ (j=1,2,\cdots,q+1).$$

Hence $F(z)$ also satisfies the condition 2^0. Consequently $F(z)\in\mathscr{F}$ as asserted.

2) $F(z)$ is a constant w_0. By the conditon 2^0, we have necessarily $|w_0|\leqslant\mu$. So if $w_0\neq 0,1$, then $F(z)\equiv w_0\in\mathscr{F}$. Bat it is also possible that $w_0=0$ or 1. In fact, consider the case $0<\mu<1$. Take a sequence of constants $w_n(n=1,2,\cdots)$ such that $w_n\neq 0$, $|w_n|\leqslant\mu$, $\lim\limits_{n\to+\infty} w_n=0$. Then the sequence of functions $f_n(z)\equiv w_n(n=1,2,\cdots)$ belong to the family $\mathscr{F}$ and the corresponding limit function $F(z)\equiv 0$. Similarly if $\mu\geqslant 1$, then we may have $F(z)\equiv 0$ or 1.

So the family $\mathscr{F}$ is not closed in Δ. However, in the case $0<\mu<1$, if we add and constant 0 to $\mathscr{F}$, we get a closed family in Δ:

$$\mathscr{F}^* = \mathscr{F}\cup(0). \tag{5.40}$$

Similarly in the case $\mu\geqslant 1$, we get a closed family in Δ:

$$\mathscr{F}^* = \mathscr{F}\cup(0,1) \tag{5.41}$$

by adding the constants 0 and 1 to $\mathscr{F}$. Consider a circle $\Gamma: |z|=r\,(0<r<1)$. Then by Theorems 4.3 and 4.2, there exists a function $f_0(z)\in\mathscr{F}^*$ such that

$$M(r,f_0) = \sup_{f\in\mathscr{F}^*} M(r,f). \tag{5.42}$$

It is interesting to see whether $f_0(z)\in\mathscr{F}$. For this, let us distinguish three cases:

a) $0<\mu<1$. In this case, the function $f(z)\equiv\mu\in\mathscr{F}$, hence $M(r,f_0)\geqslant\mu$ and $f_0(z)\in\mathscr{F}$.

b) $\mu>1$. In this case, again we have $f(z)\equiv\mu\in\mathscr{F}$, $M(r,f_0)\geqslant\mu$ and $f_0(z)\in\mathscr{F}$.

c) $\mu=1$. In this case, since the function $f(z)\equiv\mu\in\mathscr{F}^*$, we have also $M(r,f_0)\geqslant\mu$ and hence $f_0(z)\in\mathscr{F}\cup(1)$. However if $r_0\leqslant r<1$, we can always find a function $f_1(z)\in\mathscr{F}$ such that $M(r,f_1)=M(r,f_0)$. In fact, consider the function

$$g(z) = \frac{z-1}{r_0+1}.$$

Evidently $g(z)\in\mathscr{F}$ in the case $\mu=1$, and $M(r_0,g)=1$. Consequently if $r_0<r<1$, then we must have $f_0(z)\in\mathscr{F}$. On the other hand, if $r=r_0$ and $f_0(z)\equiv1$, then we take $f_1(z)=g(z)$.

Furthermore let $n\geqslant1$ be an integer. Since $\Omega(f)=f^{(n)}$ is a continuous operator defined for $f\in\mathscr{F}^*$, again by Theorem 4.3 and 4.2, there exists a function $f_n(z)\in\mathscr{F}^*$ such that

$$M(r,f_n^{(n)}) = \sup_{f\in\mathscr{F}^*} M(r,f^{(n)}), \tag{5.43}$$

where $0<r<1$ is a number. We are going to show that

$$f_n(z)\in\mathscr{F}.$$

For this, it is sufficient to show that we can find a function $f(z)\in\mathscr{F}$ such that $M(r,f^{(n)})>0$. Such a function $f(z)$ is easy to get. We need only to take

$$f(z) = \frac{\mu(z^n-1)}{r_0^n + 1}$$

for the case $0<\mu<1$, and

$$f(z) = \frac{z^n-1}{r_0^n + 1}$$

for the case $\mu\geqslant 1$.

Now again let $p_j\geqslant 0(j=1,2)$ be two integers and let $\mu>0$ be a number. Let $\mathscr{F}_1$ be the family of the functions $f(z)$ holomorphic in the circle $\Delta: |z|<1$ and satisfying the following conditions:

1′ The equations $f(z)=0$ and $f(z)=1$ have respectively at most p_1 roots and p_2 roots in Δ, with due count of order of multiplicity.

2′ $|f^{(j)}(o)|\leqslant\mu\,(j=0,1,\cdots,q;q=\min(p_1,p_2))$.

We know that the family $\mathscr{F}_1$ is normal in Δ and, more precisely, is uniformly bounded on each bounded closed set of points belonging to Δ.

Let $f_n(z)$ $(n=1,2,\cdots)$ be a sequence of functions of the family $\mathscr{F}_1$. From this sequence we can extract a subsequence $f_{n_k}(z)(k=1,2,\cdots)$ such that as $k\to+\infty$, $f_{n_k}(z)$ converges locally uniformly to a holomorphic function $F(z)$ in Δ. Distinguish two cases:

1)$F(z)$ is non-constant. Then we see easily that $F(z)\in\mathscr{F}$.

2) $F(z)$ is a constant w_0. Then $|w_0|\leqslant\mu$ and, if $w_0\neq 0,1,F(z)\equiv w_0\in\mathscr{F}$.

So from the family $\mathscr{F}_1$ we deduce a closed family $\mathscr{F}_1^*$ in Δ, by defining

$$\mathscr{F}^* = \mathscr{F}_1 \cup (0)$$

in the case $0<\mu<1$ and

$$\mathscr{F}_1^* = \mathscr{F}_1 \cup (0,1)$$

in the case $\mu\geqslant 1$.

Consider a circle $\Gamma: |z|=r(0<r<1)$. Again by Theorem 4. 3 and 4. 2, there exists a function $F_0(z)\in\mathscr{F}_1^*$ such that

$$M(r, F_0) = \sup_{f \in \mathscr{F}_1^*} M(r, f).$$

In the two cases $0 < \mu < 1$ and $\mu > 1$, the function $f(z) \equiv \mu \in \mathscr{F}_1$ and $M(r, F_0) \geqslant \mu$, $F_0(z) \in \mathscr{F}_1$. In the case $\mu = 1$, we have $M(r, F_0) \geqslant \mu$ and $F_0(z) \in \mathscr{F}_1 \cup (1)$, and we can always find a function $F_1(z) \in \mathscr{F}_1$ such that $M(r, F_1) = M(r, F_0)$, by means of the function

$$G(z) = \frac{z-1}{r+1} \in \mathscr{F}_1, \qquad M(r, G) = 1.$$

Furthermore let $n \geqslant 1$ be an integer. Then there exists a function $F_n(z) \in \mathscr{F}_1^*$ such that

$$M(r, F_n^{(n)}) = \sup_{f \in \mathscr{F}_1^*} M(r, f^{(n)}).$$

Take a number $K > 0$ such that $1/K \leqslant \mu$ and consider the function $f_1(z) = (z^n - 1)/n!\,K$. Evidently $f_1(z) \in \mathscr{F}_1$ and $M(r, f_1^{(n)}) = 1/K$. Hence $M(r, F_n^{(n)}) \geqslant 1/K$ and we have necessarily $F_n(z) \in \mathscr{F}_1$.

Definition 5. 8. Let $f(z)$ be a meromorphic function in a domain D. Let E be a bounded closed set of points belonging to D. Let $q \geqslant 1$ be an integer and $\delta > 0$ a number. We define, when $q \geqslant 2$,

$$\alpha(E, q, \delta; f) = \sup_{\sigma} \{ \min_{1 \leqslant j \leqslant q} \partial(z_j, f) \}, \tag{5.44}$$

where the sup is taken over all the systems σ: $z_j (j = 1, 2, \cdots, q)$ of q points such that $z_j \in E (j = 1, 2, \cdots, q)$ and that

$$|z_j - z_{j'}| \geqslant \delta \ (1 \leqslant j, j' \leqslant q, j \neq j').$$

When $q = 1$, we define

$$\alpha(E, q, \delta; f) = \max_{z \in E} \partial(z, f). \tag{5.45}$$

Lemma 5. 12. Let $\mathscr{F}$ be a quasi-normal family of order ν at most of meromorphic functions in a domain D, where $\nu \geqslant 0$ is an integer. Let E be a bounded closed set of points belonging to D and let $\delta > 0$ be a number. Then there is a positive number M such that for each function $f(z) \in \mathscr{F}$, we have

$$\alpha(E, \nu + 1, \delta; f) \leqslant M. \tag{5.46}$$

Proof. If $\nu = 0$, then the family $\mathscr{F}$ is normal in D, and the existence of the number M follows from (5.45) and Theorem 1.6. Consider the case $\nu \geqslant 1$. Assume that the number M does not exist. Then for each integer $n \geqslant 1$, there is a function $f_n(z) \in \mathscr{F}$ such that

$$\alpha(E, \nu + 1, \delta; f_n) > n$$

and hence there is a system $z_j^{(n)}$ $(j = 1, 2, \cdots, \nu + 1)$ of points such that $z_j^{(n)} \in E$ $(j = 1, 2, \cdots, \nu + 1)$, that

$$|z_j^{(n)} - z_{j'}^{(n)}| \geqslant \delta \quad (1 \leqslant j, j' \leqslant \nu + 1, j \neq j')$$

and that

$$\min_{1 \leqslant j \leqslant \nu + 1} \partial(z_j^{(n)}, f_n) > n. \tag{5.47}$$

From the sequence $f_n(z)$ $(n = 1, 2, \cdots)$ we can extract a subsequence $f_{n_k}(z)$ $(k = 1, 2, \cdots)$ which is a C_1-sequence in D and has at most ν nonC_0-points in D.

Consider the sequences of points $z_j^{(n_k)}$ $(k = 1, 2, \cdots)$ $(j = 1, 2, \cdots, \nu + 1)$. By Lemma 3.2, we can find an increasing sequence of positive integers k_p $(p = 1, 2, \cdots)$ such that for each $1 \leqslant j \leqslant \nu + 1$, we have, in setting $m_p = n_{k_p}$,

$$\lim_{p \to +\infty} z_j^{(m_p)} = \zeta_j, \quad \zeta_j \in E \tag{5.48}$$

and

$$|\zeta_j-\zeta_{j'}|\geqslant\delta \quad (1\leqslant j,j'\leqslant\nu+1,j\neq j').$$

The sequence $S: f_{m_p}(z)(p=1,2,\cdots)$ is also a C_1-sequence in D and has at most ν nonC_0-points in D. Among the $\nu+1$ points $\zeta_j(j=1,2,\cdots,\nu+1)$, there is at least one ζ_j which is a C_0-point of the sequence S. Consequently we can find a circle Γ : $|z-\zeta_j|<r$ belonging to D, such that S is a C_0-sequence in Γ. Accordingly S considered as a family is normal in Γ. Hence by Theorem 1.6, there is a number A >0 such that for each $p\geqslant 1$ we have

$$\partial(z,f_{m_p})\leqslant A \tag{5.49}$$

in the circle Γ' : $|z-\zeta_j|<r/2$. But from (5.47) we have

$$\partial(z_j^{(m_p)},f_{m_p})>m_p. \tag{5.50}$$

By (5.48) and (5.50), we can get a positive integer p' such that

$$z_j^{(m_{p'})}\in\Gamma',\quad \partial(z_j^{(m_{p'})},f_{m_{p'}})>A.$$

This is incompatible with (5.49).

Definition 5.9. Let f(z) be a meromorphic function in a domain D. Let $0<\eta<1/2$ be a number and $p\geqslant 0$ an integer. We say that D is a (η,p)-filling domain of f(z), if there do not exist three values $a_j\in\hat{\mathbf{C}}$ $(j=1,2,3)$ such that

$$|a_j,a_{j'}|>\eta\ (1\leqslant j,j'\leqslant 3,j\neq j')$$

and that each of the equations

$$f(z)=a_j \quad (j=1,2,3)$$

has at most p distinct roots in D.

Evidently if D is a (η,p)-filling domain of f(z), then for every value $a\in\hat{\mathbf{C}}$, the equation $f(z)=a$ has at least $p+1$ distinct roots in D, except at most values a which belong to two spherical circles $|a,a_0|\leqslant\eta$ and $|a,a_0'|\leqslant\eta$.

Theorem 5. 10. Given a domain D, a bounded closed set E of points belonging to D, two numbers $\delta>0$ and $0<\eta<1/2$ and an integer $p\geqslant0$, we can find a positive number $M(D,E,\delta,\eta,p)$ depending only on D,E,δ,η,p having the following property: If $f(z)$ is a meromorphic function in D such that

$$\alpha(E,p+1,\delta;f)>M(D,E,\delta,\eta,p),$$

then D is a (η,p)-filling domain of $f(z)$.

This theorem is a generalization of Theorem 3. 10.

Proof. Let $\mathscr{F}$ be the family of the functions $f(z)$ such that $f(z)$ is meromorphic in D and that D is not a (η,p)-filling domain of $f(z)$. By Theorem 5. 4, the family $\mathscr{F}$ is quasi-normal of order p at most in D. Next by Lemma 5. 12, there is a positive number M such that for each function $f(z)\in\mathscr{F}$, we have

$$\alpha(E,p+1,\delta;f)\leqslant M.$$

Evidently this number M has the required property.

Definition 5. 10. Let $f(z)$ be a meromorphic function in a domain D. Let $m_j\geqslant2$ $(j=1,2,\cdots,q;3\leqslant q\leqslant5)$ have the same meaning given in Corollary 5. 5. Let $0<\eta<1/q$ be a number and $p_j\geqslant0$ $(j=1,2,\cdots,q)$ q integers. We say that D is a $(m_1,m_2,\cdots,m_q;\eta;p_1,p_2,\cdots,p_q)$-filling domain of $f(z)$, if there do not exist q values $a_j\in\hat{C}$ $(j=1,2,\cdots,q)$ such that

$$|a_j,a_{j'}|>\eta\ (1\leqslant j,j'\leqslant q,j\neq j') \tag{5. 51}$$

and that for each $1\leqslant j\leqslant q$, the equation $f(z)=a_j$ has in D at most p_j distinct roots whose order of multiplicity is less than m_j.

Evidently if D is a $(m_1,m_2,\cdots,m_q;\eta;p_1,p_2,\cdots,p_q)$-filling domain of $f(z)$, then for any q values $a_j\in\hat{C}$ $(j=1,2,\cdots,q)$ satisfying the conditon (5. 51), there is an integer $1\leqslant j\leqslant q$ such that the equation $f(z)=a_j$ has in D at least p_j+1 distinct roots whose order of multiplicity is less than m_j.

Definition 5.11. Let f(z) be a meromorphic function in a domain D. Let $S \subset \hat{C}$ be a set of points and $m \geqslant 2$ an integer or $m = +\infty$. Let $\nu \geqslant 1$ be an integer. We say that the values of the function f(z) cover S with order of multiplicity less than m, strictly at least ν times, if for each value $a \in S$, the equation $f(z) = a$ has in D at least ν distinct roots whose order of multiplicity is less than m. In the particular case $m = +\infty$, this means simply that the values of f(z) cover S strictly at least ν times.

Theorem 5.11. Given a domain D, a bounded closed set E of points belonging to D, a number $\delta > 0$ and $m_j \geqslant 2$ $(j = 1, 2, \cdots, q; 3 \leqslant q \leqslant 5)$, $0 < \eta < 1/q$, $p_j \geqslant 0$ $(j = 1, 2, \cdots, q)$ having the same meaning as in Definition 5.10, we can find a positive number

$$\mu = \mu(D, E, \delta; m_1, m_2, \cdots, m_q; \eta; p_1, p_2, \cdots, p_q)$$

depending only on $D, E, \delta; m_j (j = 1, 2, \cdots, q); \eta; p_j (j = 1, 2, \cdots, q)$ having the following property: If f(z) is a meromorphic function in D such that

$$\alpha(E, P + 1, \delta; f) > \mu, \quad P = \sum_{j=1}^{q} p_j,$$

then D is a $(m_1, m_2, \cdots, m_q; \eta; p_1, p_2, \cdots, p_q)$-filling domain of f(z). Consequently for any q sets $S_j \in \hat{C}$ $(j = 1, 2, \cdots, q)$ of points such that

$$d(s_j, s_{j'}) > \eta \quad (1 \leqslant j, j' \leqslant q, j \neq j'),$$

there is an integer $1 \leqslant j \leqslant q$ such that the values of the function f(z) cover S_j with order of multiplicity less than m_j, strictly at least $p_j + 1$ times.

This theorem generalizes Theorems 3.14 and 3.15.

Proof. Let $\mathscr{F}$ be the family of functions f(z) such that f(z) is meromorphic in D and that D is not a $(m_1, m_2, \cdots, m_q; \eta; p_1, p_2, \cdots, p_q)$-filling domain of f (z). By Corollary 5.5, $\mathscr{F}$ is quasi-normal of order P at most in D. Next by Lemma 5.12, there is a positve number μ such that for each function $f(z) \in \mathscr{F}$,

we have

$$\alpha(E, P+1, \delta; f) \leqslant \mu.$$

Evidently this number μ has the required property.

The second part of the conclusion of Theorem 5. 11 is an immediate consequence of the first part of that conclusion.

Theorem 5. 12. Given a domain D, a bounded closed set E of points belonging to D, a number $\delta>0$ and M, $\eta(0<\eta<M), \nu\geqslant 1, p\geqslant 0, q\geqslant 0$ having the same meaning as in Corollary 5. 7, we can find a positive number

$$K = K(D, E, \delta; M, \eta; \nu; p, q)$$

depending only on $D, E, \delta; M, \eta; \nu; p, q$ having the following property: If f(z) is a meromorphic function in D such that

$$\alpha(E, p+q+1, \delta; f) > K,$$

then either the values of f(z) cover strictly at least $p+1$ times the domain $|w|<M$ or the values of $f^{(\nu)}(z)$ cover strictly at least $q+1$ times the domain $\eta<|w|<M$.

This theorem is a generalization of Theorem 3. 18.

Proof. Let $\mathscr{F}$ be the family of functions f(z) satisfying the conditons 1^0 and 2^0 in Corollary 5. 7, where δ is replaced by η. Then $\mathscr{F}$ is quasi-normal of order $p+q$ at most in D. Next by Lemma 5. 12, there is a positive number K such that for each function $f(z)\in\mathscr{F}$, we have

$$\alpha(E, p+q+1, \delta; f) \leqslant K.$$

This number K has the required property.

Finally we give an application of Theorem 5. 6.

Theorem 5. 13. Given a domain D, a bounded closed set E of points belonging to D, two numbers $\delta>0$, $L>0$ and an integer $k\geqslant1$, we can find a positive number $\rho(D,E,\delta;L,k)$ depending only on $D,E,\delta;L,k$ having the following property: If f(z) is a holomorphic function in D such that

$$\alpha(E,k,\delta;f)>\rho(D,E,\delta;L,k),$$

then f(z) satisfies the conditon $\Gamma(L,k)$ in D.

Proof. Let $\mathscr{F}$ be the family of the functions f(z) which are holomorphic in D and do not satisfy the condtion $\Gamma(L,k)$. By Theorem 5.6, $\mathscr{F}$ is quasi-normal of order k-1 at most in D. Next by Lemma 5.12, there is a positive number ρ such that for each function $f(z)\in\mathscr{F}$, we have

$$\alpha(E,k,\delta;f)\leqslant\rho.$$

This number ρ has the required property.

Finally we prove the following theorem:

Theorem 5.14. Let $\mathscr{F}$ be a family of meromorphic functions in a domain D and $\nu\geqslant0$ an integer. Then in order that $\mathscr{F}$ be quasi-normal of order ν at most in D, it is necessary and sufficient that for any bounded closed set E of points belonging to D and any number $\delta>0$, there is a positive number M such that for each function $f(z)\in\mathscr{F}$, we have

$$\alpha(E,\nu+1,\delta;f)\leqslant M.$$

In this theorem, when $\nu\geqslant1$, it is tacitly assumed that there exist $\nu+1$ points $z_j\in E$ $(j=1,2,\cdots,\nu+1)$ such that

$$|z_j-z_{j'}|\geqslant\delta\ (1\leqslant j,j'\leqslant q,j\neq j').$$

Proof. By Lemma 5.12 the condiotn is necessary. Conversely assume that the conditon is satisfied and distinguish two cases:

1) $\nu=0$. In this case, by Definition 5.8,

$$\alpha(E, \nu+1, \delta; f) = \max_{z\in E} \partial(z, f),$$

and hence by Theorem 1.6, the family $\mathscr{F}$ is normal in D.

2) $\nu\geqslant 1$. Let $S: f_n(z)$ $(n=1,2,\cdots)$ be a sequence of functions of the family $\mathscr{F}$ and suppose that we can not extract from S a subsequence S' of which the number of nonC_0-points in D does not exceed ν. Consider the sequence σ: $\partial(z, f_n)$ $(n=1,2,\cdots)$. σ can not be locally uniformly bounded in D, because otherwise S would be a normal family in D, by Theorem 1.6, and hence from S we can extract a subsequence S' which is a C_0-sequence in D, incompatible with the assumption. Consequently, it is easy to see that there are a point $\zeta_1\in D$ and a subsequence $S_1: f_{n_k}(z)$ $(k=1,2,\cdots)$ of S such that for any circle $\overline{\Gamma}_1$; $|z-\zeta_1|\leqslant r$ belonging to D, we have

$$\lim_{k\to+\infty} \max_{z\in\overline{\Gamma}_1} \partial(z, f_{n_k}) = +\infty.$$

For the sake of convenience, we say simply that the sequence S_1 has the property (P) at the point ζ_1. Next we see immediately that the sequence $\sigma_1: \partial(z, f_{n_k})$ $(k=1,2,\cdots)$ can not be locally uniformly bounded in the domain $D_1=D-(\zeta_1)$. Hence, for the same reason as above, there are a point $\zeta_2\in D_1$ and a subsequence $S_2: f_{m_p}(z)$ $(p=1,2,\cdots)$ of S_1 such that S_2 has the property (P) at ζ_2. Evidently S_2 also has the property (P) at ζ_1. Consider the domain $D_2=D-(\zeta_1,\zeta_2)$. Either the sequence $\sigma_2: \partial(z, f_{m_p})$ $(p=1,2,\cdots)$ is locally uniformly bounded in D_2, then we can extract from S_2 a subsequence $S_3: f_{l_q}(z)$ $(q=1,2,\cdots)$ which has only two nonC_0-points ζ_1, ζ_2 in D. Or there are a point $\zeta_3\in D_2$ and a subsequence $S_3: f_{l_q}(z)$ $(q=1,2,\cdots)$ of S_2 such that S_3 has the property (P) at ζ_3. S_3 also has the property (P) at ζ_1 and at ζ_2. Continuing in this way, the process can not stop at the μ-th step with $\mu\leqslant\nu$, because otherwise there would be a subsequence S' of S, such that S' has only μ nonC_0-points ζ_j $(j=1,2,\cdots,\mu; \mu\leqslant\nu)$ in D, incompatible with the assumption.

Consequently we can find a subsequence $S_*: f_{\lambda_t}(z)$ $(t=1,2,\cdots)$ of S, such that S_* has the property (P) at each of $\nu+1$ distinct points ζ_j $(j=1,2,\cdots,\nu+1)$ in D. Let $\overline{\Gamma}_j$: $|z-\zeta_j|\leqslant\rho_j$ $(j=1,2,\cdots,\nu+1)$ be $\nu+1$ mutually disjoint circles belonging to D. Define $E=\bigcup_{j=1}^{\nu+1}\overline{\Gamma}_j$ and let $\delta>0$ be a number such that for any $\nu+1$ points $z_j\in$

$\overline{\Gamma}_j(j=1,2,\cdots,\nu+1)$, we have

$$|z_j - z_{j'}| \geqslant \delta \ (1 \leqslant j, j' \leqslant \nu + 1, j \neq j').$$

Now given arbitrarily a positive number M, since

$$\lim_{t \to +\infty} \max_{z \in \overline{\Gamma}_j} \partial(z, f_{\lambda_t}) = +\infty \ (j = 1, 2, \cdots, \nu + 1),$$

we can find a positive integer T such that

$$\max_{z \in \overline{\Gamma}_j} \partial(z, f_{\lambda_T}) > M \ (j = 1, 2, \cdots, \nu + 1).$$

Let $z_j \in \overline{\Gamma}_j$ be such that

$$\partial(z_j, f_{\lambda_T}) = \max_{z \in \overline{\Gamma}_j} \partial(z, f_{\lambda_T}).$$

Then

$$\min_{1 \leqslant j \leqslant \nu+1} \partial(z_j, f_{\lambda_T}) > M$$

and by Definition 5. 8, we have, à fortiori,

$$\alpha(E, \nu + 1, \delta; f_{\lambda_T}) > M.$$

This result contradicts the assumption that the conditon in Theorem 5. 14 is satisfied.

5. 4. QUASI-NORMAL FAMILIES OF MEROMORPHIC FUNCTIONS OF FINITE TOTAL ORDER

In what follows, we first prove some preliminary theorems and finally introduce the notion of quasi-normal families of meromorphic functions of finite total order.

Lemma 5. 13. Let $S: f_n(z)(n=1,2,\cdots)$ be a C_0-sequence of meromorphic functions in a domain D, such that the limit function $f(z)$ of S, with respect to the spherical distance, is a meromorphic function in D. Let E be a bounded closed set of points belonging to D. Assume that $f(z)$ is finite in E. Then we can find a positive integer N such that when $n \geqslant N$, $f_n(z)$ is finite in E and as $n \to +\infty$, $f_n(z)$ converges uniformly to $f(z)$ in E.

In a particular case, this lemma has been proved in the proof of Lemma 5. 8. (see also the proof of Theorem 1. 1) In the general case, it is proved in the same way. However for the sake of completeness we give a proof.

First of all, by Theorem 1. 2, as $n \to +\infty$, $f_n(z)$ converges locally uniformly to $f(z)$ in D, with respect to the spherical distance. Consequently to each point $z_0 \in E$ corresponds a circle $\Gamma(z_0): |z-z_0|<r$ belonging to D, such that as $n \to +\infty$, $f_n(z)$ converges uniformly to $f(z)$ in $\Gamma(z_0)$, with respect to the spherical distance. By the finite covering theorem, there exist a finite number of points $z_j (j=1,2,\cdots,m)$ such that

$$E \subset \bigcup_{j=1}^{m} \Gamma(z_j).$$

Given arbitrarily a number $\varepsilon>0$. Then for each $j (1 \leqslant j \leqslant m)$, there is a positive integer N_j such that when $n \geqslant N_j$, we have

$$|f_n(z), f(z)| < \varepsilon \tag{5. 52}$$

in $\Gamma(z_j)$. Let $N = \max_{1 \leqslant j \leqslant m} N_j$. Then when $n \geqslant N$, the inequality (5. 52) holds in E. Hence as $n \to +\infty$, $f_n(z)$ converges uniformly to $f(z)$ in E, with respect to the spherical distance.

Next since $f(z)$ is finite in E, $|f(z)|$ is bounded in E. Let $A>0$ be a number such that $|f(z)|<A$ in E and set $B=A+1$. Let $\mu=\mu(A,B)$ be the positive number defined in Lemma 1. 2, and N_0 be a positive integer such that when $n \geqslant N_0$, we have

$$|f_n(z), f(z)| < \mu$$

in E. Consequently by Lemma 1. 2, we have $|f_n(z)|<B$ in E. By the formula

(1.4), we have, for $n \geqslant N_0$, $z \in E$,

$$|f_n(z)-f(z)| = \{1+|f_n(z)|^2\}^{\frac{1}{2}}\{1+|f(z)|^2\}^{\frac{1}{2}}|f_n(z),f(z)|$$
$$\leqslant (1+B^2)^{\frac{1}{2}}(1+A^2)^{\frac{1}{2}}|f_n(z),f(z)|$$

and hence as $n \to +\infty$, $f_n(z)$ converges uniformly to $f(z)$ in E.

Lemma 5.14. Let $S: f_n(z)$ $(n=1,2,\cdots)$ be a sequence of meromorphic functions in a circle $\Gamma: |z-z_0|<r$ satisfying the following conditions:

1^0 S is a C_0-sequence in the domain $d: 0<|z-z_0|<r$.

2^0 The limit function $f(z)$ of S in d, with respect to the spherical distance, is a non-constant meromorphic function.

3^0 There exist a value $w_0 \in \hat{C}$ and two positive integers p and N such that, when $n \geqslant N$, the equation $f_n(z)=w_0$ has at most p roots in Γ (with due count of order of multiplicity).

Then the function $f(z)$ is meromorphic in Γ, provided that at z_0 the value $f(z_0)$ is suitably defined.

Proof. Distinguish three cases:

1) $w_0=\infty$. In this case, we first show that the function $f(z)$ has at most a finite number of poles in d. In fact, assume that $z_1 \in d$ is a pole of $f(z)$. Then by Theorem 1.1, we can find a circle $\gamma: |z-z_1|<\rho$ belonging to d and a positive integer n_0 such that the functions $1/f_n(z)$ $(n \geqslant n_0)$ and $1/f(z)$ are holomorphic in γ and that as $n \to +\infty$, $1/f_n(z)$ converges uniformly to $1/f(z)$ in γ. Since z_1 is a zero of $1/f(z)$, it follows that, when n is sufficiently large, $1/f_n(z)$ has at least one zero in the neighborhood of z_1, by Rouché's theorem, and hence $f_n(z)$ has at least one pole in that neighborhood. It is then clear that $f(z)$ has at most p distinct poles in d.

Thus we can get a number r_0 $(0<r_0<r)$ such that $f(z)$ is finite for $0<|z-z_0| \leqslant r_0$. Consider an integer $k \geqslant 1$ and the region $\bar{d}_k: r_0/(k+1) \leqslant |z-z_0| \leqslant r_0$. By Lemma 5.13, when $n \geqslant N_k$ $(N_k \geqslant N)$ the function $f_n(z)$ is finite in $\bar{d}_k$. Let n_k $(k=1,2,\cdots)$ be the increasing sequence of positive integers defined as follows:

$$n_1 = N_1, \quad n_k = N_k + n_{k-1} \quad (k=2,3,\cdots).$$

Evidently for each $k \geqslant 1$, the function $f_{n_k}(z)$ is finite in $\bar{d}_k$. By hypothesis, the function $f_{n_k}(z)$ has at most p poles in the circle $|z-z_0|<r_0/(k+1)$. Let $P_k(z)$ be the polynomial of degree p defined as follows:

a) If $f_{n_k}(z)$ has p poles $\zeta_j(k)(j=1,2,\cdots,p)$ in the circle $|z-z_0|<r_0/(k+1)$, then $P_k(z)=\prod_{j=1}^{p}(z-\zeta_j(k))$.

b) If $f_{n_k}(z)$ has $q(1\leqslant q<p)$ poles $\zeta_j(k)$ $(j=1,2,\cdots,q)$ in the circle $|z-z_0|<r_0/(k+1)$, then $P_k(z)=(z-z_0)^{p-q}\prod_{j=1}^{q}(z-\zeta_j(k))$.

c) If $f_{n_k}(z)$ has no pole in the circle $|z-z_0|<r_0/(k+1)$, then $P_k(z)=(z-z_0)^p$.

The sequence of functions

$$g_k(z) = f_{n_k}(z)P_k(z) \quad (k = 1,2,\cdots)$$

are holomorphic in the circle $|z-z_0|\leqslant r_0$, and as $k\to+\infty$, $g_k(z)$ converges uniformly to $f(z)(z-z_0)^p$ on the circle $|z-z_0|=r_0$. Hence as $k\to+\infty$, $g_k(z)$ converges uniformly to a holomorphic function $g(z)$ in the circle $|z-z_0|<r_0$. Noting that for each point z of the domain $0<|z-z_0|<r_0$, we can find a positive integer K_z such that for $k\geqslant K_z$, $f_{n_k}(z)\neq\infty$ and as $k\to+\infty$, $f_{n_k}(z)$ converges to $f(z)$, so we have

$$g(z) = f(z)(z-z_0)^p$$

in the domain $0<|z-z_0|<r_0$. It follows that in this domain we have

$$f(z) = \frac{g(z)}{(z-z_0)^p}$$

and the function $f(z)$ is meromorphic in the circle $\Gamma: |z-z_0|<r$.

2) $w_0=0$. In this case, the sequence $1/f_n(z)$ $(n\geqslant N)$ is a C_0-sequence in the domain d and its limit function in d, with respect to the spherical distance, is the function $1/f(z)$. Since $1/f_n(z)$ has at most p poles in Γ, hence by the result in the case 1), there is a meromorphic function $F(z)$ in Γ, such that

$$\frac{1}{f(z)} = F(z)$$

in d. Consequently in d we have

$$f(z) = \frac{1}{F(z)},$$

where $1/F(z)$ is a meromorphic function in Γ.

3) $w_0 \neq 0, \infty$. In this case, by Lemma 5.2, the sequence $f_n(z)-w_0 (n=1,2,\cdots)$ is a C_0-sequence in the domain d, and it is easy to see that its limit function in d, with respect to the spherical distance, is the function $f(z)-w_0$. Since for $n \geqslant N$, the function $f_n(z)-w_0$ has at most p zeros in Γ, hence by the result in case 2), we have in d,

$$f(z)-w_0 = F_1(z),$$

where $F_1(z)$ is a meromorphic function in Γ. Consequently in d we have

$$f(z) = F_1(z) + w_0,$$

where $F_1(z)+w_0$ is a meromorphic function in Γ.

Definition 5.12. Let $S: f_n(z) (n=1,2,\cdots)$ be a sequence of meromorphic functions in a circle Γ: $|z-z_0|<r$. We say that S satisfies Montel's conditon at z_0, if there are two values $w_j \in \hat{C} (j=1,2; w_1 \neq w_2)$ and two positive integers p and N such that when $n \geqslant N$, each of the two equations $f_n(z)=w_j (j=1,2)$ has at most p roots in Γ, with due count of order of multiplicity.

Definition 5.13. Let $S: f_n(z) (n=1,2,\cdots)$ be a sequence of meromorphic functions in a circle Γ: $|z-z_0|<r$ satisfying the following conditions:

1^0 S is a C_0-sequence in the domain d: $0<|z-z_0|<r$.

2^0 z_0 is a nonC_0-point of S which is irreducible with respect to z_0.

We say that S is completely reduced at z_0, if we can find a value $w_0 \in \hat{C}$ and an integer $\mu \geqslant 1$ such that for any value $w \neq w_0$ and any number r' $(0<r'<r)$, there exist a number $\rho (0<\rho<r')$ and a positive integer N such that when $n \geqslant N$, we have

$$\nu(f_n, w, \rho) = \mu,$$

where in general we denote by $\nu(f, w, \rho)$ the number of the roots (with due count of order of multiplicity) of the equation $f(z)=w$ in the circle $|z-z_0|<\rho$. Let σ be the set of the positive integers μ having the following property: With μ we can associate a value w_0 such that w_0 and μ satisfy the condition stated above. The smallest positive integer $\mu = \mu(S, z_0)$ of the set σ is called the order of the sequence S at z_0.

Theorem 5.15. Let $S: f_n(z)$ $(n=1,2,\cdots)$ be a sequence of meromorphic functions in a circle $\Gamma: |z-z_0|<r$ satisfying the condtions 1^0 and 2^0 in Definition 5.13 and Montel's condition at z_0. Then from S we can extract a subsequence S': $f_{n_k}(z)$ $(k=1,2,\cdots)$ which is completely reduced at z_0 and the order $\mu(S', z_0)$ of S' at z_0 does not exceed p, where p is the positive integer occuring in Definition 5.12.

For the proof of this theorem, we denote by $f(z)$ the limit function of the sequence S in the domain $d: 0<|z-z_0|<r$, with respect to the spherical distance and distinguish three cases:

Case 1. $f(z)$ is a non-constant meromorphic function in d.

Case 2. $f(z)$ is a finite constant a in d.

Case 3. $f(z)$ is the constant ∞ in d.

Proof of Theorem 5.15 in case 1.

Lemma 5.15. Let $S: f_n(z)$ $(n=1,2,\cdots)$ be a sequence of meromorphic functions in a circle $\Gamma: |z-z_0|<r$ satisfying the conditions 1^0 and 2^0 in Lemma 5.14 and Montel's conditon at z_0. Furthermore $f(z)$ being the limit function of S in the domain $d: 0<|z-z_0|<r$, with respect to the spherical distance, which, by Lemma 5.14, is meromorphic in Γ, assume $f(z_0)\neq\infty$. Then the following assertions hold:

1) For any value $w\neq f(z_0)$, ∞ and any number r' $(0<r'<r)$, we can find a number ρ $(0<\rho<r')$ and a positve integer N such that, when $n\geq N$, we have

$$\nu(f_n, w, \rho) = \nu(f_n, \infty, \rho).$$

2) For any value $w \neq f(z_0)$, we can find a number $\rho(0<\rho<r)$ and a positive integer N such that, when $n \geqslant N$, we have

$$\nu(f_n, w, \rho) \leqslant p,$$

where p is the positive integer occuring in Definition 5.12.

3) For any value $w \in \hat{\mathbf{C}}$, we can find a number ρ $(0<\rho<r)$ and a positive integer N such that, when $n \geqslant N$, we have

$$\nu(f_n, w, \rho) \leqslant p + m,$$

where m denotes the order of the zero z_0 of the function $f(z)-f(z_0)$.

Proof. First of all, since $f(z_0) \neq \infty$, we can find a circle $|z-z_0|<\rho_0(0<\rho_0<r)$ in which $f(z)$ is finite. To prove 1), consider a value $w \neq f(z_0)$, ∞ and a number r' $(0<r'<r)$. Take a number $\rho(0<\rho<\min(\rho_0, r'))$ such that $f(z) \neq w$ for $|z-z_0| \leqslant \rho$. By Lemma 5.13, when n is sufficiently large, $f_n(z)$ is finite on the circle $|z-z_0|=\rho$ and as $n \to +\infty$, $f_n(z)$ converges uniformly to $f(z)$ on the circle $|z-z_0|=\rho$. Consequently by the identity

$$f_n(z)-w = \{f_n(z)-f(z)\} + \{f(z)-w\}$$

and Rouché's theorem, there is a positive integer N such that, when $n \geqslant N$, we have

$$\nu(f_n, w, \rho)-\nu(f_n, \infty, \rho) = 0.$$

By 1), in order to prove 2), it is sufficient to show that we can find a number $\rho_1(0<\rho_1<r)$ and a positive integer N_1 such that, when $n \geqslant N_1$, we have

$$\nu(f_n, \infty, \rho_1) \leqslant p. \tag{5.53}$$

For this we make use of the fact that the sequence S satisfies Montel's condition at z_0. Let $w_j(j=1,2)$ be the two values in Definition 5.12. If one of the values w_j $(j=1,2)$ is ∞, evidently ρ_1 and N_1 can be found. Otherwise, one at least of w_j

$(j=1,2)$, for example $w_1 \neq f(z_0)$. Then in 1) we may take $w=w_1$, hence ρ_1 and N_1 can also be found.

Finally by (5. 53), in order to prove 3), it is sufficient to show that we can find a number $\rho(0<\rho<\rho_1)$ and a positve integer N such that, when $n \geqslant N$, we have

$$\nu(f_n, f(z_0), \rho) = \nu(f_n, \infty, \rho) + m.$$

In fact, take a number $\rho(0<\rho<\min(\rho_0, \rho_1))$, such that $f(z) \neq f(z_0)$ for $0<|z-z_0| \leqslant \rho$. Then by the identity

$$f_n(z)-f(z_0) = \{f_n(z)-f(z)\} + \{f(z)-f(z_0)\}$$

and Rouché's theorem, we can find a positive integer N such that, when $n \geqslant N$, we have

$$\nu(f_n, f(z_0), \rho)-\nu(f_n, \infty, \rho) = m.$$

Now we prove Theorem 5. 15 in case 1. In this case we assume that the limit function $f(z)$ of the sequence S in the domain $d: 0<|z-z_0|<r$, with respect to the spherical distance, is a non-constant meromorphic function. By Lemma 5. 14, $f(z)$ is meromorphic in the circle $\Gamma: |z-z_0|<r$. Distinguish two cases:

A) $f(z_0) \neq \infty$. In this case, Lemma 5. 15 is applicable. Hence by 2) in Lemma 5. 15, we can find a number $\rho_1(0<\rho_1<r)$ and a positive integer N_1 such that, when $n \geqslant N_1$, we have

$$\nu(f_n, \infty, \rho_1) \leqslant p.$$

Next take a number $r_0(0<r_0<\rho_1)$ such that $f(z)$ is finite for $|z-z_0| \leqslant r_0$. Then by Lemma 5. 13 we can get a positive integer $N_2 \geqslant N_1$, such that, when $n \geqslant N_2$, $f_n(z)$ is finite on the circle $|z-z_0|=r_0$ and as $n \to +\infty$, $f_n(z)$ converges uniformly to $f(z)$ on the circle $|z-z_0|=r_0$. Then we can get a positive integer $N_3 \geqslant N_2$, such that, when $n \geqslant N_3$, the function $f_n(z)$ has pole in the circle $|z-z_0|<r_0$. This is because that the point z_0 is a nonC_0-point of the sequence S which is irreducible with respect to z_0. Consider an integer λ $(1 \leqslant \lambda \leqslant p)$ and denote by $I(\lambda)$ the set of

the positive integers n such that $n \geqslant N_3$ and that

$$\nu(f_n, \infty, r_0) = \lambda.$$

Evidently there exists an integer $\lambda (1 \leqslant \lambda \leqslant p)$ for which the set $I(\lambda)$ contains an infinite number of integers. Let μ be the smallest of such integers λ. Arranging the set $I(\mu)$ of integers as an increasing sequence $n_k (k=1,2,\cdots)$, then

$$\nu(f_{n_k}, \infty, r_0) = \mu \ (k = 1,2,\cdots).$$

We are going to show that the sequence $S' : f_{n_k}(z)$ $(k=1,2,\cdots)$ satisfies the required condition in Theorem 5. 15. In fact, consider a value $w \neq f(z_0)$, ∞ and a number r' $(0 < r' < r)$. By 1) in Lemma 5. 15, we can find a number ρ $(0 < \rho < \min(r_0, r'))$ and a positive integer N such that, when $n \geqslant N$, we have

$$\nu(f_n, w, \rho) = \nu(f_n, \infty, \rho).$$

Next let N' be a positive integer such that when $n \geqslant N'$, $f_n(z)$ is finite in the region $\rho \leqslant |z - z_0| \leqslant r_0$. Let K be a positive integer such that when $k \geqslant K$, we have $n_k \geqslant \max(N, N')$. Then when $k \geqslant K$, we have

$$\nu(f_{n_k}, w, \rho) = \nu(f_{n_k}, \infty, \rho) = \nu(f_{n_k}, \infty, r_0) = \mu.$$

This proves our assertion. Here the value w_0 in Definition 5. 13 is $f(z_0)$.

B) $f(z_0) = \infty$. In this case, first note that since the sequence S is irreducible with respect to z_0, we can find a positive integer n_0 such that $f_n(z) \not\equiv 0$ for $n \geqslant n_0$. Consider the sequence $S_1 : 1/f_n(z)$ $(n \geqslant n_0)$ of meromorphic functions in the circle $\Gamma : |z - z_0| < r$. It is easy to see that the sequence S_1 also satisfies the conditons 1^0 and 2^0 in Definition 5. 13 and Montel's condition at z_0. Moreover the limit function of S_1 in the domain $d : 0 < |z - z_0| < r$, with respect to the spherical distance, is the function $1/f(z)$ which is meromorphic in Γ and its value at z_0 is 0. Consequently by the result obtained in case A), we can find an integer μ $(1 \leqslant \mu \leqslant p)$ and a sequence $S_1' : 1/f_{n_k}(z)$ $(k=1,2,\cdots)$ such that for any value $w \neq 0$ and any number r' $(0 < r' < r)$, there exist a number $\rho (0 < \rho < r')$ and a positive integer K such that, when $k \geqslant K$, we have

$$\nu(\frac{1}{f_{n_k}}, w, \rho) = \mu.$$

From this we deduce that for any value $w \neq \infty$ and any number r' $(0 < r' < r)$, there exist a number $\rho(0 < \rho < r')$ and a positive integer K such that when $k \geqslant K$, we have

$$\nu(f_{n_k}, w, \rho) = \mu.$$

Hence the sequence S': $f_{n_k}(z)$ $(k = 1, 2, \cdots)$ satisfies the required condition in Theorem 5. 15. Here the value w_0 in Definition 5. 13 is $\infty = f(z_0)$.

Proof of Theorem 5. 15 in case 2.

Lemma 5. 16. Let $S: f_n(z)$ $(n = 1, 2, \cdots)$ be a sequence of meromorphic functions in a circle $\Gamma: |z-z_0| < r$ satisfying the following conditions:

1^0 S is a C_0-sequence in the domain $d: 0 < |z-z_0| < r$.

2^0 The limit function $f(z)$ of S in d, with respect to the spherical distance, is a finite constant a.

3^0 S satisfies Montel's condition at z_0.

Then the following assertions hold:

1) For any value $w \neq a, \infty$ and any number $\rho(0 < \rho < r)$, we can find a positive integer N such that, when $n \geqslant N$, we have

$$\nu(f_n, w, \rho) = \nu(f_n, \infty, \rho).$$

2) For any value $w \neq a$ and any number $\rho(0 < \rho < r)$, we can find a positive integer N such that, when $n \geqslant N$, we have

$$\nu(f_n, w, \rho) \leqslant p,$$

where p is the positive integer occuring in Definition 5. 12.

Proof. To prove 1), consider a value $w \neq a, \infty$ and a number $\rho(0 < \rho < r)$. By Lemma 5. 13, we can find a positive integer N_0 such that when $n \geqslant N_0$, the function $f_n(z)$ is finite on the circle $|z-z_0|=\rho$ and as $n \to +\infty$, $f_n(z)$ converges uniformly to a on the circle $|z-z_0|=\rho$. Then by the identity

$$f_n(z)-w = \{f_n(z)-a\} + (a-w)$$

and Rouché's theorem, we can find a positive integer $N \geqslant N_0$, such that, when $n \geqslant N$, we have

$$\nu(f_n, w, \rho)-\nu(f_n, \infty, \rho) = 0.$$

2)can be deduced from 1) and the conditon 3^0, as in the proof of 2) in Lemma 5. 15.

Now we prove Theorem 5. 15 in case 2. In this case, we assume that the limit function f(z) of the sequence S in the domain $d: 0<|z-z_0|<r$, with respect to the spherical distance, is a finite constant a. So we can apply Lemma 5. 16. Take a number $r_0 (0<r_0<r)$. By 2) in Lemma 5. 16, we can find a positive integer N_1 such that, when $n \geqslant N_1$, we have

$$\nu(f_n, \infty, r_0) \leqslant p.$$

Next, by Lemma 5. 13, we can find a positive integer $N_2 \geqslant N_1$, such that, when $n \geqslant N_2$, the function $f_n(z)$ is finite on the circle $|z-z_0|=r_0$ and as $n \to +\infty$, $f_n(z)$ converges uniformly to the constant a on the circle $|z-z_0|=r_0$. Then, we can get a positive integer $N_3 \geqslant N_2$, such that, when $n \geqslant N_3$, the function $f_n(z)$ has pole in the circle $|z-z_0|<r_0$. Let $\mu (1 \leqslant \mu \leqslant p)$ be the smallest integer such that the set I(μ) of the integers n for which $n \geqslant N_3$ and

$$\nu(f_n, \infty, r_0) = \mu$$

contains an infinite number of integers. Arranging the set I(μ) as an increasing sequence $n_k (k=1,2,\cdots)$, we have

$$\nu(f_n, \infty, r_0) = \mu \ (k = 1,2,\cdots).$$

Then, basing upon Lemma 5. 16, we see as in the proof of Theorem 5. 15 in case

1, A), that the sequence S': $f_{n_k}(z)$ $(k=1,2,\cdots)$ satisfies the required condition in Theorem 5. 15. Here the value w_0 in Definition 5. 13 is a.

Finally let us prove Theorem 5. 15 in case 3. In this case, we assume that the limit function f(z) of the sequence S in the domain $d: 0<|z-z_0|<r$, with respect to the spherical distance, is the constant ∞. as in the proof of Theorem 5. 15 in case 1, B), let n_0 be a positive integer such that $f_n(z)\not\equiv 0$ for $n\geqslant n_0$ and consider the sequence $S_1: 1/f_n(z)$ $(n\geqslant n_0)$ of meromorphic functions in the circle $\Gamma: |z-z_0|<r$, which also satisfies the conditions 1^0 and 2^0 in Definition 5. 13 and Montel's condition at z_0. Moreover the limit function of S_1 in the domain d, with respect to the spherical distance, is the constant 0. Consequently by the result obtained in the proof of Theorem 5. 15 in case 2, we can get a sequence S_1': $1/f_{n_k}(z)$ $(k=1,2,\cdots)$ such that for any value $w\neq 0$ and any number r' $(0<r'<r)$ there exist a number $\rho(0<\rho<r)$ and a positive integer K such that, when $k\geqslant K$, we have

$$\nu(\frac{1}{f_{n_k}}, w, \rho) = \mu.$$

From this we deduce that the sequence S': $f_{n_k}(z)$ $(k=1,2,\cdots)$ satisfies the required condition in Theorem 5. 15. Here the value w_0 in Definition 5. 13 is ∞.

The proof of Theorem 5. 15 is now complete.

Definition 5. 14. Let $S: f_n(z)$ $(n=1,2,\cdots)$ be an irreducible C_1-seuqence of meromorphic functions in a domain D. (see Definition 5. 6) Let E be the set of nonC_0-points of S in D. If at each point $z_0\in E$, S is completely reduced, we say that S is completely reduced in D. Then to each point $z_0\in E$ correspond a value $w_0=w_0(S,z_0)\in\dot{C}$ and an integer $\mu=\mu(S,z_0)\geqslant 1$ which is the order of S at z_0, such that w_0 and μ satisfy the condition stated in Definition 5. 13. The sum

$$\Omega(S,D) = \sum_{z_0\in E} \mu(S,z_0),$$

where the summation is taken over all points $z_0\in E$, is called the total order of S in D. We have

$$1\leqslant \Omega(S,D)\leqslant +\infty.$$

Definition 5. 15. Let $S: f_n(z)$ $(n=1,2,\cdots)$ be an irreducible C_1-sequence of meromorphic functions in a domain D. Let E be the set of nonC_0-points of S in D. We say that S satisfies Montel's condition in D, if S satisfies Montel's condition at each point $z_0 \in E$, namely there exist a circle $\Gamma: |z-z_0|<r$ belonging to D, two values $w_j \in \hat{C}$ $(j=1,2,; w_1 \neq w_2)$ and two positive integers p and N satisfying the condition stated in Definition 5. 12.

Theorem 5. 16. Let $S: f_n(z)$ $(n=1,2,\cdots)$ be an irreducible C_1-sequence of meromorphic functions in a domain D. If S satisfies Montel's condition in D, then we can extract from S a subsequence $S': f_{n_k}(z)$ $(k=1,2,\cdots)$ which is completely reduced in D.

Proof. Let E be the set of nonC_0-points of S in D and distinguish two cases:

1) E consists of a finite number of points $z_j (j=1,2,\cdots,m)$. Consider first the point z_1. By Theorem 5. 15 we can extract from S a subsequence S_1 which is completely reduced at z_1. Next from S_1 we can extract a subsequence S_2 which is completely reduced at z_2, and so on. Finally from S_{m-1} which is completely reduced at z_{m-1}, we can extract a subsequence S_m which is completely reduced at z_m. The sequence S_m is completely reduced at each of the points $z_j (j=1,2,\cdots,m)$ and S_m is a subsequence of S. Hence S_m satisfies the required condition in Theorem 5. 16.

2) E consists of an infinite number of points $z_j (j=1,2,\cdots)$. As in case 1), we first extract from S a subsequence S_1 which is completely reduced at z_1; then from S_1 we extract a subsequence S_2 which is completely reduced at z_2, and so on. Continuing in this way we get a sequence of sequences $S_j (j=1,2,\cdots)$. In general, S_j is completely reduced at z_j and S_{j+1} is a subsequence of S_j. Let S' be the diagonal sequence obtained from the sequences $S_j (j=1,2,\cdots)$. Then S' is a subsequence of the sequence S. Moreover for any j, the j-th term and all the following terms of S' belong to the sequence S_j. Consequently S' is completely reduced at z_j. Hence the sequence S' satisfies the required condition in Theorem 5.16.

Definition 5.16. For simplicity, a C_1-sequence $S: f_n(z)$ $(n=1,2,\cdots)$ of meromorphic functions in a domain D is said to be completely reduced in D, if S satisfies one of the following two conditions:

1^0 S is irreducible and completely reduced in D according to Definition 5.14.

2^0 S has no nonC_0-point in D, in other words, S is a C_0-sequence in D. Then the total order of S in D is defined to be 0, $\Omega(S,D)=0$.

Definition 5.17. Let $\mathscr{F}$ be a family of meromorphic functions in a domain D. If there exists an integer $\tau \geqslant 0$ such that from every sequence S: $f_n(z)$ $(n=1,2,\cdots)$ of functions of $\mathscr{F}$, we can extract a subsequence S': $f_{n_k}(z)$ $(k=1,2,\cdots)$ which is a completely reduced C_1-sequence in D with $\Omega(S',D) \leqslant \tau$, then we say that $\mathscr{F}$ is quasi-normal of finite total order in D. More precisely we say that $\mathscr{F}$ is quasi-normal of total order τ at most in D.

In particular if $\tau=0$, then we have necessarily $\Omega(S',D)=0$, which implies that S' is a C_0-sequence in D and $\mathscr{F}$ is normal in D.

As an example of quasi-normal families of meromorphic functions of finite total order, we prove the following theorem:

Theorem 5.17. Let D be a domain, $a_j \in \hat{C}$ $(j=1,2,3)$ three distinct values and $p_j \geqslant 0$ $(j=1,2,3)$ three integers such that $p_1 \leqslant p_2 \leqslant p_3$. Let $\mathscr{F}$ be the family of the functions $f(z)$ meromorphic in D and such that for each $1 \leqslant j \leqslant 3$, the equation

$$f(z) = a_j$$

has at most p_j roots in D, with due count of order of multiplicity. Then the family $\mathscr{F}$ is quasi-normal of total order p_2 at most in D.

Proof. By Corollary 5.9, we know already that the family $\mathscr{F}$ is quasi-normal of order p_2 at most in D. If $p_2=0$, then $\mathscr{F}$ is normal in D. So from every sequence S of functions of $\mathscr{F}$, we can extract a subsequence S' which is a C_0-sequence in D and hence $\Omega(S',D)=p_2=0$.

Assume $p_2>0$ and let S be a sequence of functions of $\mathscr{F}$. From S we can extract a subsequence S_1 which is a C_1-sequence in D and has at most p_2 nonC_0-

points in D. Assume that S_1 has nonC_0-points in D. Then by Lemma 5.5, we can extract from S_1 a subsequence S_2 which is either a C_0-sequence in D or an irreducible C_1-sequence in D. Consider the latter case and denote S_2 by $f_n(z)$ $(n=1, 2, \cdots)$. Let E be the set of nonC_0-points of S_2 in D. E consists of a finite number of points $z_j (j=1,2,\cdots,m; 1\leqslant m\leqslant p_2)$. In the domain $D_1=D-E$, S_2 is a C_0-sequence and hence has a limit function $F(z)$ defined in D_1, with respect to the spherical distance. For each point z_j let $\Gamma_j: |z-z_j|<r_j$ be a circle belonging to D such that S_2 is a C_0-sequence in the domain $d_j: 0<|z-z_j|<r_j$ and that the circles Γ_j $(j=1,2,\cdots,m)$ are mutually disjoint. Evidently S_2 satisfies Montel's condition at each point $z_j (1\leqslant j\leqslant m)$. Consequently by Theorem 5.16, we can extract from S_2 a subsequence $S': f_{n_k}(z)$ $(k=1,2,\cdots)$ which is completely reduced in D. It remains to show that

$$\Omega(S', D)\leqslant p_2. \tag{5.54}$$

Distinguish two cases:

α) $F(z)$ is a non-constant meromorphic function in D_1. Consider a point $z_j (1 \leqslant j\leqslant m)$. It is convenient to introduce the notation $\nu_j(f, w, \rho)$ which denotes the number of the roots of the equation $f(z)=w$ in the circle $|z-z_j|<\rho$, with due count of order of multiplicity. From the proof of Theorem 5.15 in case 1, we know that there is an integer $\mu_j\geqslant 1$ such that for any value $w\neq F(z_j)$ and any number r' $(0<r'<r_j)$, there exist a number $\rho(0<\rho<r')$ and a positive integer K such that when $k\geqslant K$, we have

$$\nu_j(f_{n_k}, w, \rho) = \mu_j. \tag{5.55}$$

Basing upon this fact, we are going to show that we can find a number $\rho_j (0<\rho_j< r_j)$ and a positive integer K_j such that when $k\geqslant K_j$, we have

$$\mu_j\leqslant \nu_j(f_{n_k}, a_1, \rho_j). \tag{5.56}$$

Evidently this implies

$$\Omega(S', D)\leqslant \sum_{j=1}^{m}\mu_j\leqslant \sum_{j=1}^{m}\nu_j(f_{n_k}, a_1, \rho_j)\leqslant p_1,$$

where $k \geqslant \max_{1\leqslant j\leqslant m} K_j$.

To prove (5. 56), assume first that $a_1 \neq F(z_j)$. Then the existence of ρ_j and K_j follows immediately from (5. 55). Next assuem $a_1 = F(z_j)$. Then we distinguish two cases:

1) $a_1 = F(z_j) \neq \infty$. Take a number r' $(0<r'<r_j)$ such that on the circle Υ: $|z-z_j| = r'$, $F(z) \neq a_1, \infty$. Accordingly we have

$$|F(z)-a_1| \geqslant h, \; h > 0$$

on the circle Υ. By Lemma 5. 13, we can find a positve integer K' such that when $k \geqslant K'$, $f_{n_k}(z)$ is finite on Υ and as $k \to +\infty$, $f_{n_k}(z)$ converges uniformly to $F(z)$ on Υ. Consequently when $k \geqslant K''$ $(K'' \geqslant K')$, we have

$$|f_{n_k}(z)-a_1| > h/2$$

on Υ. Let w be a value such that $0<|w-a_1|<h/2$. Then from the identity

$$f_{n_k}(z)-w = \{f_{n_k}(z)-a_1\} + (a_1-w)$$

and Rouché's theorem, we see that when $k \geqslant K''$, we have

$$\nu_j(f_{n_k}, w, r') = \nu_j(f_{n_k}, a_1, r'). \tag{5. 57}$$

Since $w \neq F(z_j)$, there exist a number $\rho(0<\rho<r')$ and a positive integer $K \geqslant K''$, such that when $k \geqslant K$ we have (5. 55). It follows from (5. 55) and (5. 57), that when $k \geqslant K$, we have

$$\mu_j = \nu_j(f_{n_k}, w, \rho) \leqslant \nu_j(f_{n_k}, w, r') = \nu_j(f_{n_k}, a_1, r').$$

2) $a_1 = F(z_j) = \infty$. Take a number r' $(0<r'<r_j)$ such that $F(z) \neq 0, \infty$ in the region $0<|z-z_j| \leqslant r'$. Accordingly we have

$$|F(z)| \geqslant l, \; l > 0$$

on the circle $\Gamma: |z-z_j|=r'$. When $k\geqslant K'$, $f_{n_k}(z)$ is finite on Γ and as $k\to+\infty$, $f_{n_k}(z)$ converges uniformly to $F(z)$ on Γ. Then from the identity

$$f_{n_k}(z) = \{f_{n_k}(z)-F(z)\} + F(z)$$

and Rouché's theorem, when $k\geqslant K''$ $(K''\geqslant K')$ we have

$$\nu_j(f_{n_k},o,r') = \nu_j(f_{n_k},\infty,r')-n_0,$$

where n_0 is the order of the pole z_j of $F(z)$. On the other hand, since $0\neq F(z_j)$, there exist ρ $(0<\rho<r')$ and $K\geqslant K''$, such that when $k\geqslant K$, we have

$$\nu_j(f_{n_k},o,\rho) = \mu_j.$$

Consequently when $k\geqslant K$, we have

$$\mu_j = \nu_j(f_{n_k},o,\rho) \leqslant \nu_j(f_{n_k},o,r') < \nu_j(f_{n_k},a_1,r').$$

β) $F(z)$ is a constant c (finite or infinite) in D_1. In this case, from the proof of Theorem 5.15 in case 2 and case 3, we know that for each point $z_j(1\leqslant j\leqslant m)$ there is an integer $\mu_j\geqslant 1$ such that for any value $w\neq c$ and any number r' $(0<r'<r_j)$ there exist a number ρ $(0<\rho<r')$ and a positive integer K such that when $k\geqslant K$, we have (5.55). At least one of the values a_1 and a_2 is different from c, for example $a_2\neq c$. Then for each $1\leqslant j\leqslant m$, we can get $\rho_j(0<\rho_j<r_j)$ and K_j such that when $k\geqslant K_j$, we have

$$\mu_j = \nu_j(f_{n_k},a_2,\rho_j).$$

Consequently we have

$$\Omega(S',D) \leqslant p_2.$$

The proof of Theorem 5.17 is now complete.

Theorem 5.17 can be generalized as follows:

Theorem 5.18. Let D be a domain, $0<\delta<1$ a number and $p_j\geqslant 0$ ($j=1,2,3$) three integers such that $p_1\leqslant p_2\leqslant p_3$. Let $\mathscr{F}$ be the family of the functions $f(z)$ satisfying the following conditions:

1° $f(z)$ is meromorphic in D.

2° There are three values $a_j(f)\in\hat{\mathbf{C}}$ ($j=1,2,3$) such that

$$|a_j(f),a_{j'}(f)|>\delta\ (1\leqslant j,j'\leqslant 3, j\neq j')$$

and that for each $1\leqslant j\leqslant 3$, the equation

$$f(z)=a_j(f)$$

has at most p_j roots in D, with due count of order of multiplicity.
Then the family $\mathscr{F}$ is quasi-normal of total order p_2 at most in D.

Proof. By Theorem 5.4, the family $\mathscr{F}$ is quasi-normal of order p_2 at most in D. Assume $p_2>0$ and let S be a sequence of functions of $\mathscr{F}$. As in the proof of Theorem 5.17, it is sufficient to consider the case that from S we can extract a subsequence S_2 which is an irreducible C_1-sequence in D. Again denote S_2 by $f_n(z)$ ($n=1,2,\cdots$) and let E be the set of nonC_0-points of S_2 in D. E consists of a finite number of points z_j ($j=1,2,\cdots,m;1\leqslant m\leqslant p_2$). S_2 is a C_0-sequence in the domain $D_1=D-E$. Let $F(z)$ be the limit function of S_2 in D_1, with respect to the spherical distance. Let Γ_j: $|z-z_j|<r_j$ ($j=1,2,\cdots,m$) be mutually disjoint circles belonging to D.

Set $a_i^{(n)}=a_i(f_n)$ ($n=1,2,\cdots$) ($i=1,2,3$). By Lemma 3.2, we can find an increasing sequence of positive integers n_p ($p=1,2,\cdots$) such that for each $1\leqslant i\leqslant 3$, the sequence $a_i^{(n_p)}$ ($p=1,2,\cdots$) satisfies one of the conditions 1° and 2° in Lemma 3.2. We have

$$\lim_{p\to+\infty}a_i^{(n_p)}=A_i\quad(A_i\in\hat{\mathbf{C}})\ (i=1,2,3),$$

$$|A_i,A_{i'}|\geqslant\delta\ (1\leqslant i,i'\leqslant 3, i\neq i').$$

Consider the sequence S_3: $f_{n_p}(z)$ ($p=1,2,\cdots$) and distinguish three cases:

α) $F(z)$ is a non-constant meromorphic function in D_1. Consider a circle Γ_j. Let r_j' $(0<r_j'<r_j)$ be a number such that on the circle Υ_j: $|z-z_j|=r_j'$ we have $F(z) \neq A_1, A_2, A_3, \infty$. We are going to show that we can find a positive integer P such that when $p \geqslant P$, we have

$$\nu_j(f_{n_p}, A_1, r_j') = \nu_j(f_{n_p}, a_1^{(n_p)}, r_j'). \tag{5.58}$$

For this, distinguish three cases:

1) $a_1^{(n_p)} \neq \infty$ $(p=1,2,\cdots)$, $A_1 \neq \infty$. Let

$$h = \min_{z \in \Upsilon_j} |F(z) - A_1| > 0.$$

By Lemma 5.13, we can find a positive integer P' such that when $p \geqslant P'$, $f_{n_p}(z)$ is finite on Υ_j and as $p \to +\infty$, $f_{n_p}(z)$ converges uniformly to $F(z)$ on Υ_j. Hence when $p \geqslant P''$ $(P'' \geqslant P')$, we have

$$|f_{n_p}(z) - A_1| > \frac{h}{2}$$

on Υ_j. Next when $p \geqslant P$ $(P \geqslant P'')$, we have

$$|a_1^{(n_p)} - A_1| < \frac{h}{2}.$$

Then when $p \geqslant P$, we have (5.58).

2) $a_1^{(n_p)} \neq \infty$ $(p=1,2,\cdots)$, $A_1 = \infty$. Let

$$M = \max_{z \in \Upsilon_j} |F(z)|.$$

Then when $p \geqslant P_1$, we have

$$|f_{n_p}(z)| < M + 1$$

on Υ_j. Next when $p \geqslant P$ $(P \geqslant P_1)$, we have

$$|a_1^{(n_p)}| > M + 1.$$

Then when $p \geqslant P$, we have

$$\nu_j(f_{n_p}, a_1^{(n_p)}, r_j') = \nu_j(f_{n_p}, \infty, r_j'),$$

that is (5.58).

3) $a_1^{(n_p)} = \infty$ $(p = 1, 2, \cdots)$, $A_1 = \infty$.

Then (5.58) is evident.

Thus (5.58) is proved. Evidently (5.58) is also true, if we replace in it A_1 by A_2 or A_3 and $a_1^{(n_p)}$ by $a_2^{(n_p)}$ or $a_3^{(n_p)}$, provided that the integer P is suitably chosen. Consequently we can find a positive integer P such that when $p \geqslant P$, we have at the same time the equalities

$$\nu_j(f_{n_p}, A_i, \Gamma_j') = \nu_j(f_{n_p}, a_i^{(n_p)}, \Gamma_j') \quad (j = 1, 2, \cdots, m; i = 1, 2, 3),$$

which imply that when $p \geqslant P$, we have

$$\sum_{j=1}^{m} \nu_j(f_{n_p}, A_i, \Gamma_j') \leqslant p, \quad (i = 1, 2, 3). \tag{5.59}$$

β) $F(z)$ is a finite constant c in D_1. Then two at least of the values $A_i (i = 1, 2, 3)$ are distinct from c, for example $c \neq A_2, A_3$. Consider a circle Γ_j. Take a number $r_j' (0 < r_j' < r_j)$ and consider the circle $\gamma_j: |z - z_j| = r_j'$. By Lemma 5.13, we can find a positive integer P' such that when $p \geqslant P'$, $f_{n_p}(z)$ is finite on γ_j and as $p \to +\infty$, $f_{n_p}(z)$ converges uniformly to c on γ_j. As for (5.58), we can prove, by distinguishing three cases, that we can find a positive integer P such that when $p \geqslant P$, we have at the same time the equalities

$$\nu_j(f_{n_p}, A_i, r_j') = \nu_j(f_{n_p}, a_i^{(n_p)}, r_j') \quad (j = 1, 2, \cdots, m; i = 2, 3),$$

which imply that when $p \geqslant P$, we have

$$\sum_{j=1}^{m} \nu_j(f_{n_p}, A_i, r_j') \leqslant p, \quad (i = 2, 3). \tag{5.60}$$

γ) $F(z)$ is the constant ∞ in D_1. Then at least two of the values $A_i(i=1,2,3)$ are finite, for example $\infty \neq A_1$, A_2. Consider a circle Γ_j. Let ρ_1, ρ_2 be two numbers such that $0<\rho_1<\rho_2<r_j$. Then in the domain $\rho_1<|z-z_j|<\rho_2$, each function $f_{n_p}(z)$ has at most a finite number of poles. Hence the set of the poles of the functions $f_{n_p}(z)$ $(p=1,2,\cdots)$ in this domain is at most enumerable. It follows that there is a number $r_j'(\rho_1<r_j'<\rho_2)$ such that the functions $f_{n_p}(z)$ $(p=1,2,\cdots)$ are all finite on the circle $\gamma_j: |z-z_j|=r_j'$. As $p \to +\infty$, $f_{n_p}(z)$ converges uniformly to ∞ on the circle γ_j. Then we see easily that we can find a positive integer P such that when $p \geqslant P$, we have at the same time the equalities

$$\nu_j(f_{n_p}, A_i, r_j') = \nu_j(f_{n_p}, a_i^{(n_p)}, r_j') \quad (j=1,2,\cdots,m; i=1,2),$$

which imply that when $p \geqslant P$, we have

$$\sum_{j=1}^{m} \nu_j(f_{n_p}, A_i, r_j') \leqslant p_i \quad (i=1,2). \tag{5.61}$$

Finally from the results (5.59), (5.60) and (5.61) obtained respectively in the cases α), β) and γ), we see that in each of these cases, we can extract from the sequence $S_3: f_{n_p}(z)$ $(p=1,2,\cdots)$ a subsequence $S': f_{m_k}(z)(k=1,2,\cdots)$ which is completely reduced in D, with

$$\Omega(S', D) \leqslant p_2.$$

The method of proof is that used in the proof of Theorem 5.17. This completes the proof of Theorem 5.18.

5.5. APPLICATIONS

In what follows, we prove some theorems on the limitation of the number of the roots of an equation of the form $f(z)=a$ in a domain Δ, where $f(z)$ is a function belonging to a quasi-normal family $\mathscr{F}$ satisfying certain conditions of meromorphic functions in a domain D, a is a constant and Δ is a bounded domain such that $\bar{\Delta} \subset D$. For this we need the following lemmas.

Lemma 5.17. Let $S: f_n(z)$ $(n=1,2,\cdots)$ be a sequence of meromorphic functions in a circle $\Gamma: |z-z_0|<r$ such that S is a C_0-sequence in the domain $d: 0<|z-z_0|<r$, and that S satisfies Montel's condition at z_0. Let $f(z)$ be the limit function of S in d, with respect to the spherical distance. Then the following assertions hold:

1^0 Assume that $f(z)$ is a non-constant meromorphic function in d and $f(z_0)=\infty$. (By Lemma 5.14, $f(z)$ is meromorphic in Γ) Then for any value $w\in\hat{\mathbf{C}}$, we can find a number $\rho(0<\rho<r)$ and a positive integer N, such that, when $n\geqslant N$, we have

$$\nu(f_n, w, \rho) \leqslant p + m,$$

where p is the positive integer occuring in Definition 5.12 and m is the order of the pole z_0 of $f(z)$.

2^0 Assume that $f(z)$ is the constant ∞ in d. Then for any value $w\neq\infty$ and any number $\rho(0<\rho<r)$, we can find a positive integer N such that, when $n\geqslant N$, we have

$$\nu(f_n, w, \rho) \leqslant p.$$

Here $\nu(f_n, w, \rho)$ denotes as usual the number of the roots (with due count of order of multiplicity) of the equation $f_n(z)=w$ in the circle $|z-z_0|<\rho$.

Proof. Let n_0 be a positive integer such that $f_n(z)\not\equiv 0$ for $n\geqslant n_0$ and consider the sequence $S_1: 1/f_n(z)\,(n\geqslant n_0)$ of meromorphic funtions in the circle Γ. By applying 3) in Lemma 5.15 to S_1, we deduce 1^0. On the other hand, by applying 2) in Lemma 5.16 to S_1, we deduce 2^0.

Lemma 5.18. Let $S: f_n(z)\ (n=1,2,\cdots)$ be a sequence of meromorphic functions in a circle $\Gamma: |z-z_0|<r$, such that S is a C_0-sequence in the domain $d: 0<|z-z_0|<r$ and that S satisfies Montel's condtion at z_0. Assume that the limit function $f(z)$ of S in d, with respect to the spherical distance, is a non-constant meromorphic function in d. Then for any sequence of values $w_n\in\mathbf{C}\,(n=1,2,\cdots)$ such that

$$\lim_{n\to+\infty} w_n = w,\ w\in\mathbf{C},$$

we can find a number $\rho(0<\rho<r)$ and a positive integer N, such that, when $n \geqslant N$, we have

$$\nu(f_n, w_n, \rho) \leqslant p + m,$$

where p is the positive integer occuring in Definition 5. 12 and m denotes the order of the zero z_0 of the function $f(z)-f(z_0)$, if $f(z_0) \neq \infty$, and denotes the order of the pole z_0 of $f(z)$, if $f(z_0)=\infty$.

Proof. First of all, by 3) in Lemma 5. 15 and 1^0 in Lemma 5. 17, we can find a number $\rho_0(0<\rho_0<r)$ and a positive integer N_0, such that, when $n \geqslant N_0$, we have

$$\nu(f_n, w, \rho_0) \leqslant p + m.$$

Next take a number $\rho(0<\rho<\rho_0)$ such that on the circle $\Gamma: |z-z_0|=\rho$ we have $f(z) \neq \infty$, w. Set

$$h = \min_{z \in \Gamma} |f(z)-w| > 0.$$

By Lemma 5. 13, when $n \geqslant N_1 (N_1 \geqslant N_0)$, $f_n(z)$ is finite on Γ and as $n \to +\infty$, $f_n(z)$ converges uniformly to $f(z)$ on Γ. Then when $n \geqslant N_2 (N_2 \geqslant N_1)$, we have

$$|f_n(z)-w| > \frac{h}{2}$$

on Γ. Finally from the identity

$$f_n(z)-w_n = \{f_n(z)-w\} + (w-w_n)$$

and Rouché's theorem, we see that when $n \geqslant N$ $(N \geqslant N_2)$, we have

$$\nu(f_n, w_n, \rho) = \nu(f_n, w, \rho) \leqslant \nu(f_n, w, \rho_0) \leqslant p + m.$$

Lemma 5.19. Let $S: f_n(z)$ $(n=1,2,\cdots)$ be a sequence of meromorphic functions in a circle $\Gamma: |z-z_0|<r$, such that S is a C_0-sequence in the domain $d: 0<|z-z_0|<r$ and that S satisfies Montel's condition at z_0. Assume that the limit function $f(z)$ of S in d, with respect to the spherical distance, is a constant c. Then the following assertions hold:

1^0 If $c\neq\infty$, then for any sequence of values $w_n\in \mathbf{C}$ $(n=1,2,\cdots)$ such that

$$\lim_{n\to+\infty} w_n = w,\ w\in \hat{\mathbf{C}},\ w\neq c$$

and any number $\rho(0<\rho<r)$, we can find a positive integer $\mathbf{N}$ such that, when $n\geqslant \mathbf{N}$, we have

$$\nu(f_n, w_n, \rho)\leqslant p. \tag{5.62}$$

For any number $\rho(0<\rho<r)$ we can also find a positive integer $\mathbf{N}$ such that, when $n\geqslant \mathbf{N}$, we have

$$\nu(f_n, \infty, \rho)\leqslant p, \tag{5.63}$$

where p is the positive integer occuring in Definition 5.12.

2^0 If $c=\infty$, then for any sequence of values $w_n\in \mathbf{C}$ $(n=1,2,\cdots)$ such that

$$\lim_{n\to+\infty} w_n = w,\ w\in \mathbf{C},$$

we can find a number $\rho(0<\rho<r)$ and a positive integer N, such that, when $n\geqslant$ N, we have

$$\nu(f_n, w_n, \rho)\leqslant p. \tag{5.64}$$

Proof of 1^0. Assume first $w\neq\infty$. By 2) in Lemma 5.16, for any number ρ $(0<\rho<r)$, we can find N_0 such that, when $n\geqslant N_0$, we have

$$\nu(f_n, w, \rho) \leqslant p.$$

By Lemma 5.13, when $n \geqslant N_1 (N_1 \geqslant N_0)$, $f_n(z)$ is finite on the circle $\gamma: |z-z_0| = \rho$ and as $n \to +\infty$, $f_n(z)$ converges uniformly to the constant c on γ. Then from the identity

$$f_n(z) - w_n = \{f_n(z) - w\} + (w - w_n)$$

and Rouché's theorem, we see that, when $n \geqslant N$ $(N \geqslant N_1)$, we have

$$\nu(f_n, w_n, \rho) = \nu(f_n, w, \rho) \leqslant p.$$

Next assume $w = \infty$. Then again by 2) in Lemma 5.16, for any number $\rho (0 < \rho < r)$, we can find N_0 such that when $n \geqslant N_0$, we have

$$\nu(f_n, \infty, \rho) \leqslant p.$$

Next when $n \geqslant N$ $(N \geqslant N_0)$, we have

$$|f_n(z)| < |w_n|$$

on the circle $|z-z_0| = \rho$. Then when $n \geqslant N$, we have

$$\nu(f_n, w_n, \rho) = \nu(f_n, \infty, \rho) \leqslant p.$$

Thus in each case we have (5.62). (5.63) follows from 2) in Lemma 5.16.

Proof of 2^0. First of all, by 2^0 in Lemma 5.17, for any number $\rho_0 (0 < \rho_0 < r)$, we can find N_0 such that, when $n \geqslant N_0$, we have

$$\nu(f_n, w, \rho_0) \leqslant p.$$

Next we can find a number $\rho (0 < \rho < \rho_0)$ such that on the circle $\gamma: |z-z_0| = \rho$ the functions $f_n(z)$ $(n = 1, 2, \cdots)$ have no pole. Since $c = \infty$, it is easy to see that on γ, as $n \to +\infty$, $f_n(z)$ converges uniformly to ∞. Then from the identity

$$f_n(z)-w_n = \{f_n(z)-w\} + (w-w_n)$$

and Rouché's theorem, we see that, when $n \geqslant N(N \geqslant N_0)$, we have

$$\nu(f_n, w_n, \rho) = \nu(f_n, w, \rho) \leqslant \nu(f_n, w, \rho_0) \leqslant p.$$

This proves (5.64).

Remark 5.1. Let $S: f_n(z)$ $(n=1,2,\cdots)$ be a sequence of meromorphic functions in a circle $\Gamma: |z-z_0|<r$. Assume that S is uniformly convergent in Γ with respect to the spherical distance and let $F(z)$ be the limit function of S in Γ, with respect to the spherical distance. By Theorem 1.1, if $F(z_0) \neq \infty$, then we can find a circle $\Gamma_0: |z-z_0|<r_0(0<r_0 \leqslant r)$ and a positive integer n_0 such that the functions $f_n(z)$ $(n \geqslant n_0)$ and $F(z)$ are holomorphic in Γ_0, and as $n \to +\infty$, $f_n(z)$ converges uniformly to $F(z)$ in Γ_0. In a smaller cirlce $\Gamma_0': |z-z_0|<r_0'(0<r_0'<r_0)$, the functions $f_n(z)$ $(n \geqslant n_0)$ are uniformly bounded. Hence the sequence S satisfies Montel's condtion at z_0. On the other hand, if $F(z_0)=\infty$, then by Theorem 1.1, we can find a circle $\Gamma_0: |z-z_0|<r_0(0<r_0 \leqslant r)$ and a positive integer n_0 such that the functions $1/f_n(z)$ $(n \geqslant n_0)$ and $1/F(z)$ are holomorphic in Γ_0, and as $n \to +\infty$, $1/f_n(z)$ converges uniformly to $1/F(z)$ in Γ_0. In a smaller circle $\Gamma_0': |z-z_0|< r_0'(0<r_0'<r_0)$, the functions $1/f_n(z)$ $(n \geqslant n_0)$ are uniformly bounded. Hence again the sequence S satisfies Montel's condition at z_0. Consequently the conclusions in Lemmas 5.18 and 5.19 hold for the sequence S.

Definition 5.18. Let $\mathscr{F}$ be a family of meromorphic functions in a domain D. We say that $\mathscr{F}$ satisfies the condition (M) in D, if from every sequence $S: f_n(z)$ $(n=1,2,\cdots)$ of functions of $\mathscr{F}$, we can extract a subsequence $S': f_{n_k}(z)$ $(k=1,2,\cdots)$ satisfying one of the following two conditions:

1^0 S' is a C_0-sequence in D.

2^0 S' is an irreducible C_1-sequence in D and satisfies Montel's condition in D. (See Definition 5.15)

Definition 5.19. Let $f(z)$ be a meromorphic function in a domain Δ and $w \in \hat{C}$ a value. We denote by $n(f, w, \Delta)$ the number of the roots of the equation $f(z)=w$ in Δ, with due count of order of multiplicity.

Theorem 5.19. Let $\mathscr{F}$ be a family of meromorphic functions in a domain D. Assume that $\mathscr{F}$ satisfies the condition (M) in D. Then for any bounded domain Δ such that $\bar{\Delta}=\Delta+\partial\Delta\subset D$ and any number $0<\varepsilon<1$, we can find a positive integer $\nu(\Delta,\varepsilon)$ having the following property: If $f(z)$ is a function of $\mathscr{F}$, then for every value $w\in\hat{C}$, we have

$$n(f,w,\Delta)\leqslant\nu(\Delta,\varepsilon), \tag{5.65}$$

except at most the values w which belong to a spherical circle $|w,w_0(f)|<\varepsilon$.

Proof. Assume on the contrary that there does not exist a positive integer ν having the required property. Then to each positive integer n correspond a function $f_n(z)\in\mathscr{F}$ and two values $w_i^{(n)}(i=1,2)$ such that

$$|w_1^{(n)},w_2^{(n)}|\geqslant\varepsilon \tag{5.66}$$

and that

$$n(f_n,w_i^{(n)},\Delta)>n\ (i=1,2). \tag{5.67}$$

From the sequence $S:f_n(z)$ $(n=1,2,\cdots)$ we can extract a subsequence $S':f_{n_k}(z)$ $(k=1,2,\cdots)$ satisfying one of the conditions 1^0 and 2^0 in Definition 5.18. Consider the two sequences $w_i^{(n_k)}(k=1,2,\cdots)$ $(i=1,2)$. By Lemma 3.2, we can find an increasing sequence of positive integers $k_p(p=1,2,\cdots)$ such that for each $1\leqslant i\leqslant 2$, the sequence $w_i^{(m_p)}(p=1,2,\cdots;m_p=n_{k_p})$ satisfies one of the conditons 1^0 and 2^0 in Lemma 3.2. We have

$$\lim_{p\to+\infty}w_i^{(m_p)}=W_i(i=1,2),\ |W_1,W_2|\geqslant\varepsilon. \tag{5.68}$$

Evidently the sequence $S'':f_{m_p}(z)$ $(p=1,2,\cdots)$ also satisfies one of the conditions 1^0 and 2^0 in Definition 5.18. First consider the case that S'' satisfies the condition 2^0 in Definition 5.18. Let E be the set of nonC_0-points of S'' in D and let $F(z)$ be the limit function of S'' in the domain $D_1=D-E$, with respect to the spherical distance. Assume first that $F(z)$ is a non-constant meromorphic function in

D_1. One at least of the two values $W_i(i=1,2)$ is finite, for example $W_1 \in \mathbf{C}$. Then $w_1^{(m_p)} \in \mathbf{C}(p=1,2,\cdots)$. Consider a point $z_0 \in \bar{\Delta}$. By the condition 2^0 in Definition 5.18 and the remark 5.1, we see that the sequence S'' satisfies Montel's condition at z_0. Consequently by Lemma 5.18, we can find a circle $\Gamma(z_0): |z-z_0| < \rho(z_0)$ belonging to D, a positive integer $P(z_0)$ and a positive integer $q(z_0)$, such that, when $p \geqslant P(z_0)$, we have

$$\nu(f_{m_p}, w_1^{(m_p)}, \rho(z_0)) \leqslant q(z_0).$$

By the finite covering theorem, there are a finite number of points $z_j \in \bar{\Delta}$ $(j=1, 2,\cdots,l)$ such that

$$\bar{\Delta} \subset \bigcup_{j=1}^{l} \Gamma(z_j).$$

Set

$$P = \max_{1 \leqslant j \leqslant l} P(z_j), \quad q = \sum_{j=1}^{l} q(z_j).$$

Then when $p \geqslant P$, we have

$$n(f_{m_p}, w_1^{(m_p)}, \Delta) \leqslant q.$$

On the other hand, by (5.67), we have

$$n(f_{m_p}, w_1^{(m_p)}, \Delta) > m_p (p = 1,2,\cdots).$$

So we get a contradiction.

Now assume that $F(z)$ is a constant c in D_1. Then by Lemma 5.19, we also get a contradiction.

In the above we have assumed that S'' satisfies the condition 2^0 in Definition 5.18. However the above proof is also valid in the case that S'' satisfies the condition 1^0 in Definition 5.18, the only difference is that now the limit function $F(z)$ of S'' is defined in D instead of D_1.

Remark 5.2. Let $\mathscr{F}$ be a family of meromorphic functions in a domain D.

Evidently if $\mathscr{F}$ satisfies one of the following conditions, then $\mathscr{F}$ satisfies the condition (M) in D.

1^0 $\mathscr{F}$ is a normal family in D.

2^0 From every sequence S of functions of $\mathscr{F}$, we can extract a subsequence S′ which is a completely reduced C_1-sequence in D. (See Definition 5.16).

À fortiori, if $\mathscr{F}$ is quasi-normal of finite total order in D (See Definition 5.17), then $\mathscr{F}$ satisfies the condition (M) in D. In particular the family $\mathscr{F}$ defined in Theorem 5.18 satisfies the condition (M) in D.

Theorem 5.20. Let $\mathscr{F}$ be a family of meromorphic functions in a domain D and let $a \in \hat{C}$ be a value. Assume that from every sequence S of functions of $\mathscr{F}$, we can extract a subsequence S′ such that S′ satisfies one of the conditons 1^0 and 2^0 in Definition 5.18 and that the limit function of S′, with respect to the spherical distance, is not the constant a. Then for any bounded domain Δ such that $\overline{\Delta} = \Delta + \partial\Delta \subset D$, we can find a positive integer $\nu(\Delta)$ such that for every function $f(z) \in \mathscr{F}$, we have

$$n(f, a, \Delta) \leqslant \nu(\Delta).$$

Proof. Assume on the contrary that there does not exist a positive integer ν having the required property. Then for each positive integer n, there is a function $f_n(z) \in \mathscr{F}$ such that

$$n(f_n, a, \Delta) > n. \tag{5.69}$$

From the sequence S: $f_n(z)$ $(n=1,2,\cdots)$ we can extract a subsequence S′ : $f_{n_k}(z)$ $(k=1,2,\cdots)$ satisfying one of the conditions 1^0 and 2^0 in Definition 5.18. Consider first the case that S′ satisfies the condition 2^0 in Definition 5.18. Let E be the set of nonC_0-points of S′ in D and let F(z) be the limit function of S′ in the domain $D_1 = D - E$, with respect to the spherical distance. By hypothesis, F(z) is not the constant a in D_1. Distinguish two cases:

1) F(z) is a non-constant meromorphic function in D_1. Then on the one hand, by the condition 2^0 in Definition 5.18 and Remark 5.1, and on the other hand, by 3) in Lemma 5.15 and 1^0 in Lemma 5.17, we see that to each point z_0

$\in \bar{\Delta}$ correspond a circle $\Gamma(z_0): |z-z_0|<\rho(z_0)$, a positive integer $K(z_0)$ and a positive integer $q(z_0)$, such that when $k \geq K(z_0)$, we have

$$\nu(f_{n_k}, a, \rho(z_0)) \leq q(z_0).$$

By the finite covering theorem, there are a finite number of points $z_j \in \bar{\Delta}$ $(j=1, 2, \cdots, l)$ such that

$$\bar{\Delta} \subset \bigcup_{j=1}^{l} \Gamma(z_j).$$

Set

$$K = \max_{1 \leq j \leq l} K(z_j), \quad q = \sum_{j=1}^{l} q(z_j).$$

Then when $k \geq K$, we have

$$n(f_{n_k}, a, \Delta) \leq q.$$

But from (5.69), we have

$$n(f_{n_k}, a, \Delta) > n_k \quad (k = 1, 2, \cdots).$$

So we get a contradiction.

2) $F(z)$ is a constant c in D_1. Since $a \neq c$, by 2) in Lemma 5.16 and 2^0 in Lemma 5.17, we also get a contradiction in the same way as above.

We have assumed that the sequence S' satisfies the condition 2^0 in Definition 5.18. The case that S' satisfies the condition 1^0 in Definition 5.18 can be treated in the same way.

Corollary 5.13. Let $\mathscr{F}$ be a normal family of meromorphic functions in a domain D and let $a \in \hat{\mathbf{C}}$ be a value. Assume that we can find a bounded closed set σ of points belonging to D and a number $0<\eta<1$, such that for every function $f(z) \in \mathscr{F}$ we have

$$\max_{z \in \sigma} |f(z), a| > \eta. \tag{5.70}$$

Then the conclusion in Theorem 5.20 holds.

This corollary contains, as a particular case, a theorem of Montel [26] in which the set σ is a single point $z_0 \in D$.

Proof. Let $S: f_n(z)$ $(n=1,2,\cdots)$ be a sequence of functions of $\mathscr{F}$. Then from S we can extract a subsequence $S': f_{n_k}(z)$ $(k=1,2,\cdots)$ which is a C_0-sequence in D. Let $F(z)$ be the limit function of S' in D, with respect to the spherical distance. It remains to show that $F(z)$ is not the constant a. In fact, assume on the contrary that $F(z) \equiv a$. Since as $k \to +\infty$, $f_{n_k}(z)$ converges uniformly to $F(z)$ in σ, with respect to the spherical distance, we can find a positive integer K such that when $k \geqslant K$, we have

$$|f_{n_k}(z), a| < \eta$$

in σ, incompatible with (5.70).

Corollary 5.13 can be generalized as follows:

Corollary 5.14. Let $\mathscr{F}$ be a family of meromorphic functions in a domain D and assume that $\mathscr{F}$ is quasi-normal of total order τ at most in D. Let $a \in \hat{\mathbb{C}}$ be a value. Suppose that we can find a bounded closed set σ of points belonging to D and two numbers $\delta > 0$, $0 < \eta < 1$, such that for every function $f(z) \in \mathscr{F}$, there are $\tau+1$ points $z_j(f) \in \sigma$ $(j=1,2,\cdots,\tau+1)$ satisfying the conditions

$$|z_j(f) - z_{j'}(f)| \geqslant \delta \ (1 \leqslant j, j' \leqslant \tau+1, j \neq j'),$$
$$|f(z_j), a| > \eta \ (z_j = z_j(f)) \ (j = 1,2,\cdots,\tau+1). \tag{5.71}$$

Then the conclusion in Theorem 5.20 holds.

In this corollary it is tacitly assumed that there exist $\tau+1$ points $z_j \in \sigma$ $(j=1,2,\cdots,\tau+1)$ such that

$$|z_j - z_{j'}| \geqslant \delta \ (1 \leqslant j, j' \leqslant \tau+1; j \neq j').$$

Proof. If $\tau=0$, then $\mathscr{F}$ is normal in D, hence the conclusion follows from Corollary 5.13. So we assume $\tau\geqslant 1$. Let $S: f_n(z)$ $(n=1,2,\cdots)$ be a sequence of functions of $\mathscr{F}$. Then from S we can extract a subsequence S' : $f_{n_k}(z)$ $(k=1,2,\cdots)$ which is a completely reduced C_1-sequence in D with $\Omega(S',D)\leqslant\tau$. (See Definition 5.17) Evidently S' satisfies one of the condtions 1^0 and 2^0 in Definition 5.18. (See Definition 5.16) So it remains to show that the limit function $F(z)$ of S', with respect to the spherical distance, is not the constant a. For this, we distinguish two cases:

1) S' is a C_0-sequence in D. In this case, from the condition (5.71) and the proof of Corollary 5.13, we see immediately that $F(z)$ is not the constant a.

2) S' is an irreducible C_1-sequence which is completely reduced in D. In this case, let E be the set of nonC_0-points of S'. E consists of a finite number of points $\zeta_i (i=1,2,\cdots,l; 1\leqslant l\leqslant\tau)$. Let $F(z)$ be the limit function of S' in the domain $D_1=D-E$, with respect to the spherical distance. We are going to show that $F(z)$ is not the constant a. Set $z_j^{(k)}=z_j(f_{n_k})$ $(k=1,2,\cdots)$ $(j=1,2,\cdots,\tau+1)$. By Lemma 3.2, we can find an increasing sequence of positive integers $k_p (p=1,2,\cdots)$ such that for each $1\leqslant j\leqslant\tau+1$, we have

$$\lim_{p\to+\infty} z_j^{(k_p)} = Z_j (Z_j \in \sigma), \ |z_j, z_{j'}| \geqslant \delta \ (1\leqslant j,j' \leqslant \tau+1; j\neq j'). \tag{5.72}$$

Consider the sequence $S'': f_{m_p}(z)$ $(p=1,2,\cdots; m_p=n_{k_p})$. Of course S'' is also a C_0-sequence in D_1 and the limit function of S'' in D_1, with respect to the spherical distance, is also $F(z)$. At least one of the points $Z_j (j=1,2,\cdots,\tau+1)$ belongs to D_1, for example $Z_1\in D_1$. Then we can find a circle Γ : $|z-Z_1|<\rho$ belonging to D_1, such that as $p\to+\infty$, $f_{m_p}(z)$ converges uniformly to $F(z)$, with respect to the spherical distance, in Γ. Hence there is a positive integer P such that when $p\geqslant P$, we have

$$|f_{m_p}(z), F(z)| < \eta$$

in Γ. Consequently, if $F(z)$ is the constant a, then by (5.72), we have

$$|f_{m_p}(z_1^{(k_p)}), a| < \eta, \tag{5.73}$$

when p is sufficiently large. Since $z_1^{(k_p)}=z_1(f_{m_p})$, (5.73) contradicts (5.71).

Finally we prove the following theorem [9]:

Theorem 5.21. Let $S:f_n(z)$ $(n=1,2,\cdots)$ be a sequence of meromorphic functions in a circle $\Gamma:|z-z_0|<r$ satisfying the following conditions:

1^0 S is a C_0-sequence in the domain $d:0<|z-z_0|<r$ and the limit function f(z) of S in d, with respect to the spherical distance, is a non-constant meromorphic function in Γ.

2^0 z_0 is a nonC_0-point of S which is irreducible with respect to z_0.

Then for each number $\rho(0<\rho<r)$ we can find a positive integer N such that, when $n\geqslant N$, the function $f_n(z)$ takes in the circle $|z-z_0|<\rho$ every value $w\in\hat{C}$.

Proof. For the sake of convenience, here we also use the symbol $\nu(f,w,\rho)$ defined in Definition 5.13. Distinguish two cases:

1) $f(z_0)\neq\infty$. Taking a number $\rho_0(0<\rho_0<r)$ such that $f(z)$ is finite in the circle $|z-z_0|<\rho_0$ and $f(z)\neq f(z_0)$ in the domain $0<|z-z_0|<\rho_0$, let ρ be a number such that $0<\rho<\rho_0$. By Lemma 5.13, there is a positive integer n_0 such that, when $n\geqslant n_0$, $f_n(z)$ is finite on the circle $|z-z_0|=\rho$ and as $n\to+\infty$, $f_n(z)$ converges uniformly to $f(z)$ on the circle $|z-z_0|=\rho$. By the condition 2^0 in Theorem 5.21, we can find a positive integer $N_1(N_1\geqslant n_0)$ such that, when $n\geqslant N_1$, we have

$$\nu(f_n,\infty,\rho)>0.$$

Set

$$h=\min_{z\in\gamma}|f(z)-f(z_0)|>0,$$

where γ denotes the circle $|z-z_0|=\rho$. Let $N_1'\geqslant N_1$ be an integer such that, when $n\geqslant N_1'$, we have

$$|f_n(z)-f(z)|<\frac{h}{2}$$

on γ. Then by the identity

$$f_n(z)-w = \{f(z)-f(z_0)\} + g_n(z),$$

where

$$g_n(z) = \{f_n(z)-f(z)\} + \{f(z_0)-w\},$$

and Rouché's theorem, we see that if $n \geqslant N_1'$, then for each point w of the circle $|w-f(z_0)| < h/2$, we have

$$\nu(f_n, w, \rho) = \nu(f_n, \infty, \rho) + m,$$

where m denotes the order of the zero z_0 of $f(z)-f(z_0)$.

Next taking a number ρ' $(0 < \rho' < \rho)$ such that

$$|f(z)-f(z_0)| < \frac{h}{4}$$

in the circle $|z-z_0| \leqslant \rho'$, let n_0' be a positive integer such that, when $n \geqslant n_0'$, $f_n(z)$ is finite on the circle $|z-z_0| = \rho'$ and as $n \to +\infty$, $f_n(z)$ converges uniformly to f (z) on the circle $|z-z_0| = \rho'$. Let $N_2 (N_2 \geqslant n_0')$ be a positive integer such that, when $n \geqslant N_2$, we have

$$\nu(f_n, \infty, \rho') > 0.$$

Then let $N_2' \geqslant N_2$ be a positive integer such that, when $n \geqslant N_2'$, we have

$$|f_n(z)-f(z)| < \frac{h}{4}$$

on the circle $|z-z_0| = \rho'$. Then by the identity

$$f_n(z)-w = \{f(z_0)-w\} + \varphi_n(z),$$

where

$$\varphi_n(z) = \{f_n(z)-f(z)\} + \{f(z)-f(z_0)\},$$

and Rouché's theorem, if $n \geqslant N_2'$, then for each point w of the region $h/2 \leqslant |w-f(z_0)| < +\infty$, we have

$$\nu(f_n, w, \rho') = \nu(f_n, \infty, \rho').$$

Finally let $N = \max(N_1', N_2')$, then when $n \geqslant N$, the function $f_n(z)$ takes in the circle $|z-z_0| < \rho$ every value $w \in \hat{\mathbf{C}}$.

2) $f(z_0) = \infty$. Evidently there is a positive integer n_0 such that $f_n(z) \not\equiv 0$ for $n \geqslant n_0$. Then the sequence of meromorphic functions $F_n(z) = 1/f_n(z)$ $(n \geqslant n_0)$ and the function $F(z) = 1/f(z)$ satisfy the conditions 1^0 and 2^0 in Theorem 5.21, with $F(z_0) = 0$. Consequently we can apply the result obtained in case 1) to $F_n(z)$ $(n \geqslant n_0)$ and $F(z)$ and then we see that the conclusion of Theorem 5.21 also holds in the present case.

6
FURTHER APPLICATIONS

6. 1. UNIFORM CONVERGENCE OF SEQUENCES OF MEROMORPHIC FUNCTIONS

The theorem of Stieltjes-Vitali[33][37] was generalized by Montel [26] as follows:

Theorem 6. 1. Let $S: f_n(z)$ $(n=1,2,\cdots)$ be a sequence of holomorphic functions in a domain D satisfying the following conditions:

1^0 S belongs to a quasi-normal family $\mathscr{F}$ of holomorphic functions in D.

2^0 There exists a set E of points of D such that E has at least one point of accumulation in D, and that at each point of E, S converges to a finite limit. Then S converges locally uniformly to a holomorphic function in D.

Proof. First let us show that we can extract from S a subsequence S' which converges locally uniformly to a holomorphic function $F(z)$ in D. In fact, by the condition 1^0, we can extract from S a subsequence S' which is a C_1-sequence in D. Let σ be the set of nonC_0-points of S' in D. Then S' is a C_0-sequence in the domain $G=D-\sigma$. By the condition 2^0, E has at least one point of accumulation z_0 in D. Let $\bar{\Gamma}: |z-z_0|\leqslant r$ be a circle belonging to D. Since σ has no point of accumulation in D, $\bar{\Gamma}$ contains at most a finite number of points of σ, but an infinite number of points of E. Hence there is a point $z_1 \in E \cap G$. By the condition 2^0, at the point z_1, S' converges to a finite limit. By Corollary 1. 1, S' converges locally uniformly to a holomorphic function $F(z)$ in G. Consider a point $z_1 \in \sigma$. Let $\bar{\Gamma}: |z-z_1|\leqslant\rho$ be a circle belonging to D such that the region $0<|z-z_1|\leqslant\rho$ belongs to G. Since S' converges uniformly to $F(z)$ on the circle $|z-z_1|=\rho$, we see that $F(z)$ is holomorphic in the circle $|z-z_1|<\rho$, provided that $F(z_1)$ is suitably defined, and then by the principle of maximum modulus, S′ converges uniformly to F(z) in the circle $\overline{\Gamma}$. Consequently F(z) is holomorphic in D and S′ converges locally uniformly to F(z) in D.

It remains to show that the sequence S converges locally uniformly to the function F(z) in D. In fact, let α be any bounded closed set of points belonging to D. Assume that S does not converge uniformly to F(z) on α. Then we can find a number $\varepsilon_0>0$ and an increasing sequence of positive integers $n_k (k=1,2,\cdots)$ such that, for each $k\geqslant 1$, we have

$$\max_{z\in\alpha}|f_{n_k}(z)-F(z)|\geqslant\varepsilon_0. \tag{6.1}$$

Consider the sequence of functions $S_1: g_k(z)=f_{n_k}(z)$ $(k=1,2,\cdots)$. Since S_1 also satisfies the conditions 1^0 and 2^0, we can also extract from S_1 a subsequence S_1': $g_{k_\lambda}(z)$ $(\lambda=1,2,\cdots)$ converging locally uniformly to a holomorphic function $F_1(z)$ in D. There is then a positive integer λ_0 such that, when $\lambda\geqslant\lambda_0$, we have

$$\max_{z\in\alpha}|g_{k_\lambda}(z)-F_1(z)|<\varepsilon_0. \tag{6.2}$$

But on the set E we have

$$F(z)=F_1(z) \tag{6.3}$$

which holds in D, by the condition 2^0. The inequalities (6.1) and (6.2) are incompatible. So we get a contradiction. Hence S converges uniformly to F(z) on α. This proves Theorem 6.1.

Corollary 6.1. Let $S: f_n(z)$ $(n=1,2,\cdots)$ be a sequence of holomorphic functions in a domain D such that S is locally uniformly bounded in D and that S satisfies the condition 2^0 in Theorem 6.1. Then S converges locally uniformly to a holomorphic function in D.

This is the theorem of Stieltjes-Vitali. It follows immediately from Theorem 6.1, because, by Corollary 1.3, the sequence S considered as a family of holomorphic functions in D, is normal in D, hence quasi-normal in D.

For the case of sequences of meromorphic functions, we have the following theorem:

Theorem 6. 2. Let $S: f_n(z)(n=1,2,\cdots)$ be a sequence of meromorphic functions in a domaiin D satisfying the following conditions:

1^0 S belongs to a normal family $\mathscr{F}$ of meromorphic functions in D.

2^0 There exists a set E of points of D such that E has at least one point of accumulation in D, and that at each point of E, S is convergent with respect to the spherical distance. Then S is a C_0-sequence in D.

Proof. By the condition 1^0 in Theorem 6. 2, we can extract from S a subsequence S′ which is a C_0-sequence in D. Let F(z) be the limit function defined in D of S′, with respect to the spherical distance. Evidently it is sufficient to show that the sequence S converges locally uniformly to F(z) in D, with respect to the spherical distance. For this, let α be any bounded closed set of points belonging to D. Assume that S does not converge uniformly to F(z) on α, with respect to the spherical distance. Then we can find a number $\varepsilon_0 > 0$ and an increasing sequence of positive integer $n_k(k=1,2,\cdots)$ such that for each $k\geqslant 1$, we have

$$\sup_{z\in\alpha} |f_{n_k}(z), F(z)| \geqslant \varepsilon_0. \tag{6. 4}$$

Next from the sequence of functions $S_1: g_k(z)=f_{n_k}(z)$ $(k=1,2,\cdots)$ we can extract a subsequence $S_1': g_{k_\lambda}(z)$ $(\lambda=1,2,\cdots)$ which is a C_0-sequence in D. Let $F_1(z)$ be the limit function defined in D of S_1', with respect to the spherical distance. By Theorem 1. 2, S_1' converges locally uniformly to $F_1(z)$ in D, with respect to the sphercial distance. Consequently we can find a positive integer λ_0 such that when $\lambda\geqslant\lambda_0$, we have

$$\sup_{z\in\alpha} |g_{k_\lambda}(z), F_1(z)| < \varepsilon_0. \tag{6. 5}$$

(See the proof of Lemma 5. 13) By the condition 2^0 in Theorem 6. 2, on the set E, we have

$$F(z) = F_1(z). \tag{6. 6}$$

But by Theorem 1.3, each of the functions $F(z)$ and $F_1(z)$ is a meromorphic function in D or the constant ∞. Hence (6.6) holds in D. The inequalities (6.4) and (6.5) being incompatible, so we get a contradiction. Hence S converges uniformly to $F(z)$ on α, with respect to the spherical distance. This proves Theorem 6.2.

Unlike the case of sequences of holomorphic functions, Theorem 6.2 is not true, if in it we replace the condition 1^0 by that S belongs to a quasi-normal family $\mathscr{F}$ of meromorphic functions in D. To see this, consider the sequence

$$S: f_n(z) = \frac{1}{nz} \quad (n = 1,2,\cdots) \tag{6.7}$$

of meromorphic functions in the complex plane **C**. S has only one nonC_0-point $z = 0$. So S is a C_1-sequence in **C**, and S considered as a family of meromorphic functions in **C**, is quasi-normal in **C**.

However under suitable additional conditions, Theorem 6.2 can be extended to the case that the sequence S belongs to a quasi-normal family $\mathscr{F}$ of meromorphic functions in D. For this we need to give first a definition.

Definition 6.1. Let $S: f_n(z)$ $(n=1,2,\cdots)$ be a sequence of meromorphic functions in a domain D and z_0 a point of D. We say that z_0 is an exceptional point of S, if the following condition is satisfied: Given arbitrarily a circle $\Gamma: |z-z_0|< r$ belonging to D, a value $a \in \hat{C}$ and a positive integer **N**, there is always an integer $n \geqslant N$ such that the function $f_n(z)$ takes the value a in Γ.

By Theorem 1.1, it is easy to see that if z_0 is an exceptional point of S, then z_0 is a nonC_0-point of S.

Theorem 6.3. Let $S: f_n(z)$ $(n=1,2,\cdots)$ be a sequence of meromorphic functions in a domain D satisfying the following conditions:

1^0 S belongs to a quasi-normal family $\mathscr{F}$ of meromorphic functions in D.

2^0 S has no exceptional point in D.

3^0 There exists a set E of points of D such that E has at least one point of accumulation in D, and that at each point $z \in E$, S converges, with respect to the spherical distance, to a limit $\lambda(z)$ such that for every value $a \in \hat{C}$, the set $\{z | z \in E, \lambda(z)=a\}$ has no point of accumulation in D.

Then S is a C_0-sequence in D and the limit function of S, with respect to the spherical distance, is a non-constant meromorphic function in D.

Proof. Let $z_0 \in D$ be a point of accumulation of E. We are going to show that z_0 is a C_0-point of the sequence S. By the condition 2^0, there exist a circle Γ: $|z-z_0|<r$ belonging to D, a value a and a positive integer N, such that for each $n \geqslant N$, the function $f_n(z)$ does not take the value a in Γ. Set

$$F_n(z)=\begin{cases} f_n(z) \text{ if } a=\infty, \\ \dfrac{1}{f_n(z)-a} \text{ if } a \neq \infty \end{cases} (n \geqslant N). \tag{6.8}$$

Then by the condition 1^0 in Theorem 6.3 and Lemmas 5.2 and 5.3, we see that the sequence $F_n(z)(n \geqslant N)$ belongs to a quasi-normal family of holomorphic functions in Γ. By the condition 3^0, there is a number $\rho(0<\rho<r)$ such that the domain ω: $0<|z-z_0|<\rho$ contains no point of the set $\{z|z \in E, \lambda(z)=a\}$. Consequently on the set $E_1=E \cap \omega$, we have $\lambda(z) \neq a$. So at each point $z \in E_1$, $F_n(z)$ converges to a finite limit, as $n \to +\infty$. Hence we can apply Theorem 6.1 to the sequence of holomorphic functions $F_n(z)$ $(n \geqslant N)$ in the circle Γ and conclude that as $n \to +\infty$, $F_n(z)$ converges locally uniformly to a holomorphic function in Γ. Then from (8) we see that the sequence S is a C_0-sequence in Γ and à fortiori, z_0 is a C_0-point of S.

Next we are going to show that each point of D is a C_0-point of S. Consider a point $z_1 \in D(z_1 \neq z_0)$. Join z_0 and z_1 by a polygonal line $Z(t)$ $(0 \leqslant t \leqslant 1)$ lying in D with $Z(0)=z_0$ and $Z(1)=z_1$. Since z_0 is a C_0-point of S, there is a number $\tau(0<\tau<1)$ such that each point of the polygonal line $Z(t)$ $(0 \leqslant t \leqslant \tau)$ is a C_0-point of S. Let τ_* be the least upper bound of the set of such numbers τ. We have $0<\tau_* \leqslant 1$. Then each point of the polygonal line $Z(t)(0 \leqslant t<\tau_*)$ is a C_0-point of S. To each t $(0 \leqslant t<\tau_*)$ corresponds a circle Γ_t: $|z-Z(t)|<r_t$ belonging to D, such that S is a C_0-sequence in Γ_t, where $\Gamma_0=\Gamma$. Consider the set

$$d=\bigcup_{0 \leqslant t<\tau_*} \Gamma_t.$$

It is easy to see that d is a subdomain of D and S is a C_0-sequence in d. Let $f(z)$ be the limit function of S in d, with respect to the spherical distance. On the set

E_1, we have $f(z)=\lambda(z)$. By the condition 3^0, $\lambda(z)$ is non-constant on E_1. Consequently $f(z)$ is a non-constant meromorphic function in d.

Consider the point $z_\cdot=Z(\tau_\cdot)$. Again by the condition 2^0, there exist a circle $\Gamma_\cdot: |z-z_\cdot|<r_\cdot$ belonging to D, a value $a_\cdot$ and a positive integer $N_\cdot$, such that for each $n\geqslant N_\cdot$, the function $f_n(z)$ does not take the value $a_\cdot$ in $\Gamma_\cdot$. Evidently we can find a line segment $L:Z(t)$ $(t_1\leqslant t\leqslant t_2, 0<t_1<t_2<\tau_\cdot)$ belonging to $\Gamma_\cdot$. Since $f(z)$ is non-constant in d and $L\subset d$, the set $\{z|z\in L, f(z)=a_\cdot\}$ consists of at most a finite number of points. Consequently as for the point z_0, we conclude that the point $z_\cdot$ is a C_0-point of S. It follows that we can not have $\tau_\cdot<1$. Therefore $\tau_\cdot=1$ and the point $z_1=Z(1)$ is a C_0-point of S.

We have shown that S is a C_0-sequence in D. Let $f(z)$ be the limit function of S defined in D, with respect to the spherical distance. By the condition 3^0, $f(z)$ is non-constant on the set E. Hence $f(z)$ is a non-constant meromorphic function in D.

Consider again the sequence S of meromorphic functions in **C** defined by (6.7). Since the functions $f_n(z)$ $(n=1,2,\cdots)$ of S do not take the value o in **C**, S has no exceptional point. On the other hand, at each point $z\in \mathbf{C}$, S converges to a limit $f(z)$, with respect to the spherical distance. We have

$$f(0)=\infty,\ f(z)=0 \ for\ z\neq 0.$$

So the sequence S does not satisfy the condition 3^0 in Theorem 6. 3.

As an application of Theorem 6. 2, we prove the following theorem:

Theorem 6. 4. Let $S:f_n(z)$ $(n=1,2,\cdots)$ be a sequence of meromorphic functions in a domain D. If at each point of D, the sequence S is convergent with respect to the spherical distance, then the set of the C_0-points of S is everywhere dense in D.

For the proof of this theorem we need the following lemma:

Lemma 6. 1. Let $f_j(z)$ $(j=1,2)$ be two meromorphic functions in a domain D. Then the function $\varphi(z)=|f_1(z),f_2(z)|$ is continuous in D.

Proof. Consider a point $z_0\in D$. We have

$$|f_1(z),f_2(z)| \leqslant |f_1(z),f_1(z_0)| + |f_1(z_0),f_2(z_0)| + |f_2(z_0),f_2(z)|$$
$$|f_1(z_0),f_2(z_0)| \leqslant |f_1(z_0),f_1(z)| + |f_1(z),f_2(z)| + |f_2(z),f_2(z_0)|$$

and hence

$$|\varphi(z) - \varphi(z_0)| \leqslant |f_1(z),f_1(z_0)| + |f_2(z),f_2(z_0)|.$$

By Lemma 1.4, $f_j(z)$ $(j=1,2)$ are continuous in D with respect to the spherical distance. It follows that

$$\lim_{z\to z_0}\varphi(z) = \varphi(z_0)$$

and the function $\varphi(z)$ is continuous at z_0.

Now let us prove Theorem 6.4. Consider a circle Γ: $|z-z_0|<r$ belonging to D. We are going to show that there exist a circle Γ' : $|z-z_0'|<r'$ and a positive integer N, such that $\Gamma'\subset\Gamma$ and that for $m\geqslant N$, $n\geqslant N$ we have

$$|f_m(z),f_n(z)| \leqslant \frac{1}{4} \tag{6.9}$$

in Γ'. In fact, assume, on the contrary, that such Γ' and N do not exist. Then we can find $m_1\geqslant 1$, $n_1\geqslant 1$ and a point $z_1\in\Gamma$, such that

$$|f_{m_1}(z_1),f_{n_1}(z_1)| > \frac{1}{4}.$$

By Lemma 6.1, the function $|f_{m_1}(z),f_{n_1}(z)|$ is continuous at z_1, hence we can find a circle Γ_1: $|z-z_1|<r_1$ such that $\overline{\Gamma}_1\subset\Gamma$ ($\overline{\Gamma}_1=(|z-z_1|\leqslant r_1)$) and that

$$|f_{m_1}(z),f_{n_1}(z)| > \frac{1}{4}$$

in $\overline{\Gamma}_1$. Then we can find $m_2\geqslant 2$, $n_2\geqslant 2$ and a point $z_2\in\Gamma_1$, such that

$$|f_{m_2}(z_2), f_{n_2}(z_2)| > \frac{1}{4}.$$

Again we can find a circle Γ_2: $|z-z_2|<r_2$ such that $\overline{\Gamma}_2 \subset \Gamma_1$ and that

$$|f_{m_2}(z), f_{n_2}(z)| > \frac{1}{4}$$

in $\overline{\Gamma}_2$. In this way we find successively a sequence of circles Γ_p: $|z-z_p|<r_p (p=1,2,\cdots)$ and a sequence of pairs of integers $m_p, n_p (p=1,2,\cdots)$, such that $\overline{\Gamma}_1 \subset \Gamma$ and for $p \geqslant 1$, we have

$$\overline{\Gamma}_{p+1} \subset \Gamma_p, \ m_p \geqslant p, \ n_p \geqslant p \tag{6.10}$$

and

$$|f_{m_p}(z), f_{n_p}(z)| > \frac{1}{4}$$

in $\overline{\Gamma}_p$.

Let z_* be a limiting point of the sequence of points $z_p (p=1,2,\cdots)$. Then we see easily that $z_* \in \overline{\Gamma}_p (p=1,2,\cdots)$ and hence

$$|f_{m_p}(z_*), f_{n_p}(z_*)| > \frac{1}{4} \ (p = 1,2,\cdots). \tag{6.11}$$

(6.10) and (6.11) are incompatible with the condition that the sequence $S: f_n(z)$ $(n=1,2,\cdots)$ is convergent at z_*, by hypothesis.

This contradiction proves that there exist a circle $\Gamma' \subset \Gamma$ and a positive integer N, such that for $m \geqslant N$, $n \geqslant N$, the inequality (6.9) holds in Γ', as asserted. Let $\gamma: |z-z_0'|<\rho \ (0<\rho<r')$ be a circle such that

$$|f_N(z), f_N(z_0)| < \frac{1}{4}$$

in γ. Then for $m \geqslant N$ and $z \in \gamma$, we have

$$|f_m(z), f_N(z_0)| \leqslant |f_m(z), f_N(z)| + |f_N(z), f_N(z_0)|$$
$$< \frac{1}{4} + \frac{1}{4} = \frac{1}{2}.$$

Thus for $m \geqslant N$, the function $f_m(z)$ does not take in Υ the values w belonging to the region $|w, f_N(z_0)| \geqslant 1/2$ and hence the family $f_m(z)$ $(m \geqslant N)$ is normal in Υ by Theorem 3. 1. Evidently the family $f_n(z)$ $(n \geqslant 1)$ is also normal in Υ. Since by hypothesis, the sequence $S: f_n(z)$ $(n=1,2,\cdots)$ is convergent at each point of Υ with respect to the spherical distance, it follows from Theorem 6. 2 that S is a C_0-sequence in Υ. Since $\Upsilon \subset \Gamma' \subset \Gamma$ and $\Gamma \subset D$ is arbitrary, the set of C_0-points of S is therefore everywhere dense in D.

From Theorem 6. 4 we deduce immediately the following theorem of Osgood [29]:

Corollary 6. 2. Let $S: f_n(z)$ $(n=1,2,\cdots)$ be a sequence of holomorphic functions in a domain D. If at each point of D, the sequence S converges to a finite limit, then we can find a domain $D_1 \subset D$ such that as $n \to +\infty$, $f_n(z)$ converges uniformly to a holomorphic function $F(z)$ in D_1.

Proof. By Theorem 6. 4 the set of C_0-points of S is everywhere dence in D. Let $z_0 \in D$ be a C_0-points of S. Then we can find a circle $\Gamma: |z-z_0|<r$ belonging to D, such that S is a C_0-sequence in Γ. Then by Corollary 1. 1, as $n \to +\infty$ $f_n(z)$ converges locally uniformly to a holomorphic function $F(z)$ in Γ. A fortiori, as $n \to +\infty$, $f_n(z)$ converges uniformly to $F(z)$ in a circle $\Gamma': |z-z_0|<r_0'(0<r_0'<r_0)$.

In Theorems 6. 1, 6. 2 and 6. 3, we have always assumed that the set E has at least one point of accumulation in the domain D. The case where the points of accumulation of the set E all lie on the boundary of the domain D was first treated by Blaschke [1] for bounded holomorphic functions in the unit circle. We are going to prove a more general theorem [8] which contains that of Blaschke as a particular case.

Lemma 6. 2. Let $f(z)$ be a holomorphic function in the circle $C: |z|<1$. Assume that $f(z)$ is nonidentically equal to zero and has an infinite number of zeros in C, arranged in a sequence:

$$z_1, z_2, \cdots, z_n, \cdots \tag{6.12}$$

such that each zero appears as many times as its order of multiplicity and that

$$|z_n| \leqslant |z_{n+1}| \ (n = 1, 2, \cdots). \tag{6.13}$$

If the function m(r,f) (see Appendix B) is bounded for $0<r<1$, then the series

$$\sum_{n=1}^{\infty}(1 - |z_n|) \tag{6.14}$$

is convergent.

Proof. Without loss of generality, we may assume $f(0)\neq 0$. In fact, if $z=0$ is a zero of order k of f(z), then we have

$$f(z) = z^k g(z),\ g(0) \neq 0,$$

where g(z) is a holomorphic function in C, whose zeros are precisely those of f(z) and with the same orders of multiplicity, except $z=0$. Moreover by hypothesis, for $0<r<1$,

$$m(r,f) \leqslant A \tag{6.15}$$

where A is a positive constant, and we have

$$m(r,g) \leqslant m(r,f) + k\log\frac{1}{r} \leqslant A + k\log\frac{1}{r}.$$

Since m(r,g) is a nondecreasing function of r, we have for $0<r<r'<1$,

$$m(r,g) \leqslant A + k\log\frac{1}{r'}.$$

Keeping r fixed and letting $r'\to 1$, we get

$$m(r,g) \leqslant A.$$

So the function $m(r,g)$ is also bounded for $0<r<1$.

Now set

$$r_n = |z_n| \quad (n = 1,2,\cdots) \tag{6.16}$$

and consider a number $r(r_1<r<1)$. By (6.15) and Jensen formula

$$\log|f(0)| = \frac{1}{2\pi}\int_0^{2\pi}\log|f(re^{i\theta})|d\theta - \sum_{j=1}^{p}\log\frac{r}{r_j},$$

where p is a positive integer such that $r_p<r\leqslant r_{p+1}$, we find

$$\frac{r_1}{r}\frac{r_2}{r}\cdots\frac{r_p}{r} \geqslant e^{-A}|f(0)|. \tag{6.17}$$

This inequality remains valid, if we replace p by any other positive integer n. In fact, if $n>p$, then r_j/r $(j=p+1,p+2,\cdots,n)$ are greater or equal to 1. On the other hand, if $n<p$, then r_j/r $(j=n+1,n+2,\cdots,p)$ are less or equal to 1. Consequently

$$r_1r_2\cdots r_n \geqslant r^ne^{-A}|f(0)|. \tag{6.18}$$

Keeping n fixed and letting $r\to 1$, we get

$$r_1r_2\cdots r_n \geqslant e^{-A}|f(0)|$$

or

$$\frac{1}{r_1r_2\cdots r_n} \leqslant \frac{e^A}{|f(0)|}.$$

Since

$$e^{1-r_n} < \frac{1}{1-(1-r_n)} \quad (n = 1,2,\cdots),$$

we have

$$e^{\sum_{j=1}^{n}(1-r_j)} < \frac{e^A}{|f(0)|},$$

which implies the convergence of the series

$$\sum_{n=1}^{\infty}(1-r_n).$$

Lemma 6. 3. Let $\mathscr{F}$ be a family of holomorphic functions in the circle C: $|z|<1$. If there is a positive number A such that for each function $f(z)\in\mathscr{F}$, the inequality

$$m(r,f) \leqslant A \tag{6. 19}$$

holds for $0<r<1$, then the family $\mathscr{F}$ is normal in C. (see [26],[36])

Proof. Consider a function $f(z)\in\mathscr{F}$ and two numbers r, R $(0<r<R<1)$. By Lemma 5 in Appendix B and (6. 19), we have

$$\log M(r,f) \leqslant \frac{R+r}{R-r}m(R,f) \leqslant A\,\frac{R+r}{R-r}$$

and hence

$$|f(z)| \leqslant A\,\frac{R+r}{R-r}$$

for $|z|\leqslant r$. This shows that the family $\mathscr{F}$ is uniformly bounded in the circle $|z|\leqslant r$. Since r is arbitrary, $\mathscr{F}$ is normal in C, by Corollary 1. 3.

Lemma 6. 4. Let $S: f_n(z)$ $(n=1,2,\cdots)$ be a sequence of holomorphic functions in the circle $C: |z|<1$. If S converges locally uniformly to a holomorphic function $f(z)$ in C, then for each number r $(0<r<1)$, we have

$$\lim_{n\to+\infty} m(r,f_n) = m(r,f). \tag{6.20}$$

Proof. Consider a number $r(0<r<1)$. By the hypothesis, we see as usual that S converges uniformly to $f(z)$ on the circle $|z|=r$. Then it is easy to see that

$$\lim_{n\to+\infty} \overset{+}{\log} |f_n(re^{i\theta})| = \overset{+}{\log} |f(re^{i\theta})|$$

uniformly in the interval $0\leqslant\theta\leqslant 2\pi$. This implies (6. 20).

Theorem 6. 5. Let $S: f_n(z)$ $(n=1,2,\cdots)$ be a sequence of holomorphic functions in the circle $C: |z|<1$ satisfying the following conditions:

1^0 There is a positive number A such that for each $n\geqslant 1$, the inequality

$$m(r,f_n) \leqslant A \tag{6.21}$$

holds for $0<r<1$.

2^0 There is a sequence of distinct points $\zeta_j (j=1,2,\cdots)$ of C such that the series

$$\sum_{j=1}^{\infty}(1-|\zeta_j|) \tag{6.22}$$

is divergent and that for each $j\geqslant 1$, S converges at ζ_j to a finite limit.
Then S converges locally uniformly to a holomorphic function in C.

Proof. By Lemma 6. 3 and the condition 1^0, S constitutes a normal family in C. Hence we can extract from S a subsequence S′ which converges locally uniformly to a holomorphic function $F(z)$ in C. If S does not converge locally uniformly to the function $F(z)$ in C, then as in the proof of Theorem 6. 1, we can find another subsequence S_1' of S, such taht S_1' converges locally uniformly to a holomorphic function $F_1(z)$ in C, which is not identically equal to $F(z)$. Consider the function $f(z)=F(z)-F_1(z)$ which is holomorphic and nonidentically equal to zero in C, and of which, by the condition 2^0, the points $\zeta_j (j=1,2,\cdots)$ are zeros. Moreover by the condition 1^0 and Lemma 6. 4, we have

$$m(r,F) \leqslant A, \quad m(r,F_1) \leqslant A$$

and hence

$$m(r,f) \leqslant 2A + \log 2$$

for $0<r<1$. Consequently by Lemma 6. 2, the series (22) is convergent, incompatible with the condition 2^0. Hence S converges locally uniformly to the function $F(z)$ in C.

Corollary 6. 3. If a sequence $S: f_n(z)$ $(n=1,2,\cdots)$ of holomorphic functions in the circle $C: |z|<1$ is uniformly bounded in C and satisfies the condition 2^0 in Theorem 6. 5, then S converges locally uniformly to a holomorphic function in C.

This is Blaschke's theorem mentioned above.

Theorem 6. 5. can be extended to the case where the constant A in the condition 1^0 is replaced by an unbounded function of r. We first prove the following lemma: [8]

Lemma 6. 5. Let $f(z)$ be a holomorphic function in the circle $C: |z|<1$. Assume that $f(z)$ is nonidentically equal to zero and has an infinite number of zeros in C, arranged in a sequence:

$$z_1, z_2, \cdots, z_n, \cdots \tag{6. 23}$$

in the same way as in Lemma 6. 2. Let $U(x)$ be a positive function defined for $x>1$ and tending to infinity with x. If for $r_0<r<1$ $(0\leqslant r_0<1)$ we have

$$m(r,f) \leqslant AU\left(\frac{1}{1-r}\right), \tag{6.24}$$

where A is a positive constant, then the series

$$\sum_{n=1}^{\infty}\left\{\frac{1-|z_n|}{U\left(\frac{\alpha}{1-|z_n|}\right)}\right\}^{1+\varepsilon} \tag{6.25}$$

converges for any numbers $\alpha>1$ and $\varepsilon>0$.

Proof. As in the proof of Lemma 6.2, we may assume $f(0)\neq 0$. In fact, if $z=0$ is a zero of order k of $f(z)$, then we have

$$f(z) = z^k g(z),\ g(0) \neq 0.$$

Take a number $r_0'(r_0<r_0'<1)$ such that for $r_0'<r<1$ we have

$$U\left(\frac{1}{1-r}\right) < 1.$$

Then for $r_0'<r<1$, we have

$$m(r,g) \leqslant m(r,f) + k\log\frac{1}{r} < AU\left(\frac{1}{1-r}\right) + k\log\frac{1}{r_0'}$$
$$< A'U\left(\frac{1}{1-r}\right),$$

where $A' = A + k\log\frac{1}{r_0'}$.

Now set

$$r_n = |z_n|\ (n = 1,2,\cdots)$$

and consider a number $r(\max(r_0,r_1)<r<1)$. By (24) and Jensen formula, we see as in the proof of Lemma 6.2, that for any positive integer $n\geqslant 1$, we have

$$\frac{r_1}{r}\frac{r_2}{r}\cdots\frac{r_n}{r} \geqslant e^{-AU(\frac{1}{1-r})}|f(0)|,$$

and à fortiori,

$$(\frac{r_n}{r})^n \geqslant e^{-AU(\frac{1}{1-r})}|f(0)|. \tag{6.26}$$

Replacing r_0, if necessary, by a greater number, we may assume that for $r_0<r<1$, we have

$$U(\frac{1}{1-r}) > 1. \tag{6.27}$$

Then from (6.26) we get

$$\log\frac{r}{r_n} \leqslant \frac{K}{n}U(\frac{1}{1-r}), \tag{6.28}$$

where $K=A+\overset{+}{\log}\frac{1}{|f(0)|}$. Observing that if $0<a<1$, $0<b<1$ and $a<b$, then

$$b-a<\log\frac{1-a}{1-b}.$$

Hence if $1-r_n>1-r$, then (28) yields

$$1-r_n<1-r+\frac{K}{n}U(\frac{1}{1-r}). \tag{6.29}$$

This inequality is then valid for any $n\geqslant 1$ and $\max(r_0, r_1)<r<1$. $\alpha>1$ being a number, let us determine r by the relation

$$1-r=\frac{1-r_n}{\alpha}. \tag{6.30}$$

Then when n is sufficiently large, we have $\max(r_0, r_1) < r < 1$, hence, by (6. 29), we have

$$\frac{1-r_n}{U\left(\frac{\alpha}{1-r_n}\right)} < \frac{K'}{n},$$

where $K' = \alpha K/(\alpha-1)$. This inequality implies the convergence of the series (6. 25).

By the same method of proof of Theorem 6. 5, we deduce from Lemma 6. 5 the following theorem:

Theorem 6. 6. Let $S: f_n(z)$ $(n=1,2,\cdots)$ be a sequence of holomorphic functions in the circle $C: |z|<1$ satisfying the following conditions:

1^0 There exist a positive function $U(x)$ defined for $x>1$ with $\lim\limits_{x\to+\infty} U(x) = +\infty$ and a number $r_0 (0 \leqslant r_0 < 1)$ such that for each $n \geqslant 1$, the inequality

$$m(r, f_n) \leqslant AU\left(\frac{1}{1-r}\right) \tag{6. 31}$$

holds for $r_0 < r < 1$, where A is a positive constant independent of n.

2^0 There are a sequence of distinct points $\zeta_j (j=1,2,\cdots)$ of C and two numbers $\alpha > 1, \varepsilon > 0$, such that the series

$$\sum_{j=1}^{\infty}\left\{\frac{1-|\zeta_j|}{U\left(\frac{\alpha}{1-|\zeta_j|}\right)}\right\}^{1+\varepsilon} \tag{6. 32}$$

is divergent and that for each $j \geqslant 1$, S converges at ζ_j to a finite limit.

Then S converges locally uniformly to a holomorphic function in C.

In particular if we take $U(x) = x^\sigma (\sigma > 0)$, the inequality (6. 31) becomes

$$m(r, f_n) \leqslant \frac{A}{(1-r)^\sigma} \tag{6. 33}$$

and the series (6. 32), neglecting a constant factor, becomes

$$\sum_{j=1}^{\infty}(1-|\zeta_j|)^{\sigma+1+\varepsilon'}.$$

In this way we deduce from Theorem 6.6 a corollary which is more general than a generalization of Blaschke's theorem due to Montel [26] who, instead of the condition (6.33), assumes

$$\log M(r,f_n)<\frac{A}{(1-r)^{\sigma}}.$$

Next we are going to show that in Lemma 6.5 we can remove the constant $\alpha>1$ in the series (6.25), provided that the function $U(x)$ satisfies the following condition of normal growth: To each number $\varepsilon>0$ corresponds a numbr $x_\varepsilon\geqslant 1$ such that for $x>x_\varepsilon$ we have

$$U(x')\leqslant U(x)^{1+\varepsilon}\ with\ x'=x\left(1+\frac{1}{U(x)^{\varepsilon}}\right). \tag{6.34}$$

In fact, given $\varepsilon>0$, take a number $\delta<\varepsilon/2$ and then a positive integer n_0 such that for $n\geqslant n_0$ we have

$$r_n>max(r_0,r_1),\ \frac{1}{1-r_n}>x_\delta. \tag{6.35}$$

Consider an integer $n\geqslant n_0$ and determine r by the relation

$$\frac{1}{1-r}=\frac{1}{1-r_n}\left\{1+\frac{1}{U(\frac{1}{1-r_n})^{\delta}}\right\}, \tag{6.36}$$

which implies $r_n<r<1$. We can then apply (6.29) and get

$$1-r_n<\frac{1-r_n}{1+\frac{1}{U(\frac{1}{1-r_n})^{\delta}}}+\frac{K}{n}U\left\{\frac{1}{1-r_n}\left[1+\frac{1}{U(\frac{1}{1-r_n})^{\delta}}\right]\right\}.$$

Then by the condition (6. 34) and (6. 35), we have

$$\frac{1-r_n}{U(\frac{1}{1-r_n})^{\delta}+1}<\frac{K}{n}U(\frac{1}{1-r_n})^{1+\delta}$$

and

$$\frac{1-r_n}{U(\frac{1}{1-r_n})^{1+2\delta}}<\frac{2K}{n}.$$

A fortiori,

$$\left\{\frac{1-r_n}{U(\frac{1}{1-r_n})}\right\}^{1+2\delta}<\frac{2K}{n}$$

and

$$\left\{\frac{1-r_n}{U(\frac{1}{1-r_n})}\right\}^{1+\varepsilon}<\frac{K'}{n^{\lambda}},$$

where $K'>0$ is a constant and $\lambda=(1+\varepsilon)/(1+2\delta)>1$. The series

$$\sum_{n=1}^{\infty}\left\{\frac{1-r_n}{U(\frac{1}{1-r_n})}\right\}^{1+\varepsilon}$$

is therefore convergent. This proves our assertion.

Correspondingly in Theorem 6. 6. , we can remove the constant $\alpha>1$ in the series (6. 32), provided that the function $U(x)$ satisfies the condition (6. 34) of normal growth.

6.2. GENERALIZATIONS OF BLOCH'S THEOREM AND SOME THEOREMS OF VALIRON

The method of normal families is very useful for the study of the domains covered by the values of meromorphic functions. In the preceding chapters, we have already used that method in the proof of some particular covering theorems and in dealing with filling circles or filling domains. In this paragraph, we shall prove some general theorems in the case of holomorphic functions.

Let us first recall Definition 2. 1 and the following theorem of Bloch (see Appendix A):

Theorem 6. 7. Let $w=f(z)$ be a holomorphic function in the circle $C: |z| < 1$ such that $f(0)=0$, $|f'(0)| \geqslant 1$. Then there is a circle $|w-w_0|<\rho$ which is a simply covered image domain of $f(z)$ for C and whose radius ρ is greater than an absolute positive constant b.

An immediate consequence of this theorem is the following corollary:

Corollary 6. 4. Let $w=f(z)$ be a holomorphic function in a circle $\Gamma: |z-z_0|<r$ and let $\delta>0$ be a number. Assume that there does not exist a circle $|w-w_0|<\rho$ which is a simply covered image domain of $f(z)$ for Γ and whose radius ρ is greater than δ, then

$$|f'(z_0)| < \frac{\delta}{br}, \tag{6. 37}$$

where b is the positive constant in Theorem 6. 7.

Proof. Consider the holomorphic function

$$\varphi(\zeta) = \frac{b}{\delta}\{f(z_0 + r\zeta) - f(z_0)\}$$

in the circle $C: |\zeta|<1$, with

$$\varphi(0) = 0, \ \varphi'(0) = \frac{br}{\delta} f'(z_0).$$

Suppose that the inequality (6. 37) does not hold. Then we have $|\varphi'(0)|\geqslant 1$. Consequently by Theorem 6. 7, there is a circle $|Z-Z_0|<R$ which is a simply covered image domain of $\varphi(\zeta)$ for C, with

$$R > b. \tag{6. 38}$$

Then we see easily that the circle $|w-w_0|<\rho$, with

$$w_0 = f(z_0) + \frac{\delta}{b}Z_0, \ \rho = \frac{\delta}{b}R,$$

is a simply covered image domain of the function $f(z)$ for Γ. But (6. 38) implies $\rho>\delta$, incompatible with the hypothesis of Corollary 6. 4.

Theorem 6. 8. Let $\mathscr{F}$ be a family of holomorphic functions in a domain D. Assume that there does not exist a sequence of functions $f_n(z)$ $(n=1,2,\cdots)$ of the family $\mathscr{F}$, which converges locally uniformly to a finite constant or ∞ in D. Then there is a constant $\beta>0$ such that for each function $f(z)\in\mathscr{F}$, there exists a circle $|w-w_0|<\rho$ with $\rho>\beta$, which is a simply covered image domain of f (z) for D.

Proof. Suppose that such a constant β does not exist. Then to each positive integer $n\geqslant 1$ corresponds a function $f_n(z)\in\mathscr{F}$, such that we can not find a circle $|w-w_0|<\rho$ with $\rho>1/n$, which is a simply covered image domain of $f_n(z)$ for D. We are going to show that the sequence of functions $f'_n(z)$ $(n=1,2,\cdots)$ converges locally uniformly to zero in D. In fact, consider a circle $\Gamma: |z-z_0|<r$ such that the circle $|z-z_0|<2r$ belongs to D. Let z' be a point of Γ and consider the circle $\Gamma': |z-z'|<r$. Let $n\geqslant 1$ be an integer. By Corollary 6. 4, we have

$$|f'_n(z')| < \frac{1}{nbr}.$$

This shows that the sequence of functions $f'_n(z)$ $(n=1,2,\cdots)$ converges uniformly to zero in Γ, and hence proves our assertion.

Next we are going to show that we can extract from the sequence $f_n(z)$ $(n=$

$1,2,\cdots$), a subsequence $f_{n_k}(z)$ $(k=1,2,\cdots)$ which converges locally uniformly to a finite constant or ∞ in D. In fact, consider a point $z_0\in D$ and distinguish two cases:

Case 1. The sequence $f_n(z_0)$ $(n=1,2,\cdots)$ is bounded. Then we can extract from this sequence, a subsequence $f_{n_k}(z_0)$ $(k=1,2,\cdots)$ which converges to a finite limit l. Consider a circle $\Gamma: |z-z_0|<r$ such that its closure $\overline{\Gamma}: |z-z_0|\leqslant r$ belongs to D. By the result obtained above, the sequence $f'_{n_k}(z)$ $(k=1,2,\cdots)$ converges uniformly to zero in $\overline{\Gamma}$. Evidently in the forminla

$$f_{n_k}(z)-f_{n_k}(z_0)=\int_{z_0}^{z}f'_{n_k}(\zeta)d\zeta, \tag{6.39}$$

where the integral is taken along the segment joining z_0 to z, the right member of (6.39) tends uniformly to zero in Γ, as $k\to+\infty$. Then we see that in Γ, the sequence $f_{n_k}(z)$ $(k=1,2,\cdots)$ converges uniformly to l. Now consider a point $z'_0\in$ D, $z'_0\neq z_0$, and join z_0, z'_0 by a polygonal line λ lying in D. Let $\delta>0$ be a number such that for each point $a\in\lambda$, the circle $|z-a|\leqslant\delta$ belongs to D. Next take in the sense from z_0 to z'_0, a finite number of points $z_j (j=0,1,\cdots,m)$ such that

$$|z_{j+1}-z_j|<\delta\ (j=0,1,\cdots,m-1),\ z_m=z'_0. \tag{6.40}$$

By the result just obtained, the sequence $f_{n_k}(z)$ $(k=1,2,\cdots)$ converges uniformly to l in the circle $\Gamma_0: |z-z_0|<\delta$. À fortiori, by (6.40), the sequence $f_{n_k}(z_1)$ $(k=1,2,\cdots)$ converges to l. So in the circle $\Gamma_1: |z-z_1|<\delta$, the sequence $f_{n_k}(z)$ $(k=1,2,\cdots)$ converges uniformly to l. Continuing succesively in this way, we see finally that in the cricle $\Gamma_m: |z-z'_0|<\delta$, the sequence $f_{n_k}(z)$ $(k=1,2,\cdots)$ converges uniformly to l. This proves that the sequence $f_{n_k}(z)$ $(k=1,2,\cdots)$ converges locally uniformly to the constant l in D.

Case 2. The sequence $f_n(z_0)$ $(n=1,2,\cdots)$ is unbounded. Then we can extract from this sequence, a subsequence $f_{n_k}(z_0)$ $(k=1,2,\cdots)$ which converges to ∞. As in case 1, we see that the sequence $f_{n_k}(z)$ $(k=1,2,\cdots)$ converges locally uniformly to ∞ in D.

Thus in both cases we can get a sequence of functions $f_{n_k}(z)$ $(k=1,2,\cdots)$ of the family $\mathscr{F}$, which converges locally uniformly to a finite constant or ∞ in D. This contradiction with the hypothesis of Theorem 6.8 proves this theorem.

The numbers $\beta > 0$ having the property in Theorem 6. 8 consititute a set whose least upper bound $\beta_0 = \beta_0(D, \mathscr{F})$ depending only on D and $\mathscr{F}$, may be called the Bloch constant with respect to D and $\mathscr{F}$.

Given a domain D, we can define in various ways a family $\mathscr{F}$ of holomorphic functions in D, satisfying the required condition in Theorem 6. 8. We give some examples.

1) Let z_1, z_2 be two points of D. Let a and $b \neq 0$ be two finite values. Define $\mathscr{F}$ to be the family of holomorphic functions f(z) in D, such that

$$f(z_1) = a, \ f'(z_2) = b.$$

2) Let $z_1, z_2 (z_1 \neq z_2)$ be two points of D. Let $a, b (a \neq b)$ be two finite values. Define $\mathscr{F}$ to be the family of holomorphic functions f(z) in D, such that

$$f(z_1) = a, \quad f(z_2) = b.$$

More generally, we can replace the two points z_1, z_2 by two bounded closed sets E_1, E_2 of points of D.

3) Let E_1, E_2 be two bounded closed sets of points of D. Let A, B be two positive numbers. Let $m \geqslant 1$ be an integer. Define $\mathscr{F}$ to be the family of holomorphic functions f(z) in D, such that

$$\min_{z \in E_1} |f(z)| \leqslant A, \ \max_{z \in E_2} |f^{(m)}(z)| \geqslant B.$$

4) Let E_1, E_2 be two bounded closed sets of points of D. Let A, B$(A < B)$ be two positive numbers. Define $\mathscr{F}$ to be the family of holomorphic functions f (z) in D, such that

$$\min_{z \in E_1} |f(z)| \leqslant A, \ \max_{z \in E_2} |f(z)| \geqslant B.$$

Now we give some generalizations of theorems of Valiron [35].

Theorem 6. 9. Given a domain D, two bounded closed sets E_1, E_2 of points belonging to D, an integer $m \geqslant 0$ and a number $0 < \varepsilon < \pi/2$, we can find a number $A=A(D,E_1,E_2,m,\varepsilon)\geqslant 0$ depending only on D,E_1,E_2,m,ε, having the following property: If $f(z)$ is a holomorphic function in D satisfying the following conditions:

1^0 $f(z)$ has at least one zero in E_1;

$2^0 \max_{z\in E_2}|f^{(m)}(z)|\geqslant 1$;

then either the values of $f(z)$ cover the domain $\Omega: |w|<1/A$ $(0<1/A\leqslant +\infty)$, or there exists a point $w_0=w_0(f)\in\Omega$ such that $f(z)$ does not take the value w_0 in D and that the values of $f(z)$ cover the set Ω-Γ, where Γ denotes the circle:

$$|w-w_0|<|w_0|\sin\frac{\varepsilon}{2}. \tag{6.41}$$

Proof. Consider the number $0<\varepsilon'<1$ defined by

$$\sin\frac{\varepsilon}{2}=\frac{\varepsilon'}{1-\varepsilon'} \tag{6.42}$$

and let $\mathscr{G}$ be the family of holomorphic functions $g(z)$ in D satisfying the following conditions:

1) $g(z)$ has at least one zero in E_1.

2) $g(z)$ does not take the value 1 in D.

3) There exists a value $a=a(g)$ such that $|a|\leqslant 1$, $|1-a|\geqslant\varepsilon'$ and that $g(z)$ does not take the value a in D. By Theorem 2. 8, the family $\mathscr{G}$ is normal in D. Consider a sequence of functions $g_n(z)$ $(n=1,2,\cdots)$ of $\mathscr{G}$. By Corollary 1. 2 and the condition 1), from this sequence we can extract a subsequence $g_{n_k}(z)$ $(k=1,2,\cdots)$ such that as $k\to+\infty$, $g_{n_k}(z)$ converges locally uniformly to a holomorphic function $\varphi(z)$ in D. by Weierstrass theorem, as $k\to+\infty$, $g_{n_k}^{(m)}(z)$ also converges locally uniformly to $\varphi^{(m)}(z)$ in D. Thus the family $\mathscr{G}_m$: $\{g^{(m)}(z)\,|\,g(z)\in\mathscr{G}\}$ is also normal in D. Then by the method of proof of Lemma 2. 1, we see that the family $\mathscr{G}_m$ is uniformly bounded in the set E_2, namely there is a positve constant α such that for each function $g(z)\in\mathscr{G}$, we have

$$\max_{z\in E_2}|g^{(m)}(z)|\leqslant\alpha.$$

Such positive constants α constitute a set σ. Let $A \geqslant 0$ be the greatest lower bound of σ. Then for each function $g(z) \in \mathscr{G}$, we have

$$\max_{z \in E_2} |g^{(m)}(z)| \leqslant A. \tag{6.43}$$

Now let f(z) be a holomorphic function in D satisfying the conditions 1^0 and 2^0 in Theorem 6.9. Assume that there is a point w' of the domain $\Omega: |w| < 1/A$, such that f(z) does not take the value w' in D. We see easily that among the set S of such points w', there is one $w_0 \in S$ such that

$$|w_0| \leqslant |w'| \tag{6.44}$$

for $w' \in S$. We are going to show that the values of f(z) cover the set Ω-Γ, where Γ is the circle (6.41). Suppose that this is untrue, so there is a point $w_1 \in \Omega$-Γ such that f(z) does not take the value w_1 in D. Since $w_1 \notin \Gamma$, we must have

$$|w_1 - w_0| \geqslant \varepsilon' |w_1|, \tag{6.45}$$

where ε' is defined by (6.42). In fact, if

$$|w_1 - w_0| < \varepsilon' |w_1|,$$

then

$$|w_1| - |w_0| < \varepsilon' |w_1|, \ \varepsilon' |w_1| < \frac{\varepsilon'}{1 - \varepsilon'} |w_0| = |w_0| \sin \frac{\varepsilon}{2}$$

and

$$|w_1 - w_0| < |w_0| \sin \frac{\varepsilon}{2},$$

incompatible with the condition $w_1 \notin \Gamma$.

Consider the function

$$g(z) = \frac{1}{w_1} f(z)$$

which evidently does not take the values 1 and w_0/w_1 in D. By (6. 44) and (6. 45), we have

$$\left|\frac{w_0}{w_1}\right| \leqslant 1, \quad \left|1 - \frac{w_0}{w_1}\right| \geqslant \varepsilon' .$$

Moreover by the condition 1^0, g(z) satisfies the condition 1). Conseqently $g(z) \in \mathscr{G}$, and by (6. 43) and the condition 2^0, we have

$$\frac{1}{|w_1|} \leqslant \max_{z \in E_2} |g^{(m)}(z)| \leqslant A ,$$

$$|w_1| \geqslant \frac{1}{A}.$$

which contradicts the condition $w_1 \in \Omega$. This proves Theorem 6. 9.

The conclusion of Theorem 6. 9 implies the following consequence: The values of the function f(z) cover a circle $|w| = \rho$ with $\rho \geqslant \min(1, \frac{1}{2A}(1 - \frac{1}{\sqrt{2}}))$.

In fact, if $A = 0$, this is obvious. If $A > 0$, we see this by distinguishing two cases: $|w_0| < 1/2A$ and $|w_0| \geqslant 1/2A$.

The above consequence is a generalization of Corollary 2. 5.

A particular case of Theorem 6. 9 is that D is the circle C: $|z| < 1$, E_1, E_2 are the point $z = 0$ and $m = 1$. In this case, we get a theorem of Valiron [35] who stated it in the following form:

Corollary 6. 5. If the function

$$Z = f(z) = z + \cdots$$

is holomorphic for $|z| < 1$, then to each positive number ε $(0 < \varepsilon < \pi/2)$ corresponds a positive number $K(\varepsilon)$ such that the domain described by Z contains all points of the circle $|Z| < K(\varepsilon)$, except perhaps points which can be enclosed in a circle which looked from $Z = 0$ is in an angle ε.

Next we give a generalization of Theorem 6. 9.

Theorem 6. 10. Given a domain D, two bounded closed sets E_1, E_2 of points belonging to D, two integers $m \geqslant 0$, $p \geqslant 1$ and a number $0 < \varepsilon < \pi/2$, we can find a number $A = A(D, E_1, E_2, m, p, \varepsilon) \geqslant 0$ depending only on $D, E_1, E_2, m, p, \varepsilon$ having the following property: If f(z) is a holomorphic function in D satisfying the following conditions:

1^0 f(z) has at least p zeros in E_1, with due count of order of multiplicity;

$2^0 \max_{z \in E_2} |f^{(m)}(z)| \geqslant 1$;

then either the values of f(z) cover the domain $\Omega: |w| < 1/A$ $(0 < 1/A \leqslant +\infty)$ at least p times (see definition 4. 8), or there exists a point $w_0 = w_0(f) \in \Omega$ such that $f(z) - w_0$ has at most $p-1$ zeros in D and that the values of f(z) cover the set Ω-Γ at least p times, where Γ denotes the circle (6. 41).

Proof. Consider the number $0 < \varepsilon' < 1$ defined by (6. 42) and let $\mathscr{G}$ be the family of holomorphic functions g(z) in D satisfying the following conditions:

$1'$) g(z) has at least p zeros in E_1.

$2'$) $g(z) - 1$ has at most $p-1$ zeros in D.

$3'$) There exists a value $a = a(g)$ such that $|a| \leqslant 1$, $|1-a| \geqslant \varepsilon'$ and that g(z)-a has at most $p-1$ zeros in D.

By Corollary 5. 10 the family $\mathscr{G}$ is quasi-normal of order $p-1$ at most in D. We are going to show that the family $\mathscr{G}$ is normal in D. In fact, let $g_n(z)$ $(n = 1, 2, \cdots)$ be a sequence of functions of $\mathscr{G}$. According to Definition 5. 3, from this sequence we can extract a subsequence $g_{n_k}(z)$ $(k = 1, 2, \cdots)$ which is a C_1-sequence in D and of which the set σ of nonC_0-points consists of at most $p-1$ points. It is sufficient to show that the sequence $g_{n_k}(z)$ $(k = 1, 2, \cdots)$ is a C_0-sequence in D. Suppose that this is untrue. Then as $k \to +\infty$, $g_{n_k}(z)$ converges locally uniformly to ∞ in the domain $D_1 = D - \sigma$. (see the beginning part of Chapter 5) By the condition $1'$) the set $E_1 \cap \sigma = (z_1, z_2, \cdots, z_q)$ must be nonempty, because if otherwise, then $E_1 \subset D_1$ and as $k \to +\infty$, $g_{n_k}(z)$ converges uniformly to ∞ in E_1, incompatible with the condition $1'$). With each point $z_j (1 \leqslant j \leqslant q)$ describe a circle $\Upsilon_j: |z - z_j| < \delta$ such that the region $0 < |z - z_j| \leqslant \delta$ belongs to D_1 and that the closures $\bar{\Upsilon}_j: |z - z_j| \leqslant \delta$ $(j = 1, 2, \cdots, q)$ are mutually disjoint. Consider the set $H = E_1 - \bigcup_{j=1}^{q} \Upsilon_j$. We see easily that H is bounded closed and $H \subset D_1$. Consequently as $k \to +\infty$, $g_{n_k}(z)$ converges uniformly to ∞ in H. So we can find a positive integer k_0 such that when $k \geqslant k_0$, $g_{n_k}(z)$ has no zero in H. It follows that if $k \geqslant k_0$ and if we denote by $\nu_{kj} \geqslant 0$ the number of the zeros of the function $g_{n_k}(z)$ in the circle Γ_j, then, by the condition 1′), we have

$$\sum_{j=1}^{q} \nu_{kj} \geqslant p. \tag{6.46}$$

On the other hand, take a positive integer k_0' such that when $k \geqslant k_0'$, we have

$$|g_{n_k}(z)| > 1 \tag{6.47}$$

on the circles $|z-z_j|=\delta$ $(j=1,2,\cdots,q)$, and therefore, by Rouché's theorem, if we denote by $\nu'_{kj} \geqslant 0$ the number of the zeros of the function $g_{n_k}(z)-1$ in the circle Γ_j, we have

$$\nu'_{kj} = \nu_{kj} \quad (j=1,2,\cdots,q).$$

Consequently if $k \geqslant \max(k_0, k_0')$, then

$$\sum_{j=1}^{q} \nu'_{kj} = \sum_{j=1}^{q} \nu_{kj} \geqslant p,$$

incompatible with the condition 2′). This contradiction proves that the sequence $g_{n_k}(z)$ $(k=1,2,\cdots)$ is a C_0-sequence in D and hence the family $\mathscr{G}$ is normal in D.

The remaining part of the proof is similar to that of Theorem 6.9. First of all, in the same way we define the number $A \geqslant 0$ satisfying the condition (6.43). Next let f(z) be a holomorphic function in D satisfying the conditions 1^0 and 2^0 in Theorem 6.10. Assume that the values of f(z) do not cover the domain $\Omega: |w| < 1/A$ at least p times. Then there is a point $w' \in \Omega$ such that the number of the zeros of the function $f(z)-w'$ in D is at most equal to $p-1$. We see easily that among the set S of such points w', there is one $w_0 \in S$ such that

$$|w_0| \leqslant |w'| \tag{6.48}$$

for $w' \in S$. In fact, let $\rho_0 = \inf_{w' \in S} |w'|$. Then we can find a sequence of points $w'_k \in S$ $(k=1,2,\cdots)$ such that $\lim_{k \to +\infty} |w'_k| = \rho_0$. The sequence of points $w'_k (k=1,2,\cdots)$ has at least a limiting point w_0 such that $|w_0| = \rho_0$. This point $w_0 \in S$ and satisfies the condition (6.48). It remains to show that the values of the function $f(z)$ cover the set Ω-Γ at least p times, where Γ denotes the circle (6.41). Suppose that this is untrue, then there is a point $w_1 \in \Omega$-Γ such that the number of the zeros of $f(z)-w_1$ in D is at most equal to $p-1$. We have seen that w_1 must satisfy the inequality (6.45). Consider the function

$$g(z) = \frac{1}{w_1} f(z),$$

for which the number of the zeros of each of the two functions $g(z)-1$ and $g(z) - \frac{w_0}{w_1}$ in D is at most equal to $p-1$. Moreover by the condition 1^0 in Theorem 6.10, $g(z)$ also satisfies the condition $1'$). Consequently $g(z) \in \mathscr{G}$ and from the condition 2^0 in Theorem 6.10 and (6.43) we get $|w_1| \geqslant 1/A$. This contradiction with the condition $w_1 \in \Omega$ proves Theorem 6.10.

The conclusion of Theorem 6.10 implies the following consequence: The values of the function $f(z)$ cover at least p times a circle $|w| = \rho$ with $\rho \geqslant \min(1, \frac{1}{2A}(1-\frac{1}{\sqrt{2}}))$.

This consequence implies in particular the following corollary: (see [35])

Corollary 6.6. Given an integer $p \geqslant 1$, we can find a positive number $\alpha = \alpha(p) > 0$ depending only on p having the following property: If $f(z)$ is a holomorphic function in the circle $|z|<1$ such that $f(z)$ has a zero of order p at $z=0$ with $|f^{(p)}(0)| \geqslant 1$, then the values of $f(z)$ cover at least p times a circle $|w| = \rho$ with $\rho \geqslant \alpha$.

In fact it is sufficient to take in Theorem 6.10, $D = (|z|<1)$, $E_1 = E_2 = (0)$, $m = p$ and $\varepsilon = \pi/4$.

A particular case of Theorem 6.10 is that D is the circle $C: |z|<1$, E_1 is a circle $|z| \leqslant \rho$ $(0<\rho<1)$, E_2 is a circle $|z| = \rho'$ $(0<\rho'<1)$ and $m=0$. In this case we get the following theorem of Valirom [35]:

Corollary 6.7. If the function $Z=f(z)$ is holomorphic for $|z|<1$, has at least p zeros in the circle $|z|<\rho(0<\rho<1)$ and has a modulus greater or equal to 1 at a point of the circle $|z|=\rho'$, then to $\varepsilon(0<\varepsilon<\pi/2)$ corresponds a number $\Omega(\varepsilon,p,\rho,\rho')$ having the following property: $f(z)$ takes at least p times the values of modulus less than $\Omega(\varepsilon,p,\rho,\rho')$, except those which belong to a circle $c(f)$ which looked from $Z=0$ is in an angle ε.

In theorem 6.10, if besides the conditions 1^0 and 2^0, we assume furthermore that $f(z)$ has at most $q(q\geqslant p)$ zeros in D, we have the following theorem:

Theorem 6.11. Given a domain D, two bounded closed sets E_1, E_2 of points belonging to D and three integers $m\geqslant 0$, $p\geqslant 1, q\geqslant p$, we can find a number $A=A(D,E_1,E_2,m,p,q)\geqslant 0$ depending only on D,E_1,E_2,m,p,q having the following property: If $f(z)$ is a holomorphic function in D satisfying the following conditions:

1^0 $f(z)$ has at least p zeros in E_1 and at most q zeros in D, with due count of order of multiplicity;

2^0 $\max\limits_{z\in E_2}|f^{(m)}(z)|\geqslant 1$;

then the values of $f(z)$ cover the domain $|w|<1/A$ $(0<1/A\leqslant+\infty)$ at least p times.

Proof. Let $\mathscr{G}$ be the family of holomorphic functions $g(z)$ in D satisfying the following conditions:

$1'$) $g(z)$ has at least p zeros in E_1 and at most q zeros in D.

$2'$) $g(z)-1$ has at most $p-1$ zeros in D.

By Corollary 5.11, the family $\mathscr{G}$ is quasi-normal of order $p-1$ at most in D. As in the proof of Theorem 6.10 we see that the family $\mathscr{G}$ is normal in D, and then as in the proof of Theorem 6.9, we define the number $A\geqslant 0$ such that for each function $g(z)\in\mathscr{G}$, we have the inequality (6.43).

It remains to show that this number A has the required property in Theorem 6.11. In fact let $f(z)$ be a holomorphic function in D satisfying the conditions 1^0 and 2^0 in Theorem 6.11. Assume, on the contrary, that there is a point w_0 of the domain $|w|<1/A$ such that the function $f(z)-w_0$ has at most $p-1$ zeros in D. Evidently $w_0\neq 0$, and the function $g(z)=\frac{1}{w_0}f(z)\in\mathscr{G}$. Then by the inequality (6.43) and the condition 2^0 in Theorem 6.11, we get

$$\frac{1}{|w_0|} \leqslant A, \ |w_0| \geqslant \frac{1}{A},$$

incompatible with $|w_0| < 1/A$.

As a particular case of Theorem 6.11, we have the following corollary: (see [35])

Corollary 6.8. Given an integer $p \geqslant 1$, we can find a positive number $A = A(p) > 0$ depending only on p having the following property: If $f(z)$ is a holomorphic function in the circle $|z| < 1$ such that $f(z)$ has a zero of order p at $z=0$ with $|f^{(p)}(0)| \geqslant 1$ and that $f(z)$ has no zero in the domain $0 < |z| < 1$, then the values of $f(z)$ cover the domain $|w| < 1/A$ at least p times.

In fact it is sufficient to take in Theorem 6.11, $D = (|z| < 1)$, $E_1 = E_2 = (0)$ and $m = q = p$.

We wish to point out that, more generally, in the Theorems 6.9, 6.10 and 6.11, we may replace the condition 2^0 by the following condition:

$$\max_{z \in E_2} |P(f, f', \cdots, f^{(m)})| \geqslant 1, \tag{6.49}$$

where $P(f, f', \cdots, f^{(m)})$ is a given homogeneous polynomial of $f, f', \cdots, f^{(m)}$ of degree d, whose coefficients are given holomorphic functions in the domain D. Correspondingly (6.43) is now replaced by

$$\max_{z \in E_2} |P(g, g', \cdots, g^{(m)})| \leqslant A \tag{6.50}$$

and the domain $|w| < 1/A$ is replaced by the domain $|w| < 1/A^{\frac{1}{d}}$.

6.3. UNIVALENT AND MULTIVALENT FUNCTIONS

The main purpose of this paragraph is to prove that some families of univalent functions or multivalent functions are closed families, so that the general theorems in Chapter 4 are applicable to these families. In this way we shall see that many extremal problems concerning these families are "theoretically solvable" in the sense that both the extremal value and the corresponding extremal function exist. However in general it is very difficult to find out the exact value of the extremal value and the explicit expression of the extremal function.

Theorem 6. 12. Let D be a domain and z_0 a point of D. Let $\mathscr{F}$ be the family of the functions $f(z)$ satisfying the following conditions:

1^0 $f(z)$ is holomorphic and univalent in D.

2^0 $f(z_0)=0$, $f'(z_0)=1$.

Then the family $\mathscr{F}$ is closed in D.

Proof. Consider the domain $D_1=D-(z_0)$. Take three distinct values a, b, c. If $f(z)\in\mathscr{F}$, then $f(z)$ takes at most once each of the values a, b, c in D, hence by Corollary 5. 9, the family $\mathscr{F}$ is quasi-normal of order 1 at most in D_1. Next since $f(z)$ does not take the values 0 and ∞ in D_1, the family $\mathscr{F}$ is normal in D_1, by Theorem 5. 2. Consequently from any sequence of functions $f_n(z)$ $(n=1,2,\cdots)$ of the family $\mathscr{F}$, we can extract a subsequence $f_{n_k}(z)$ $(k=1,2,\cdots)$ which converges locally uniformly in D_1 either to a holomorphic function $f(z)$ in D_1 or to ∞. We are going to show that the second case can not occur. In fact, consider the function

$$g_k(z)=\frac{f_{n_k}(z)}{z-z_0}$$

which is holomorphic and has no zero in the domain D, with $g_k(z_0)=1$. Let $\Gamma: |z-z_0|=r$ be a circle which and its interior belong to D. If the sequence $f_{n_k}(z)$ $(k=1,2,\cdots)$ converges locally uniformly to ∞ in D_1, then it converges uniformly to ∞ on Γ, by the finite covering theorem. Hence the sequence $g_k(z)$ $(k=1,2,\cdots)$ converges uniformly to ∞ on Γ. Consequently the sequence $1/g_k(z)$ $(k=1,2,\cdots)$ converges uniformly to zero on Γ and therefore also in its interior. But this contradicts the fact $1/g_k(z_0)=1$. Thus the second case can not occur, and hence the first case is ture. Again by the finite covering theorem, the sequence $f_{n_k}(z)$ $(k=1,2,\cdots)$ converges uniformly on the circle Γ and hence also in its interior. So we see finally that the sequence $f_{n_k}(z)$ $(k=1,2,\cdots)$ converges locally uniformly to a holomorphic function $f(z)$ in D, with

$$f(z_0)=0,\ f'(z_0)=1.$$

The second condition implies that $f(z)$ is non-constant and hence univalent in D. Consequently the function $f(z) \in \mathscr{F}$. This proves that the family $\mathscr{F}$ is closed in D.

Consider again the family $\mathscr{F}$ defined in Theorem 6. 12. On the other hand, let $\Delta_t (0<t<+\infty)$ be a domain in the w-plane, depending on the variable t and satisfying the four conditions in Theorem 4. 11. Then by Theorems 6. 12 and 4. 11, there is correspondingly a domain Ω containing the point $w=0$, such that for each function $f(z) \in \mathscr{F}$, the values of $f(z)$ cover the domain Ω. We have the following theorem:

Theorem 6. 13. Assume that the domain D has boundary points $a \neq \infty$ and that for each such boundary point a, in any circle $|z-a|<r$ there are exterior points of D. Let $\mathscr{G}$ be the family of functions $g(w)$ satisfying the following conditions:

1^0 $g(w)$ is holomorphic and univalent in Ω.

2^0 For $w \in \Omega$, we have $g(w) \in D$.

3^0 There exists a function $f(z) \in \mathscr{F}$ such that for $w \in \Omega$, we have $f\{g(w)\} = w$, where $\mathscr{F}$ is the family of functions defined in Theomrem 6. 12.

Then the family $\mathscr{G}$ is closed in Ω.

Evidently the conditions 2^0 and 3^0 signify that $g(w)$ is a branch of the inverse function of the function $f(z)$.

Proof. The condition 3^0 and the condition 2^0 in Theorem 6. 12 imply that

$$g(0) = z_0, \; g'(0) = 1.$$

Consider a sequence of functions $g_n(w)$ $(n=1,2,\cdots)$ of the family $\mathscr{G}$. By means of the sequence of functions $g_n(w)-z_0 (n=1,2,\cdots)$, we deduce from Theorem 6. 12, that we can extract from the sequence $g_n(w)$ $(n=1,2,\cdots)$ a subsequence $g_{n_k}(w)$ $(k=1,2,\cdots)$ which converges locally uniformly in Ω to a function $g_*(w)$ satisfying the condition 1^0 in Theorem 6. 13. It remains to show that $g_*(w)$ also satisfies the conditions 2^0 and 3^0 in Theorem 6. 13.

In fact, consider a point $w_0 \in \Omega$. Since

$$g_{n_k}(w_0) \in D \ (k = 1,2,\cdots), \ \lim_{k\to+\infty} g_{n_k}(w_0) = g_*(w_0),$$

the point $\zeta_0 = g_*(w_0)$ either belongs to D or is a boundary point of D. The second case can not occur, because, otherwise, the values of $g_*(w)$ cover a circle $|w-\zeta_0|<\rho$ which, by hypothesis, contains exterior points of D. Hence $g_*(w)$ satisfies the condition 2^0 in Theorem 6.13.

Furthermore to each function $g_{n_k}(w)$ corresponds a function $f_k(z)\in\mathscr{F}$ such that for $w\in\Omega$, we have

$$f_k\{g_{n_k}(w)\} = w \tag{6.51}$$

By Theorem 6.12, we can extract from the sequence $f_k(z)$ $(k=1,2,\cdots)$ a subsequence $f_{k_p}(z)$ $(p=1,2,\cdots)$ which converges locally uniformly in D to a function $f_*(z)\in\mathscr{F}$. We are going to show that for $w\in\Omega$, we have

$$f_*\{g_*(w)\} = w.$$

In fact, consider a point $w_0\in\Omega$ and the point $\zeta_0=g_*(w_0)\in D$. Let $c: |z-\zeta_0|<r$ be a circle belonging to D, such that the sequence $f_{k_p}(z)$ $(p=1,2,\cdots)$ converges uniformly to $f_*(z)$ in c. Since by (6.51), we have

$$f_{k_p}\{g_{m_p}(w_0)\} = w_0 \qquad (p = 1,2,\cdots) \ (m_p = n_{k_p}),$$

letting $p\to+\infty$ we find

$$f_*(\zeta_0) = w_0.$$

Hence $g_*(w)$ also satisfies the condition 3^0 in Theorem 6.13.

An important particular case of Theorem 6.12 is the following corollary:

Corollary 6.9. Let S be the family of functions $f(z)$ satisfying the following conditions:

1^0 $f(z)$ is holomorphic and univalent in the circle $|z|<1$.

2^0 $f(0)=0$, $f'(0)=1$.

Then the family S is closed in the circle $|z|<1$.

Applying Theorem 4.11 to the family S with $\Delta_t=(|w|<t)$ $(0<t<+\infty)$, we see that there is a circle $|w|<\rho$ such that for each function $f(z)\in S$, the values of $f(z)$ cover the circle $|w|<\rho$. We know that the largest value of ρ is 1/4. Hence, as a particular case of Theorem 6.13, we have the following corollary:

Corollary 6.10. Let $\tilde{s}$ be the family of the functions $g(w)$ satisfying the following conditions:

1^0 $g(w)$ is holomorphic and univalent in the circle $|w|<1/4$.

2^0 In the circle $|w|<1/4$, we have $|g(w)|<1$.

3^0 There exists a function $f(z)\in S$ such that in the circle $|w|<1/4$, we have $f\{g(w)\}=w$.

Then the family $\tilde{s}$ is closed in the circle $|w|<1/4$.

Consider a real continuous functional $\varphi(f)$ defined on S. By Theorem 4.2, $\varphi(f)$ has a maximum M and minimum m defined by

$$M=\sup_{f\in S}\varphi(f), \qquad m=\inf_{f\in S}\varphi(f)$$

which are respectively attained by two functions $g(z)$ and $h(z)$ of S. Thus to each real continuous functional $\varphi(f)$ defined on S, corresponds an extremal problem which consists of finding out the exact values of M and m and the explicit expressions of the two functions $g(z)$ and $h(z)$.

From Theorem 4.3, the rules 1)—5) concerning continuous operators and the remark following Definition 4.2, we can deduce many continuous functionals defined on S. We give some examples.

1) Consider a function $f(z)\in S$ and its Taylor series expansion

$$f(z)=z+a_2z^2+\cdots+a_nz^n+\cdots \tag{6.52}$$

in the circle $|z|<1$. Let $n\geqslant 2$ be an integer. By Theorem 4.3 $f^{(n)}(0)$ is a continuous functional defined on S. Then by the remark following Defination 4.2, $a_n=f^{(n)}(0)/n!$, $|a_n|$, $|a_{n+1}|-|a_n|$ and $||a_{n+1}|-|a_n||$ are continuous functionals defined on S.

2) By the rule 3), $\Omega(f)=f(z)/z$ is a continuous operator defined on S. Moreover $\Omega(f)$ has no zero in the circle $|z|<1$ with $\Omega(f)(0)=1$. It follows from

the rule 4), the operator

$$L_f(z) = log \frac{f(z)}{z}, \ L_f(0) = 0$$

is a continuous operator defined on S. Let

$$L_f(z) = log \frac{f(z)}{z} = 2\sum_{n=1}^{\infty} \gamma_n z^n$$

be the Taylor series expansion in the circle $|z|<1$. Consider on integer $n \geqslant 1$. By the rule 2), $L_f^{(n)}(z)$ is a continuous operator defined on S. Then as in example 1), we see that γ_n, $|\gamma_n|$, $|\gamma_n|^2$ and

$$\sum_{n=1}^{p} n|\gamma_n|^2 \qquad (p = 1, 2, \cdots)$$

are continuous functionals defined on S.

3) Consider a number $\lambda > 0$. By the rule 5), the operator

$$\Omega(f) = \left\{\frac{f(z)}{z}\right\}^{\lambda} = e^{\lambda log \frac{f(z)}{z}}$$

is a continuous operator defined on S. Then ~~as in~~ example 2), we see that in the Taylor series expansion of $\Omega(f)$:

$$\left\{\frac{f(z)}{z}\right\}^{\lambda} = \sum_{n=0}^{\infty} D_n(\lambda) z^n,$$

the coefficents $D_n(\lambda)$ $(n = 2, 3, \cdots)$ are continuous functionals defined on S.

Now we are going to show that the following subfamilies of S are all closed in the circle $|z|<1$.

a) The family S^* consisting of the functions $f(z) \in S$ such that $f(z)$ maps the circle $|z|<1$ onto a starlike domain with respect to the point $w=0$.

b) The family S^c consisting of the functions $f(z) \in S$ such that $f(z)$ maps the circle $|z|<1$ onto a convex domain.

c) The family S^T consisting of the functions $f(z)\in S$ such that there exists a function $g(z)\in S^c$ satisfying the inequality

$$Re\left\{\frac{f'(z)}{g'(z)}\right\}\geqslant 0$$

for $|z|<1$.

d) The family S^r consisting of the functions $f(z)\in S$ such that the coefficients $a_n(n=2,3,\cdots)$ of the Taylor series expansion (6.52) of $f(z)$ are all real.

e) The family $S^{(2)}$ consisting of the odd functions $f(z)\in S$.

f) $p\geqslant 2$ being an integer, the family $S^{(p)}$ consisting of the functions $f(z)\in S$ such that

$$f(\varepsilon_p z)=\varepsilon_p f(z)\quad \varepsilon_p=e^{\frac{2\pi i}{p}}$$

for $|z|<1$.

Proof. Consider a sequence of functions $f_n(z)\in S^*(n=1,2,\cdots)$. Since $f_n(z)\in S$ $(n=1,2,\cdots)$ and S is closed, we can extract from the sequence $f_n(z)$ $(n=1,2,\cdots)$ a subsequence $f_{n_k}(z)$ $(k=1,2,\cdots)$ such that as $k\to+\infty$, $f_{n_k}(z)$ converges locally uniformly in the circle $|z|<1$ to a function $f(z)\in S$. By a well known condition (see for example, [14]), we have

$$Re\left\{\frac{zf'_{n_k}(z)}{f_{n_k}(z)}\right\}\geqslant 0\qquad (k=1,2,\cdots)$$

for $|z|<1$. Keeping z fixed and letting $k\to+\infty$, we get

$$Re\left\{\frac{zf'(z)}{f(z)}\right\}\geqslant 0,$$

hence $f(z)\in S^*$, and S^* is closed in the circle $|z|<1$.

Similarly basing upon the condition

$$Re\left\{1+\frac{zf''(z)}{f'(z)}\right\}\geqslant 0,$$

we see that S^c is closed in the circle $|z|<1$.

Now consider a sequence of functions $f_n(z)\in S^T(n=1,2,\cdots)$. By definition, to each function $f_n(z)$ corresponds a function $g_n(z)\in S^c$ satisfying the inequality

$$Re\left\{\frac{f'_n(z)}{g'_n(z)}\right\}\geqslant 0$$

for $|z|<1$. As above, we can extract from the sequence $f_n(z)$ $(n=1,2,\cdots)$ a subsequence $f_{n_k}(z)$ $(k=1,2,\cdots)$ such that as $k\to+\infty$, $f_{n_k}(z)$ converges locally uniformly in the circle $|z|<1$ to a function $f(z)\in S$. Since S^c is closed, we can extract from the sequence $g_{n_k}(z)$ $(k=1,2,\cdots)$ a subsequence $g_{m_p}(z)$ $(p=1,2,\cdots;m_p=n_{k_p})$ such that as $p\to+\infty$, $g_{m_p}(z)$ converges locally uniformly in the circle $|z|<1$ to a function $g(z)\in S^c$. Of course $f_{m_p}(z)\to f(z)$ locally uniformly in the circle $|z|<1$. In the inequality

$$Re\left\{\frac{f'_{m_p}(z)}{g'_{m_p}(z)}\right\}\geqslant 0$$

keeping z fixed and letting $p\to+\infty$, we get

$$Re\left\{\frac{f'(z)}{g'(z)}\right\}\geqslant 0,$$

hence $f(z)\in S^T$ and S^T is closed in the circle $|z|<1$.

Similarly we prove that S^r, $S^{(2)}$ and $S^{(p)}$ are closed in the circle $|z|<1$.

Finally we point out the following fact: Let $\mathscr{F}$ be a closed subfamily of S and $\varphi(f)$ a functional defined on S. Then $\varphi(f)$ is of course also defined on $\mathscr{F}$. Furthermore if $\varphi(f)$ is continuous with respect to S, then $\varphi(f)$ is continuous with respect to $\mathscr{F}$. This fact is obvious but useful.

Theorem 6.14. Let σ be the family of the functions $\varphi(z)$ satisfying the following conditions:

1^0 $\varphi(z)$ is holomorphic in the domain $1<|z|<+\infty$ with $\lim\limits_{z\to\infty}\varphi(z)=0$.

2^0 The function $f(z)=z+\varphi(z)$ is univalent in the domain $1<|z|<+\infty$.

Then the family σ is closed in the domain $1<|z|<+\infty$.

Proof. Consider a function $\varphi(z)\in\sigma$. By the conditoin 1^0 in Theorem 6. 14, the Laurent series expansion of $\varphi(z)$ in the domain $1<|z|<+\infty$ has the following form:

$$\varphi(z)=\sum_{n=1}^{\infty}\frac{b_n}{z^n}. \tag{6. 53}$$

By the condiiton 2^0 in Theorem 6. 14 and the area theorem (see for example [14]) we have

$$\sum_{n=1}^{\infty}n\,|b_n|^2\leqslant 1,$$

and à fortiori,

$$|b_n|\leqslant 1\ (n=1,2,\cdots).$$

Consequently for $1<|z|<+\infty$, we have

$$|\varphi(z)|\leqslant\sum_{n=1}^{\infty}\frac{1}{|z|^n}=\frac{1}{|z|-1}. \tag{6. 54}$$

This inequality shows that for each number $r>1$, the family σ is uniformly bounded in the domain $r<|z|<+\infty$, hence by Corollary 1. 3, the family σ is normal in the domain $1<|z|<+\infty$. Let $\varphi_n(z)$ $(n=1,2,\cdots)$ be a sequence of functions of the family σ. Then from this sequence we can extract a subsequence $\varphi_{n_k}(z)$ $(k=1,2,\cdots)$ such that as $k\to+\infty$, $\varphi_{n_k}(z)$ converges locally uniformly in the domain $1<|z|<+\infty$ to a holomorphic function $\Phi(z)$. It remians to show that $\Phi(z)\in\sigma$. In fact, first of all, from (6. 54), we have for $1<|z|<+\infty$,

$$|\Phi(z)|\leqslant\frac{1}{|z|-1}$$

which evidently implies $\lim\limits_{z\to\infty}\Phi(z)=0$. Furthermore the sequence of functions

$$f_{n_k}(z) = z + \varphi_{n_k}(z) \ (k = 1,2,\cdots),$$

by the condition 2^0 in Theorem 6. 14, are univalent in the domain $1<|z|<+\infty$ and as $k\to+\infty$, $f_{n_k}(z)$ converges locally uniformly in the same domain to the function $F(z)=z+\Phi(z)$ which is evidnetly non-constant. Hence $F(z)$ is univalent in the domain $1<|z|<+\infty$ and $\Phi(z)\in\sigma$.

Next we are going to show that for each $n\geqslant1$, the coefficient $b_n=b_n(\varphi)$ in (6. 53) is a continuous functional defined on σ. We can see this easily by the formula

$$b_n(\varphi) = \frac{1}{2\pi i}\int_\Gamma \varphi(z) z^{n-1} dz,$$

where Γ is a circle $|z|=r$ $(1<r<+\infty)$. Evidently if $\varphi_m(z)$ $(m=1,2,\cdots)$ is a sequence of functions of the family σ, such that as $m\to+\infty$, $\varphi_m(z)$ converges locally uniformly in the domain $1<|z|<+\infty$ to a function $\Phi(z)\in\sigma$, then $\varphi_m(z)\to\Phi(z)$ uniformly on the circle Γ, hence

$$\lim_{m\to+\infty} b_n(\varphi_m) = b_n(\Phi).$$

Of course $|b_n(\varphi)|$ is also a continuous functional defined on σ.

Corollary 6. 11. Let Σ be the family of the functions $f(z)$ satisfying the following conditions:

1^0 $f(z)$ is holomorphic and univalent in the domain $1<|z|<+\infty$.

2^0 $\lim\limits_{z\to+\infty}\{f(z)-z\}=0$.

Then the family Σ is closed in the doamin $1<|z|<+\infty$.

Proof. Evidently every function $f(z)\in\Sigma$ has the following form:

$$f(z) = z + \varphi(z),\ \varphi(z)\in\sigma. \tag{6. 55}$$

Consider a sequence of functions

$$f_n(z) = z + \varphi_n(z),\ \varphi_n(z) \in \sigma\ (n = 1,2,\cdots)$$

of the family Σ. Since the family σ is closed in the domain $1<|z|<+\infty$, from the sequence $\varphi_n(z)$ $(n=1,2,\cdots)$ we can extract a subsequence $\varphi_{n_k}(z)$ $(k=1,2,\cdots)$ such that as $k\to+\infty$, $\varphi_{n_k}(z)$ converges locally uniformly in the domain $1<|z|<+\infty$ to a function $\Phi(z)\in\sigma$. Then evidently $f_{n_k}(z)\to z+\Phi(z)$ locally uniformly in the domain $1<|z|<+\infty$. Since $F(z)=z+\Phi(z)\in\Sigma$, this proves Corollary 6. 11. More precisely, we are going to show that for any number $r>1$, $f_{n_k}(z)\to F(z)$ uniformly in the region $r\leqslant|z|<+\infty$. In fact, consider a number $\varepsilon>0$. By (6. 54), we can find a number $R>r$ such that for each function $\varphi(z)\in\sigma$, we have

$$|\varphi(z)| < \frac{\varepsilon}{2}$$

in the domain $R<|z|<+\infty$. Hence for $k\geqslant1$, we have

$$|f_{n_k}(z) - F(z)| = |\varphi_{n_k}(z) - \Phi(z)| < \varepsilon$$

in the domain $R<|z|<+\infty$. On the other hand we can find an integer $K\geqslant1$ such that for $k\geqslant K$ we have

$$|f_{n_k}(z) - F(z)| < \varepsilon$$

in the region $r\leqslant|z|\leqslant R$. It follows that for $k\geqslant K$, the inequality (6. 56) holds in the region $r\leqslant|z|<+\infty$.

We have seen that for each $n\geqslant1$, the coefficient b_n in (6. 53) is a continuous functional defined on σ. Evidently b_n can also be considered as a continuous functional defined on Σ.

We shall also need the following two facts concerning Σ:

1)For any number $r>1$ and any function $f(z)\in\Sigma$, we have $|f(z)|\leqslant r+\dfrac{1}{r-1}$ on the circle $|z|=r$.

2) For any function $f(z)\in\Sigma$, we have $|f(z)|>1$ in the domain $2<|z|<+\infty$.

These two facts are immediate consequences of (6. 54) and (6. 55).

Now let us introduce the family **M** of meromorphic functions $h(\zeta)$ in the circle $|\zeta|<1$ of the following form:

$$h(\zeta)=\frac{1}{f(\frac{1}{\zeta})}\ (0<|\zeta|<1),\ f(z)\in\Sigma;\ h(0)=0. \tag{6.56}$$

We are going to show that the family **M** is closed in the circle $|\zeta|<1$. In fact consider a sequence of functions

$$h_n(\zeta)=\frac{1}{f_n(\frac{1}{\zeta})}\ (0<|\zeta|<1),\ f_n(z)\in\Sigma;\ h_n(0)=0$$

$$(n=1,2,\cdots)$$

of the family **M**. By a result just obtained in the above, from the sequence $f_n(z)$ $(n=1,2,\cdots)$ we can entract a subsequence $f_{n_k}(z)(k=1,2,\cdots)$ such that for any number $r>1$, $f_{n_k}(z)\to F(z)$ uniformly for $r\leqslant|z|<+\infty$ where $F(z)\in\Sigma$, and hence $f_{n_k}(\frac{1}{\zeta})\to F(\frac{1}{\zeta})$ uniformly for $0<|\zeta|<1/r$. Consequently $H(\zeta)\in M$ being the function defined by

$$H(\zeta)=\frac{1}{F(\frac{1}{\zeta})}\quad (0<|\zeta|<1),\quad H(0)=0,$$

we see that

$$|h_{n_k}(\zeta),H(\zeta)|\to 0 \tag{6.57}$$

uniformly for $0<|\zeta|<1/r$. On the other hand, by the above fact 2), $h_{n_k}(\zeta)$ and $H(\zeta)$ are holomorphic in the circle $|\zeta|<1/2$ and $h_{n_k}(\zeta)\to H(\zeta)$ uniformly for $|\zeta|=1/3$, hence (6.57) also holds uniformly for $|\zeta|<1/3$. This proves that as $k\to+\infty$, $h_{n_k}(z)$ converges locally uniformly to $H(\zeta)$, in the circle $|\zeta|<1$, with respect to the spherical distance. Hence **M** is closed in the circle $|\zeta|<1$ according to Definition 4.5.

Since $h(\zeta)$ defined by (6.56) is univalent in the circle $|\zeta|<1$, we have

$h'(0) \neq 0$ and $h(\zeta)$ has a simple zero at $\zeta = 0$. We can then apply Theorem 4.11 to the family M with $\Delta_t = (|v| < t)$ $(0 < t < +\infty)$, and conclude that there is a number $0 < t_0 < +\infty$ such that for each function $h(\zeta) \in M$, the values of $h(\zeta)$ cover the circle $|v| < t_0$. Then we see that to each function $h(\zeta) \in M$ corresponds a function $k(v)$ holomorphic and univalent in the circle $|v| < t_0$ such that identically

$$h\{k(v)\} = v. \tag{6.58}$$

We have

$$0 < |k(v)| < 1 \; for \; 0 < |v| < t_0; \; k(0) = 0. \tag{6.59}$$

In what follows, we shall need the following fact:

1°) Given any number $0 < \varepsilon < 1/2$, we can find a number $0 < \delta < t_0$ depending only on ε and independent of $h(\zeta)$, such that for $|v| < \delta$, we have $|k(v)| < \varepsilon$.

In fact by the above fact 1), for any function $f(z) \in \Sigma$, we have $|f(z)| \leqslant \frac{1}{\varepsilon} + \frac{\varepsilon}{1-\varepsilon}$ on the circle $|z| = \frac{1}{\varepsilon}$. Take a number δ such taht

$$0 < \delta < min\left\{t_0, \frac{1}{\frac{1}{\varepsilon} + \frac{\varepsilon}{1-\varepsilon}}\right\}.$$

Then from (6.56), we get $|h(\zeta)| > \delta$ on the circle $|\zeta| = \varepsilon$. Consequently by Rouché's theorem, if $|v| < \delta$, then the function $h(\zeta) - v$ has a zero in the circle $|\zeta| < \varepsilon$. This proves 1°.

Next from (6.56), (6,58) and (6.59), we deduce that the function

$$g(w) = \frac{1}{k\left(\frac{1}{w}\right)} \tag{6.60}$$

is holomorphic and univalent in the domain $\rho_0 < |w| < +\infty$ $(\rho_0 = 1/t_0)$ and satisfies in this domain

$$1 < |g(w)| < +\infty,\ f\{g(w)\} = w. \tag{6.61}$$

$g(w)$ is therefore a branch of the inverse function of $f(z)$.

We have the following theorem:

Theorem 6.15. Let $\widetilde{\Sigma}$ be the family of the functions $g(w)$ satisfying the following conditions:

1^0 $g(w)$ is holomorphic and univalent in the domain $\rho_0 < |w| < +\infty$ $(\rho_0 = 1/t_0)$.

2^0 In the domain $\rho_0 < |w| < +\infty$, we have $1 < |g(w)| < +\infty$.

3^0 There exists a function $f(z) \in \Sigma$ such that in the domain $\rho_0 < |w| < +\infty$ we have $f\{g(w)\} = w$.

Then the family $\widetilde{\Sigma}$ is closed in the domain $\rho_0 < |w| < +\infty$.

Proof. First consider a function $g(w) \in \widetilde{\Sigma}$. By the conditions 2^0 and 3^0 in Theorem 6.15, $g(w)$ satisfies (6.61) with $f(z) \in \Sigma$. To this function $f(z)$ corresponds a function $h(\zeta) \in M$ defined by (6.56) and then to $h(\zeta)$ corresponds a function $k(v)$ satisfying (6.58) and (6.59). By the univalence of $f(z)$, evidently we must have (6.60).

Next from (6.60) and the fact 1^{*}), we deduce immediately the following result: Given any number $r > 2$, we can find a number $\rho > \rho_0$ such that for each function $g(w) \in \widetilde{\Sigma}$, we have $|g(w)| > r$ for $\rho < |w| < +\infty$.

In fact, to the number $\varepsilon = 1/r$ corresponds a number $0 < \delta < t_0$ having the property in 1^{*}). Then the number $\rho = 1/\delta$ satisfies the above required condition.

Again consider a function $g(w) \in \widetilde{\Sigma}$. By the condition 3^0 there exists a function $f(z) \in \Sigma$:

$$f(z) = z + \varphi(z),\ \varphi(z) \in \sigma$$

such that in the domain $\rho_0 < |w| < +\infty$ we have

$$g(w) = w - \varphi\{g(w)\}. \tag{6.62}$$

Then by the above result and (6.54), given $r > 2$, we can find $\rho > \rho_0$ such that for $\rho < |w| < +\infty$, we have

$$|w| - \frac{1}{r-1} < |g(w)| < |w| + \frac{1}{r-1}. \qquad (6.63)$$

By the condition 2^0 in Theorem 6. 15 and Theorem 2. 1, the family $\widetilde{\Sigma}$ is normal in the domain $\rho_0 < |w| < +\infty$. Let

$$g_n(w) = w - \varphi_n\{g_n(w)\},\ \varphi_n(z) \in \sigma\ (n = 1,2,\cdots) \qquad (6.64)$$

be a sequence of functions of $\widetilde{\Sigma}$. From the sequence $g_n(w)$ $(n=1,2,\cdots)$ we can extract a subsequence $g_{n_\lambda}(w)$ $(\lambda=1,2,\cdots)$ such that as $\lambda\to+\infty$, $g_{n_\lambda}(w)$ converges locally uniformly in the domain $\rho_0<|w|<+\infty$ to a holomorphic function $G(w)$ or ∞. By (6. 63), the second case can not occur. it remains to show that $G(w)\in\widetilde{\Sigma}$. From (6. 63) we see that

$$|G(w)| \geqslant |w| - \frac{1}{r-1}$$

for $\rho<|w|<+\infty$. This implies that $G(w)\to\infty$, as $w\to\infty$, à fortiori $G(w)$ is non-constant and hence univalent in the domain $\rho_0<|w|<+\infty$. On the other hand from the condiiton 2^0 in Theorem 6. 15, we see that $|G(w)|\geqslant 1$. Since $G(w)$ is non-constant, we have $|G(w)|>1$, by the maximum modulus theorem. So $G(w)$ satisfies the conditions 1^0 and 2^0 in Theorem 6. 15. Since by Theorem 6. 14, the family σ is closed in the domain $1<|z|<+\infty$, from the sequence $\varphi_{n_\lambda}(z)$ $(\lambda=1,2,\cdots)$ we can extract a subsequence $\varphi_{m_\mu}(z)$ $(\mu=1,2,\cdots;m_\mu=n_{\lambda_\mu})$ such that as $\mu\to+\infty$, $\varphi_{m_\mu}(z)$ converges locally uniformly in the domain $1<|z|<+\infty$ to a function $\Phi(z)\in\sigma$. From (6. 64) we have

$$g_{m_\mu}(w) = w - \varphi_{m_\mu}\{g_{m_\mu}(w)\}\ (\mu = 1,2,\cdots). \qquad (6.65)$$

Consider a point w_0 of the domain $\rho_0<|w|<+\infty$. In (6. 65) setting $w=w_0$ and letting $\mu\to+\infty$, then $g_{m_\mu}(w_0)\to G(w_0)$, and from the inequality

$$\begin{aligned}|\varphi_{m_\mu}\{g_{m_\mu}(w_0)\} - \Phi\{G(w_0)\}| &\leqslant |\varphi_{m_\mu}\{g_{m_\mu}(w_0)\} - \Phi\{g_{m_\mu}(w_0)\}| \\ &\quad + |\Phi\{g_{m_\mu}(w_0)\} - \Phi\{G(w_0)\}|\end{aligned}$$

we see that

$$\varphi_{n_k}\{g_{m_k}(w_0)\} \to \Phi\{G(w_0)\}.$$

Consequently

$$G(w_0) = w_0 - \Phi\{G(w_0)\}$$

and in general we have

$$G(w) = w - \Phi\{G(w)\}$$

in the domain $\rho_0<|w|<+\infty$. this shows that $G(w)$ also satisfies the condition 3^0 in Theorem 6. 15. Hence $G(w)\in\widetilde{\Sigma}$ and the proof of Theorem 6. 15 is complete.

Consider a function $g(w)\in\widetilde{\Sigma}$. By (6. 63), $g(w)\to\infty$, as $w\to\infty$, and then by (6. 62), in the domain $\rho_0<|w|<+\infty$, we have

$$g(w) = w + \psi(w), \tag{6. 66}$$

where $\psi(w)$ is a holomorphic function in the domain $\rho_0<|w|<+\infty$ with $\lim_{w\to\infty}\psi(w)=0$. Consequently the Laurent series expansion of $\psi(w)$ in the domain $\rho_0<|w|<+\infty$ has the following form:

$$\psi(w) = \sum_{n=1}^{\infty}\frac{c_n}{w^n}. \tag{6. 67}$$

As for the coefficient b_n in (6. 53), we see that for each $n\geqslant 1$, the coefficient c_n in (6.67) is a continuous functional defined on $\widetilde{\Sigma}$. Of course this also holds for $|c_n|$. Note that we can also consider c_n as a functional defined on σ.

Theorem 6. 16. Let D be a domain and $z_1, z_2 (z_1\neq z_2)$ two points of D. Let $\mathscr{F}$ be the family of functions $f(z)$ satisfying the following conditions:

1^0 $f(z)$ is meromorphic and univalent in D.

2^0 $f(z_1)=0$, $f'(z_1)=1$; $f(z_2)=\infty$.

Then the family $\mathscr{F}$ is closed in D.

Proof. First let us show that if $f(z)\in\mathscr{F}$, then z_2 is a simple pole of $f(z)$. In fact, consider a circle $c: |z-z_2|<r$ interior to D, such that $f(z)$ has no zero in c. Then the function $g(z)=1/f(z)$ is holomorphic and univalent in c and $g(z_2)=0$. So we have $g'(z_2)\neq 0$ and z_2 is a simple zero of $g(z)$. Hence z_2 is a simple pole of $f(z)$.

Now consider the domain $D_* = D-(z_2)$ and the family $\mathscr{F}_*$ of the functions $f(z)$ holomorphic and univalent in D_*, such that $f(z_1)=0$, $f'(z_1)=1$. By Theorem 6.12, $\mathscr{F}_*$ is closed in D_*. Consequently from any sequence of functions $f_n(z)$ $(n=1,2,\cdots)$ of the family $\mathscr{F}$, we can extract a subsequence $f_{n_k}(z)$ $(k=1,2,\cdots)$ which converges locally uniformly in D_* to a function $F(z)\in\mathscr{F}_*$. We are going to show that z_2 is a pole of $F(z)$. In fact, consider the sequence of functions

$$g_k(z) = (z-z_2)f_{n_k}(z) \quad (k=1,2,\cdots)$$

holomorphic in D, which converges locally uniformly in D_* to $(z-z_2)F(z)$. Let $\bar{\gamma}: |z-z_2|\leqslant\rho$ be a circle which belongs to D and does not contain the point z_1. Since on the circle $|z-z_2|=\rho$, $g_k(z)$ converges uniformly to $(z-z_2)F(z)$, hence $g_k(z)$ converges uniformly to a holomorphic function $g(z)$ in the circle $\gamma: |z-z_2|<\rho$. We have

$$g(z) = (z-z_2)F(z)$$

for $0<|z-z_2|<\rho$. Furthermore since $g_k(z)$ has no zero in γ, $g(z)$ has no zero in γ. So for $0<|z-z_2|<\rho$, we have

$$F(z) = \frac{g(z)}{z-z_2}$$

and z_2 is a simple pole of $F(z)$. Consequently $F(z)\in\mathscr{F}$, provided that we define $F(z_2)=\infty$.

Since $f_{n_k}(z)\to F(z)$ uniformly on the circle $|z-z_2|=\rho$ and the functions $f_{n_k}(z)$ and $F(z)$ have no zero in the circle $\bar{\gamma}$, we see that $1/f_{n_k}(z)\to 1/F(z)$ uniformly in $\overline{\Gamma}$.

Thus the sequence $f_{n_k}(z)$ $(k=1,2,\cdots)$ converges locally uniformly in D. to $F(z)$ and the sequence $1/f_{n_k}(z)$ $(k=1,2,\cdots)$ converges uniformly to $1/F(z)$ in $\overline{\Gamma}$. This implies that the sequence $f_{n_k}(z)$ $(k=1,2,\cdots)$ converges locally uniformly in D to $F(z)$, with respect to the spherical distance. Hence the family $\mathscr{F}$ is closed in D according to Definition 4. 5.

Corollary 6. 12. Given a number $0<q<1$. let $S(q)$ be the family of functions $f(z)$ satisfying the following conditions:

1^0 $f(z)$ is meromorphic and univalent in the circle C: $|z|<1$.

2^0 $f(0)=0$, $f'(0)=1$; $f(q)=\infty$.

Then the family $S(q)$ is closed in the circle C.

Consider the domain $D_1=C-(q)$ in which the functions of the family $S(q)$ are holomorphic. À fortiori, $S(q)$ is closed in D_1 according to Definition 4. 5, and hence according to Definition 4. 1, by Lemma 4. 3. It follows that, $m\geqslant 0$ being an integer and z_0 a point of D_1, $f^{(m)}(z_0)$ and $|f^{(m)}(z_0)|$ are continuous functionals defined on $S(q)$. On the other hand, by the remark following Definition 4. 5, the family

$$\mathscr{T}(q) = \{g(z) = 1/f(z) \mid f(z) \in S(q)\}$$

is closed in C. Similarly we see that $m\geqslant 0$ being an integer and z_0 a point of the domain $C-(0)$, $\{1/f(z)\}^{(m)}_{z=z_0}$ and $|\{1/f(z)\}^{(m)}_{z=z_0}|$ are continuous functionals defined on $S(q)$.

Since each function $f(z)\in S(q)$ has a simple zero at $z=0$, we can apply Theorem 4. 11 to the family $S(q)$ with $\Delta_t=(|w|<t)$ $(0<t<+\infty)$ and conclude that there exists a circle Γ: $|w|<\rho$ $(0<\rho<+\infty)$ such that for each function $f(z)\in S(q)$, the values of $f(z)$ cover Γ. On the other hand, since each function $f(z)\in S(q)$ has a simple pole at $z=q$, we deduce from Corollary 4. 4, in taking $\Delta_t=(|w|>1/t)$ $(0<t<+\infty)$, that there exists a domain Ω: $|w|>R$ $(0<R<+\infty)$ such that for each function $f(z)\in S(q)$, the values of $f(z)$ cover Ω.

Now we treat the case of multivalent functions. First give the following definition:

Definition 6. 2. $p\geqslant 2$ being an integer, a meromorphic function $f(z)$ in a domain D is said to be p-valent in D, if the following conditions are satisfied:

1^0 For each value $w \in \hat{C}$ the equation $f(z)=w$ has at most p roots in D, with due count of order of multiplicity.

2^0 There exists at least one value $w \in \hat{C}$ such that the equation $f(z)=w$ has p roots in D.

Theorem 6.17. Let D be a domain, z_0 a point of D and $p \geqslant 2$ an integer. Let $\mathscr{F}$ be the family of the functions $f(z)$ satisfying the following conditions:

1^0 $f(z)$ is holomorphic and p-valent in D.

2^0 $f^{(j)}(z_0)=0$ $(j=0,1,\cdots,p-1)$, $f^{(p)}(z_0)=1$.

Then the family $\mathscr{F}$ is closed in D.

Proof. As in the proof of Theorem 6.12, we first show that given any sequence of functions $f_n(z)$ $(n=1,2,\cdots)$ of the family $\mathscr{F}$, we can extract a subsequence $f_{n_k}(z)$ $(k=1,2,\cdots)$ which converges locally uniformly to a holomorphic function $f(z)$ in D, with

$$f^{(j)}(z_0) = 0 \ (j = 0,1,\cdots,p-1), f^{(p)}(z_0) = 1.$$

It remains to show that the function $f(z)$ is p-valent in D. In fact consider a finite value w and assume that the function $f(z)-w$ has a zero z_1 of order m in D. Let $\Gamma: |z-z_1|=r$ be a circle such that $\bar{\Gamma}: |z-z_1| \leqslant r$ belongs to D and that $f(z)-w$ has no zero for $0<|z-z_1| \leqslant r$. Since the sequence $f_{n_k}(z)$ $(k=1,2,\cdots)$ converges uniformly to $f(z)$ on Γ, when k is sufficiently large, the function $f_{n_k}(z)-w$ has m zeros in the interior of Γ, by a known theorem. It is then clear that the function $f(z)-w$ has at most p zeros in D. On the other hand, $f(z)$ has a zero of order p at z_0. Consequently the function $f(z)$ is p-valent in D and hence belongs to the family $\mathscr{F}$. $\mathscr{F}$ is therefore closed in D.

Corollary 6.13. Given an integer $p \geqslant 2$, let $M(p)$ be the family of the functions $f(z)$ satisfying the following conditions:

1^0 $f(z)$ is holomorphic and p-valent in the circle $|z|<1$.

2^0 $f^{(j)}(0)=0$ $(j=0,1,\cdots,p-1)$, $f^{(p)}(0)=1$.

Then the family $M(p)$ is closed in the cricle $|z|<1$.

Applying Theorem 4.11 to the family $M(p)$, we see that there is a circle Γ:

$|w|<\rho(0<\rho<+\infty)$ such that for each function $f(z)\in M(p)$, the values of $f(z)$ cover at least p times the circle Γ, where ρ depends on p.

Theorem 6. 18. Let D be a domain, $z_1, z_2 (z_1 \neq z_2)$ two points of D and $p \geqslant 2$ an integer. Let $\mathscr{F}$ be the family of the functions $f(z)$ satisfying the following conditions:

1^0 $f(z)$ is meromorphic and p-valent in D.

2^0 $f^{(j)}(z_1)=0$ $(j=0,1,\cdots,p-1)$, $f^{(p)}(z_1)=1$.

3^0 z_2 is a pole of order p of $f(z)$.

Then the family $\mathscr{F}$ is closed in D.

With the help of Theorem 6. 17, the proof of Theorem 6. 18 is quite similar to that of Theorem 6. 16.

Corollary 6. 14. Given an integer $p\geqslant 2$ and a number $0<\xi<1$, let $M(p,\xi)$ be the family of the functions $f(z)$ satisfying the following conditions:

1^0 $f(z)$ is meromorphic and p-valent in the circle $|z|<1$.

2^0 $f^{(j)}(0)=0$ $(j=0,1,\cdots,p-1)$, $f^{(p)}(0)=1$.

3^0 ξ is a pole of order p of $f(z)$.

Then the family $M(p,\xi)$ is closed in the circle $|z|<1$.

Applying Theorem 4. 11 and Corollary 4. 4 to the family $M(p,\xi)$ we see that there are two numbers ρ and $R(0<\rho, R<+\infty)$ such that for each function $f(z)\in M(p,\xi)$, the values of $f(z)$ cover at least p times each of the two domains $|w|<\rho$ and $|w|>R$.

Up to now we have shown that some families of univalent functions or multivalent functions are closed families. However besides these families, we can define more general families of such functions, which are also closed. For this we give only some examples in the case of holomorphic functions.

Theorem 6. 19. Let D be a domain, E a bounded closed set of points belonging to D and $\alpha>0, \beta>0(\alpha<\beta)$ two numbers. Let $\mathscr{F}$ be the family of the functions $f(z)$ satisfying the following conditions:

1^0 $f(z)$ is holomorphic and univalent in D.

2^0 There is a point $\zeta=\zeta(f)\in E$ such that $f(\zeta)=0$ and $\alpha\leqslant|f'(\zeta)|\leqslant\beta$.

Then the family $\mathscr{F}$ is closed in D.

Proof. Let $f_n(z)$ $(n=1,2,\cdots)$ be a sequence of functions of the family $\mathscr{F}$. From the sequence of points $\zeta_n=\zeta(f_n)$ $(n=1,2,\cdots)$ we can extract a subsequence $\zeta_{n_k}=\zeta(f_{n_k})$ $(k=1,2,\cdots)$ which converges to a point $Z\in E$. Consider a circle $\bar{c}: |z-Z|\leqslant\delta$ belonging to D. Let k_0 be a positive integer such that for $k\geqslant k_0$, ζ_{n_k} belongs to the circle $c: |z-Z|<\delta$. Then the family of functions $\{f_{n_k}(z)\ (k\geqslant k_0)\}$ is normal in the domain $D_1=D-\bar{c}$. Adding the finite number of functions $f_{n_k}(z)$ $(1\leqslant k\leqslant k_0-1)$ to the family $\{f_{n_k}(z)\ (k\geqslant k_0)\}$, we get the family of functions $\{f_{n_k}(z)\ (k\geqslant 1)\}$ which is also normal in D_1. Since δ can be taken arbitrarily small so the family $\{f_{n_k}(z)\ (k\geqslant 1)\}$ is normal in the domain $D_0=D-(Z)$. Consequently we can extract from the sequence $f_{n_k}(z)$ $(k=1,2,\cdots)$ a subsequence $f_{m_\lambda}(z)$ $(\lambda=1,2,\cdots;m_\lambda=n_{k_\lambda})$ which converges locally uniformly in D_0 to a holomorphic function $F(z)$ or to ∞. We are going to show that the second case can not occur. In fact consider the sequence of functions

$$g_\lambda(z)=\frac{f_{m_\lambda}(z)}{(z-\zeta_{m_\lambda})}\ (\lambda=1,2,\cdots)$$

which are holomorphic and have no zero in D. Let $\bar{\Gamma}: |z-Z|\leqslant r$ be a circle belonging to D. Let λ_0 be an integer such that for $\lambda\geqslant\lambda_0$, $|\zeta_{m_\lambda}-Z|<r$. Then for $\lambda\geqslant\lambda_0$, we have

$$|g_\lambda(z)|\geqslant\frac{1}{2r}|f_{m_\lambda}(z)|$$

on the circle $|z-Z|=r$. Consequently if the second case occurs, then the sequence $g_\lambda(z)$ $(\lambda=1,2,\cdots)$ converges uniformly to ∞ on the circle $|z-Z|=r$, and hence also in the circle $\bar{\Gamma}$. But

$$|g_\lambda(\zeta_{m_\lambda})|=|f'_{m_\lambda}(\zeta_{m_\lambda})|\leqslant\beta,$$

so the second case can not occur. It is then clear that the sequence $f_{m_\lambda}(z)$ $(\lambda=1,2,\cdots)$ converges locally uniformly to a holomorphic function $F(z)$ in D, and it is easy to see that

$$F(Z)=0,\ \alpha\leqslant|F'(Z)|\leqslant\beta.$$

Thus $F(z)\in\mathscr{F}$ and the proof of Theorem 6.19 is complete.

Theorem 6.20. Let D be a domain, $p\geqslant 2$ an integer, $z_j(j=1,2,\cdots,m;1\leqslant m\leqslant p)$ m distinct points of D and $\alpha>0$, $\beta>0$ $(\alpha<\beta)$ two numbers. Let $\mathscr{F}$ be the family of the functions satisfying the following conditions:

1^0 $f(z)$ is holomorphic and p-valent in D.

2^0 For each $1\leqslant j\leqslant m$, z_j is a zero of order $\lambda_j=\lambda_j(f)$ of $f(z)$, with $\alpha\leqslant|f^{(\lambda_j)}(z_j)|\leqslant\beta$ and $\sum_{j=1}^{m}\lambda_j=p$.

Then the family $\mathscr{F}$ is closed in D.

Proof. Let $f_n(z)$ $(n=1,2,\cdots)$ be a sequence of functions of the family $\mathscr{F}$. It is easy to see that there is a system of posiitve integers $\lambda_j(j=1,2,\cdots,m)$ with $\sum_{j=1}^{m}\lambda_j=p$ such that for an infinite number of positive integers n, we have

$$\lambda_j(f_n)=\lambda_j\quad(j=1,2,\cdots,m).$$

So we can find a subsequence $f_{n_k}(z)$ $(k=1,2,\cdots)$ of the sequence $f_n(z)$ $(n=1,2,\cdots)$ such that

$$\lambda_j(f_{n_k})=\lambda_j\quad(j=1,2,\cdots,m).$$

The family of functions $\{f_{n_k}(z)\ (k=1,2,\cdots)\}$ is normal in the domain $D_1=D-(z_1,z_2,\cdots,z_m)$ and hence we can extract from the sequence $f_{n_k}(z)$ $(k=1,2,\cdots)$ a subsequence $f_{q_s}(z)$ $(s=1,2,\cdots;q_s=n_{k_s})$ which converges locally uniformly in D_1 to a holomorphic function $F(z)$ or to ∞. By means of the sequence of functions

$$g_s(z)=\frac{f_{q_s}(z)}{(z-z_1)^{\lambda_1}(z-z_2)^{\lambda_2}\cdots(z-z_m)^{\lambda_m}}\ (s=1,2,\cdots)$$

and the condition 2^0 in Theorem 6.20, we see, as in the proof of Theorem 6.19, that the second case can not occur. Hence the sequence $f_{q_s}(z)$ $(s=1,2,\cdots)$ converges locally uniformly in D to a holomorphic function $F(z)$. It is easy to see that $F(z)$ satisfies the condition 2^0 in Theorem 6.20 and is p-valent in **D**. Thus

$F(z) \in \mathscr{F}$ and Theorem 6.20 is proved.

7

EXTENSIONS OF SOME CRITERIONS OF NORMALITY AND QUASI—NORMALITY

In this chapter we shall give some extensions of Theorems 2.1, 2.8, 3.1, 3.9, 5.4 and Corollaries 5.9, 5.10, such that the values a, b, c and $a_j(j=1,2,3)$ in these theorems and corollaries are replaced by functions. The results thus obtained will then be applied to prove theorems on sets of functions covered, in a certain sence, by the values of holomorphic or meromrophic functions. Similar theorems concerning univalent or multivalent functions will also be proved, by a different method.

7.1. CASE OF HOLOMORPHIC FUNCTIONS

Theorem 7.1. Let D be a domain and M, δ two positive numbers. Let $\mathscr{F}$ be the family of the functions f(z) satisfying the following conditions:

1^0 f(z) is holomorphic in D.

2^0 There are two holomorphic functions φ(z) and ψ(z) in D such that

$$|\varphi(z)-\psi(z)|>\delta,\ |\varphi(z)|<M,\ |\psi(z)|<M \tag{7.1}$$

$$f(z)-\varphi(z)\neq 0,\ f(z)-\psi(z)\neq 0 \tag{7.2}$$

in D.

Then the family $\mathscr{F}$ is normal in D.

Proof. Consider a sequence of functions $f_n(z)$ $(n=1,2,\cdots)$ of the family $\mathscr{F}$. By the condition 2^0, to each $f_n(z)$ correspond two functions $\varphi_n(z)$ and $\psi_n(z)$

holomorphic and satisfying the conditions (7.1), (7.2) in D, with respect to $f_n(z)$. The sequence of functions

$$g_n(z) = \frac{f_n(z) - \varphi_n(z)}{\psi_n(z) - \varphi_n(z)} \quad (n = 1,2,\cdots)$$

are holomorphic in D and do not take the values 0 and 1. By Theorem 2.1, the family $\{g_n(z)\ (n=1,2,\cdots)\}$ is normal in D. Hence we can find a subsequence $g_{n_k}(z)$ $(k=1,2,\cdots)$ converging locally uniformly in D either to a holomorphic function or to ∞. In the second case, we deduce from the relation

$$f_{n_k}(z) = \{\psi_{n_k}(z) - \varphi_{n_k}(z)\} g_{n_k}(z) + \varphi_{n_k}(z)$$

and (7.1) that $f_{n_k}(z)$ $(k=1,2,\cdots)$ also converges locally uniformly to ∞ in D. In the first case, the same relation and (7.1) show that the sequence $f_{n_k}(z)$ $(k=1,2,\cdots)$ is locally uniformly bounded in D, hence by Corollary 1.3, we can extract from $f_{n_k}(z)$ $(k=1,2,\cdots)$ a subsequence $f_{m_s}(z)$ $(s=1,2,\cdots)$ converging locally uniformly to a holomorphic function in D.

Definition 7.1. Let $S_i (i=1,2)$ be two sets of holomorphic functions in a domain D. We say that $S_i (i=1,2)$ are locally uniformly disjoint, if to each point $z_0 \in D$ correspond a circle $|z-z_0|<r$ interior to D and two positive numbers η, M such that for any two functions $\varphi_i(z) \in S_i (i=1,2)$, we have

$$|\varphi_1(z) - \varphi_2(z)| > \eta, \ |\varphi_i(z)| < M \ (i = 1,2)$$

in the circle $|z-z_0|<r$.

Theorem 7.2. Let $\mathscr{F}$ be a family of holomorphic functions in a domain D and $S_i (i=1,2)$ two locally uniformly disjoint sets of holomorphic functions in D. Suppose that to each point $z_0 \in D$ corresponds a circle $|z-z_0|<r$ interior to D having the following property: For every function $f(z) \in \mathscr{F}$, there are two functions $\varphi_i(z) \in S_i (i=1,2)$ such that the two functions $f(z)-\varphi_i(z)$ $(i=1,2)$ have no zero in the cricle $|z-z_0|<r$. Then the family $\mathscr{F}$ is normal in D.

Proof. Consider a point $z_0 \in D$. Let $|z-z_0|<r$ be a circle having the property in Theorem 7.2. On the other hand, since $S_i(i=1,2)$ are locally uniformly disjoint, there exist a circle $|z-z_0|<r'$ and two positive numbers η, M having the property in Definition 7.1. Setting $r_0=\min(r,r')$, then by Theorem 7.1, the family $\mathscr{F}$ is normal in the circle $|z-z_0|<r_0$. Since z_0 is arbitrary, $\mathscr{F}$ is normal in D by Theorem 1.4.

Definition 7.2. Let f(z) be a holomorphic function in a domain D and S a set of holomorphic functions in D. We say that the values of the function f(z) cover in generalized sense the set S, if for each function $\varphi(z) \in S$, the function $f(z)-\varphi(z)$ has at least one zero in D.

Corollary 7.1. Let D be a domain, $E_i(i=1,2)$ two bounded closed sets of points belonging to D and $S_i(i=1,2)$ two locally uniformly disjoint sets of holomorphic functions in D. Then there exists a positive number b depending only on $D, E_i, S_i(i=1,2)$ having the following property: If f(z) is a holomorphic function in D such that

$$\min_{z \in E_1}|f(z)| \leqslant 1, \quad \max_{z \in E_2}|f(z)| \geqslant b,$$

then the values of the function f(z) cover in generalized sense at least one of the sets $S_i(i=1,2)$.

Proof. Assume that the positive number b does not exist. Then for each positive integer n, there is a holomorphic function $f_n(z)$ in D such that

$$\min_{z \in E_1}|f_n(z)| \leqslant 1, \quad \max_{z \in E_2}|f_n(z)| \geqslant n \tag{7.3}$$

and that the values of the function $f_n(z)$ do not cover in generalized sense either of the sets $S_i(i=1,2)$. Accordingly there are two functions $\varphi_{ni}(z) \in S_i(i=1,2)$ such that the two functions $f_n(z)-\varphi_{ni}(z)$ $(i=1,2)$ have no zero in D. By Theorem 7.2, the family $\{f_n(z)\ (n=1,2,\cdots)\}$ is normal in D. Consequently by Lemma 2.1 and the first inequality in (7.3), the sequence $f_n(z)$ $(n=1,2,\cdots)$ is uniformly bounded in E_2. But this contradicts the second inequality in (7.3).

Definition 7. 3. Let D be a domain in the z-plane and σ a set of points of the w-plane. We denote by $H_D(\sigma)$ the set of holomorphic functions $\varphi(z)$ in D such that the values taken by $\varphi(z)$ in D all belong to σ. In particular, each point $w\in\sigma$ considered as a constant function defined in D belongs to $H_D(\sigma)$.

Corollary 7. 2. Let D be a domain and $E_i(i=1,2)$ two bounded closed sets of points belonging to D. Then there exist three positive numbers α,δ,λ depending only on $D,E_i(i=1,2)$ having the following property: If $f(z)$ is a holomorphic function in D such that

$$\min_{z\in E_1}|f(z)|\leqslant\alpha,\ \max_{z\in E_2}|f(z)|\geqslant 1, \tag{7. 4}$$

then the values of the function $f(z)$ cover in generalized sense a set $H_D(\rho_1<|w|<\rho_2)$ with

$$\rho_1>\delta,\ \rho_2-\rho_1>\lambda. \tag{7. 5}$$

Proof. By Corollary 7. 1, corresponding to D, $E_i(i=1,2)$, $S_1=H_D(1<|w|<2)$, $S_2=H_D(3<|w|<4)$, there is a positive number b having the property stated in that corollary. Then the numbers $\alpha=1/b,\delta=1/2b$, $\lambda=1/2b$ have the required property in Corollary 7. 2. In fact, if $f(z)$ satisfies (7. 4), then $g(z)=bf(z)$ satisfies the condition

$$\min_{z\in E_1}|g(z)|\leqslant 1,\ \max_{z\in E_2}|g(z)|\geqslant b,$$

consequently the values of $g(z)$ cover in generalized sense at least one of the sets $S_i(i=1,2)$, which implles that the values of $f(z)$ cover in generalized sense at least one of the sets $H_D(1/b<|w|<2/b)$ and $H_D(3/b<|w|<4/b)$.

Theorem 2. 9. can be deduced from Corollary 7. 2 by applying the latter to the particular case where D is the circle $|z|<1$ and $E_i(i=1,2)$ are respectively the point $z=0$ and the circle $|z|=1/2$.

Theorem 7. 3. Let D be a domain, $S_i(i=1,2)$ two locally uniformly disjoint sets of holomorphic functions in D and p a positive integer. Let $\mathscr{F}$ be the family of the functions f(z) satisfying the following conditions:

1^0 f(z) is holomorphic in D.

2^0 There are two functions $\varphi_i(z)\in S_i(i=1,2)$ such that each of the two functions $f(z)-\varphi_i(z)$ $(i=1,2)$ has at most p distinct zeros in D.

Then the family $\mathscr{F}$ is quasi-nrmal in D of order p at most.

Proof. Consider a suquence of functions $f_n(z)$ $(n=1,2,\cdots)$ of the family $\mathscr{F}$. Then to each $f_n(z)$ correspond $\varphi_{ni}(z)\in S_i(i=1,2)$ such that each of $f_n(z)-\varphi_{ni}(z)$ $(i=1,2)$ has at most p distinct zeros in D. Accordingly the function

$$g_n(z)=\frac{f_n(z)-\varphi_{n1}(z)}{\varphi_{n2}-\varphi_{n1}(z)}$$

has at most p distinct zeros in D and the same is true for $g_n(z)-1$. By Corollary 5. 11, the family $\{g_n(z)\ (n=1,2,\cdots)\}$ is quasi-normal in D of order p at most. Hence from the sequence $g_n(z)$ $(n=1,2,\cdots)$ we can extract a subsequence $g_{n_k}(z)$ $(k=1,2,\cdots)$ converging locally uniformly to a holomorphic function or to ∞ in a domain $D-\sigma$, where σ is a set consisting of at most p points of D. Making use of the condition that $S_i(i=1,2)$ are locally uniformly disjoint, we see as in the proof of Theorem 7. 1, that either the sequence $f_{n_k}(z)$ $(k=1,2,\cdots)$ converges locally uniformly to ∞ in $D-\sigma$ or we can extract from the sequence $f_{n_k}(z)$ $(k=1,2,\cdots)$ a subsequence $f_{m_s}(z)$ $(s=1,2,\cdots)$ converging locally uniformly to a holomorphic function in $D-\sigma$. This proves Theorem 7. 3.

Definition 7. 4. Let f(z) be a holomorphic function in a doamin D, S a set of holomorphic functions in D and $p\geqslant 2$ an integer. We say that the values of the function f(z) cover in generalized sense the set S at least p times (strictly at least z times), if for each function $\varphi(z)\in S$, the function $f(z)-\varphi(z)$ has at least p zeros in D with due count of order of multiplicity (at least p distinct zeros in D).

Corollary 7. 3. Let D be a domain, $E_i(i=1,2)$ two bounded closed sets of points belonging to D, $S_i(i=1,2)$ two locally uniformly disjoint sets of holomorphic functions in D, $p\geqslant 2$ an integer and δ a positive number. Then there exists a positive number b depending only on $D, E_1, S_i (i=1,2)$, p and δ having the following property: If $f(z)$ is a holomorphic function in D such that

$$|f(z_j)| \leqslant 1 \quad (j = 1,2,\cdots,p) \tag{7.6}$$

for p points $z_j \in E_1 (j=1,2,\cdots,p)$ with

$$|z_j - z_l| \geqslant \delta \quad (j,l = 1,2,\cdots,p\,;j \neq l) \tag{7.7}$$

and that

$$\max_{z\in E_2} |f(z)| \geqslant b,$$

then the values of the function $f(z)$ cover in generalized sense strictly at least p times one of the sets $S_i (i=1,2)$.

Proof. Assume that the positive number b does not exist. Then for each positive integer n, there is a holomorphic function $f_n(z)$ in D such that

$$|f_n(z_j^{(n)})| \leqslant 1 \quad (j = 1,2,\cdots,p) \tag{7.8}$$

for p points $z_j^{(n)} \in E_1 (j=1,2,\cdots,p)$ with

$$|z_j^{(n)} - z_l^{(n)}| \geqslant \delta \quad (j,l = 1,2,\cdots,p\,;j \neq l), \tag{7.9}$$

that

$$\max_{z\in E_2} |f_n(z)| \geqslant n \tag{7.10}$$

and that the values of the function $f_n(z)$ do not cover in generalized sense strictly at least p times either of the sets $S_i (i=1,2)$. By Theorem 7.3, the family $\{f_n(z)\ (n=1,2,\cdots)\}$ is quasi-normal in D of order $p-1$ at most. Hence there is a subsequence $f_{n_k}(z)\ (k=1,2,\cdots)$ converging locally uniformly to a holomorphic function or to ∞ in a domain D-σ where σ is a set consisting of at most $p-1$ points of D. The first case can not happen, because by (7. 10) the sequence $f_{n_k}(z)$ $(k=1, 2,\cdots)$ is not uniformly bounded on E_2. So the sequence $f_{n_k}(z)$ $(k=1,2,\cdots)$ converges locally uniformly to ∞ in D-σ. by (7. 9) we can find an increasing sequence of positive integers $k_\lambda(\lambda=1,2,\cdots)$ such that the p sequences

$$z_j^{(m_\lambda)}(\lambda = 1,2,\cdots)\ (j = 1,2,\cdots,p),m_\lambda = n_{k_\lambda}$$

converge respectively to the limits $\zeta_j \in E_1 (j=1,2,\cdots,p)$ with

$$|\zeta_j - \zeta_l| \geqslant \delta\ (j,l = 1,2,\cdots,p\,;j \neq l).$$

(see Lemma 3. 2) One at least of the points $\zeta_j(j=1,2,\cdots,p)$, say $\zeta_{j_0} \in D-\sigma$. This leads to a contradicition in view of (7. 8).

Corollary 7. 4. Let D, $E_i(i=1,2)$, $S_i(i=1,2)$ and p have the same meaning as in Corollary 7. 3. Then there exists a positive number b depending only on $D,E_i,S_i(i=1,2)$ and p having the following property: If $f(z)$ is a holomorphic function in D such that

$$|f^{(m)}(z_0)| \leqslant 1\ (m = 0,1,\cdots,p-1)$$

for a point $z_0 \in E_1$ and that

$$\max_{z\in E_2} |f(z)| \geqslant b,$$

then the values of the function $f(z)$ cover in generalized sense at least p times one of the sets $S_i(i=1,2)$.

Proof. Assume that the positive number b does not exist. Then for each positive integer n, there is a holomorphic function $f_n(z)$ in D such that

$$|f_n^{(m)}(z_0^{(n)})| \leqslant 1\ (m = 0,1,\cdots,p-1)$$

for a point $z_0^{(n)} \in E_1$, that

$$\max_{z \in E_1} |f_n(z)| \geqslant n$$

and that the values of the function $f_n(z)$ do not cover in generalized sense at least p times either of the sets $S_i(i=1,2)$. Then as in the proof of Corollary 7. 3, we find an increasing sequence of positive integers $l_\lambda(\lambda=1,2,\cdots)$ such that the sequence $f_{l_\lambda}(z)$ $(\lambda=1,2,\cdots)$ converges locally uniformly to ∞ in D-σ, where σ is a set consisting of at most $p-1$ points of D, and that the sequence $z_0^{(l_\lambda)}(\lambda=1,2,\cdots)$ converges to a limit $\zeta_0 \in E_1$. By Definition 7. 1, we can find a circle $\overline{\Gamma}: |z-\zeta_0| \leqslant r$ belonging to D such that the sets $S_i(i=1,2)$ are uniformly bounded in $\overline{\Gamma}$ and that $\overline{\Gamma}-(\zeta_0) \subset D-\sigma$. Then on the circle $|z-\zeta_0|=r$ the sequence $f_{l_\lambda}(z)$ $(\lambda=1,2,\cdots)$ converges uniformly to ∞, and from the identity

$$f_{l_\lambda}(z) - \varphi(z) = \{f_{l_\lambda}(z) - P_\lambda(z)\} + \{P_\lambda(z) - \varphi(z)\},$$

where

$$P_\lambda(z) = \sum_{h=0}^{p-1} \frac{1}{h!} f_{l_\lambda}^{(h)}(z_0^{(l_\lambda)})(z - z_0^{(l_\lambda)})^h,$$

we see that, when λ is sufficiently large, the function $f_{l_\lambda}(z)-\varphi(z)$ has at least p zeros (with due count of order of multiplicity) in the circle $\Gamma: |z-\zeta_0|<r$ for any $\varphi(z) \in S_1 \cup S_2$. So we get a contradiciton.

By the same method we can also prove the following corollary:

Corollary 7. 5. Let D, $E_i(i=1,2)$, $S_i(i=1,2)$, p and δ have the same meaning as in Corollary 7. 3. Then there exists a positive number b depending only on D, E_i, $S_i(i=1,2)$, p and δ having the following property: If $f(z)$ is a holomorphic function in D satisfying the following conditions:

1° There are $q(2 \leqslant q \leqslant p)$ points $z_j \in E_1(j=1,2,\cdots,q)$ and q non-negative integers $\nu_j(j=1,2,\cdots,q)$ such that

$$|z_j - z_l| \geqslant \delta \ (j,l=1,2,\cdots,q; j \neq l),$$
$$|f^{(m)}(z_j)| \leqslant 1 (m=0,1,\cdots,\nu_j) \ (j=1,2,\cdots,q),$$

$$\sum_{j=1}^{q}(\nu_j+1)=p;$$

$2^0 \max_{z\in E_2}|f(z)|\geqslant b;$

then the values of the function $f(z)$ cover in generalized sense at least p times one of the sets $S_i(i=1,2)$.

Remark 7. 1. In Corollary 7. 3 and some other similar places, it is tacitly assumed that there exist p points $z_j\in E_1(j=1,2,\cdots,p)$satisfying the condition (7. 7).

Remark 7. 2. In the same way that Corollary 7. 2 is deduced from Corollary 7. 1, we can deduce respectively from Corollaries 7. 3, 7. 4 and 7. 5 three corollaries on the convering of a set of the form $H_D(\rho_1<|w|<\rho_2)$.

7. 2. CASE OF MEROMORPHIC FUNCTIONS

Definition 7. 5. Let $S_i(i=1,2)$ be two sets of meromorphic functions in a domain D. We say that $S_i(i=1,2)$ are locally uniformly disjoint, if to each point $z_0\in D$ correspond a circle $|z-z_0|<r$ belonging to D and a number δ $(0<\delta<1)$ such that for any two functions $\varphi_i(z)\in S_i(i=1,2)$, we have

$$|\varphi_1(z),\varphi_2(z)|>\delta \tag{7.11}$$

in the circle $|z-z_0|<r$.

Theorem 7. 4. Let D be a domain and $S_i(i=1,2,3)$ three sets of meromorphic functions in D, such that for $i\neq j$ $(i,j=1,2,3)$ S_i and S_j are locally uniformly disjoint and that $S_i(i=1,2)$ are normal families in D. Let $\mathscr{F}$ be the family of the functions $f(z)$ satisfying the following conditions:

1^0 $f(z)$ is meromorphic in D.

2^0 There are three functions $\varphi_i(z)\in S_i(i=1,2,3)$ such that each of the equations $f(z)=\varphi_i(z)$ $(i=1,2,3)$ has no root in D.

Then the family $\mathscr{F}$ is normal in D.

For the proof of this theorem we need the following lemma:

Lemma 7.1. Let $f_n(z)$ $(n=1,2,\cdots)$ be a C_0-sequence of meromorphic functions in a domain D and $a_n(z)$ $(n=1,2,\cdots)$ be a sequence of holomorphic functions in D converging locally uniformly to a holomorphic function in D. Then the sequence of functions $g_n(z)=f_n(z)+a_n(z)$ $(n=1,2,\cdots)$ is a C_0-sequence in D.

Proof. Let $F(z)$ be the limit function of the sequence $f_n(z)$ $(n=1,2,\cdots)$, defined in D (with respect to the spherical distance). Let $a(z)$ be the limit function of the sequence $a_n(z)$ $(n=1,2,\cdots)$, which is holomorphic in D. Consider a point $z_0\in D$ and distinguish two cases:

1) $F(z_0)\neq\infty$. Then by Theorem 1.1, we can find a circle $\Gamma_0: |z-z_0|<r_0$ belonging to D and a positive integer n_0 such that the functions $f_n(z)$ $(n\geqslant n_0)$ and $F(z)$ are holomorphic in Γ_0, and the sequence $f_n(z)$ $(n\geqslant n_0)$ converges uniformly to $F(z)$ in Γ_0. Accordingly in $\Gamma_0': |z-z_0|<r_0'(0<r_0'<r_0)$, $g_n(z)$ $(n\geqslant n_0)$ is holomorphic and converges uniformly to $F(z)+a(z)$. Hence z_0 is a C_0-point of the sequence $g_n(z)$ $(n=1,2,\cdots)$.

2) $F(z_0)=\infty$. Then by Theorem 1.1, we can find a circle $\Gamma_0: |z-z_0|<r_0$ belonging to D and a positive integer n_0 such that the functions $1/f_n(z)$ $(n\geqslant n_0)$ and $1/F(z)$ are holomorphic in Γ_0, and the sequence $1/f_n(z)$ $(n\geqslant n_0)$ converges uniformly to $1/F(z)$ in Γ_0. Writing

$$\frac{1}{g_n(z)}=\frac{1}{f_n(z)}\frac{1}{1+\frac{1}{f_n(z)}a_n(z)},$$

we see that we can find a circle $\Gamma_0': |z-z_0|<r_0'(0<r_0'<r_0)$ and a positive integer $n_0'(n_0'>n_0)$ such that in Γ_0', $1/g_n(z)$ $(n\geqslant n_0')$ is holomorphic and converges uniformly to the function

$$\frac{1}{F(z)}\frac{1}{1+\frac{1}{F(z)}a(z)}$$

which is holomorphic in Γ_0'. Hence again z_0 is a C_0-point of the sequence $g_n(z)$ $(n=1,2,\cdots)$. Since z_0 is arbitrary, this proves Lemma 7.1.

Now let us return to the proof of Theorem 7.4. Let $f_n(z)$ $(n=1,2,\cdots)$ be a sequence of functions of the family $\mathscr{F}$. Then to each $f_n(z)$ correspond three functions $a_n(z)\in S_1$, $b_n(z)\in S_2$, $c_n(z)\in S_3$ such that each of the three equations

$$f_n(z)=a_n(z),\ f_n(z)=b_n(z),\ f_n(z)=c_n(z)$$

has no root in D. Since $S_i(i=1,2)$ are normal families in D, we can find an increasing sequence of positive integers n_k $(k=1,2,\cdots)$ such that the two sequences $a_{n_k}(z)$, $b_{n_k}(z)$ $(k=1,2,\cdots)$ are C_0-sequences in D and hence have respectively limit functions $a(z)$, $b(z)$ defined in D with respect to the spherical distance. For each point $z\in D$, the two corresponding values $a(z)$, $b(z)$ must be unequal. In fact, if at a point $z_0\in D$ we have $a(z_0)=b(z_0)$, then the inequality

$$|a_n(z_0),b_n(z_0)|\leqslant|a_n(z_0),a(z_0)|+|b_n(z_0),b(z_0)|$$

yields $|a_n(z_0),b_n(z_0)|\to 0$ $(n\to+\infty)$. This contradicts the inequality $|a_n(z_0),b_n(z_0)|>\delta$ resulting from (7.11).

Now consider a point $z_0\in D$. Then one at least of the values $a(z_0)$ and $b(z_0)$ is finite, say $a(z_0)\neq\infty$. By Theorem 1.1, we can find a circle Γ_0: $|z-z_0|<r_0$ belonging to D and a positive integer k_0 such that the functions $a_{n_k}(z)$ $(k\geqslant k_0)$ and $a(z)$ are holomorphic in Γ_0 and the sequence $a_{n_k}(z)$ $(k\geqslant k_0)$ converges uniformly to $a(z)$ in Γ_0. We may assume that r_0 is small enough, so that in Γ_0, the sequence $a_{n_k}(z)$ $(k\geqslant k_0)$ is uniformly bounded:

$$|a_{n_k}(z)|\leqslant M\ (k\geqslant k_0) \tag{7.12}$$

and

$$|b_{n_k}(z),a_{n_k}(z)|>\delta,\ |c_{n_k}(z),a_{n_k}(z)|>\delta,\ |b_{n_k}(z),\ c_{n_k}(z)|>\delta, \tag{7.13}$$

where $0<\delta<1$ is independent of k.

Consider the three sequences

$$F_k(z)=\frac{1}{f_{n_k}(z)-a_{n_k}(z)}$$

$$(k\geqslant k_0)$$

$$B_k(z) = \frac{1}{b_{n_k}(z) - a_{n_k}(z)},\ C_k(z) = \frac{1}{c_{n_k}(z) - a_{n_k}(z)}.$$

$F_k(z), B_k(z), C_k(z)$ are holomorphic in Γ_0 and each of the two equations

$$F_k(z) = B_k(z),\ F_k(z) = C_k(z)$$

has no root in Γ_0. Moreover by (7. 13), we have

$$|B_k(z)| < \frac{1}{\delta},\ |C_k(z)| < \frac{1}{\delta}$$

in Γ_0, and by (7. 12), (7. 13) and Lemma 1. 1, we have

$$|B_k(z) - C_k(z)| \geqslant |B_k(z), C_k(z)| = |\frac{1}{B_k(z)}, \frac{1}{C_k(z)}|$$
$$= |b_{n_k}(z) - a_{n_k}(z),\ c_{n_k}(z) - a_{n_k}(z)| \geqslant \frac{1}{2}\frac{1}{1+M^2}\delta$$

in Γ_0. Consequently by Theorem 7. 1, the family $F_k(z)$ $(k \geqslant k_0)$ is normal in Γ_0, hence the family

$$\frac{1}{F_k(z)} = f_{n_k}(z) - a_{n_k}(z)\ (k \geqslant k_0)$$

is normal in Γ_0. Then by Lemma 7. 1, we see that the identity

$$f_{n_k}(z) = \{f_{n_k}(z) - a_{n_k}(z)\} + a_{n_k}(z)\ (k \geqslant k_0)$$

implies that the family $f_{n_k}(z)$ $(k \geqslant k_0)$ is normal in Γ_0. Again this implies that the family $f_{n_k}(z)$ $(k \geqslant 1)$ is normal in Γ_0.

Since z_0 is arbitrary, the family $f_{n_k}(z)$ $(k \geqslant 1)$ is normal in D, by Theorem 1. 4. Hence we can extract from the sequence $f_{n_k}(z)$ $(k \geqslant 1)$ a subsequence $f_{m_h}(z)$ $(h \geqslant 1)$ which is a C_0-sequence in D. This proves Theorem 7. 4.

Definition 7. 6. Let $f(z)$ be a meromorphic function in a domain D and S a set of meromorphic functions in D. We say that the values of the function $f(z)$ cover in generalized sense the set S, if for each function $\varphi(z)\in S$, the equation $f(z)=\varphi(z)$ has at least one root in D.

Definition 7. 7. Let E be a bounded closed set of points of the complex plane **C**. Consider a system of a finite number of circles $\Gamma: |z-z_j|<r_j (j=1,2,\cdots,m)$ such that

$$E\subset\bigcup_{j=1}^{m}(|z-z_j|<r_j).$$

The linear measure of E, denoted by ΛE, is defined by

$$\Lambda E=\inf_{\Gamma}(\sum_{j=1}^{m}r_j),$$

where the inf is taken over all such systems Γ of circles.

As an example, consider a continuous curve $C: z=z(t)\ (a\leqslant t\leqslant b)$, where $z(t)$ is a continuous function in the interval $a\leqslant t\leqslant b$ such that $z(a)\neq z(b)$. It can be shown that

$$\Lambda C\geqslant\frac{1}{2}|z(a)-z(b)|.$$

Corollary 7. 6. Let D be a domain, $E_i(i=1,2)$ two bounded closed sets of points belonging to D, $S_i(i=1,2,3)$ three sets of meromorphic functions in D such that S_i and S_j are locally uniformly disjoint for $i\neq j\ (i,j=1,2,3)$ and that S_i $(i=1,2)$ are normal families in D, and ε a positive number. Then there exists a positive number b depending only on D, $E_i(i=1,2)$, $S_i(i=1,2,3)$ and ε, having the following property: If $f(z)$ is a meromorphic function in D such that

$$\min_{z\in E_1}|f(z)|\leqslant 1,\ \Lambda(z|z\in E_2,\ |f(z)|\geqslant b)\geqslant\varepsilon, \tag{7.14}$$

then the values of the function f(z) cover in generalized sense at least one of the sets $S_i(i=1,2,3)$.

Proof. Assume that the positive number b does not exist. Then for each positive integer n, there is a meromorphic function $f_n(z)$ in D such that

$$\min_{z\in E_1}|f_n(z)|\leqslant 1,\ \Lambda(z\,|\,z\in E_2,\ |f_n(z)|\geqslant n)\geqslant \varepsilon \tag{7.15}$$

and that the values of the function $f_n(z)$ do not cover in generalized sense either of the sets $S_i(i=1,2,3)$. By Theorme 7.4, the family $\{f_n(z)\ (n=1,2,\cdots)\}$ is normal in D. Accordingly we can find a subsequence $f_{n_k}(z)\ (k=1,2,\cdots)$ which is a C_0-sequence in D. Let F(z) be the limit function of the sequence $f_{n_k}(z)\ (k=1,2,\cdots)$, defined in D with respect to the spherical distance. In view of (7.15), F(z) can not be the constant ∞, so F(z) is a meromorphic function in D, by Theorem 1.3. Let $\zeta_j(j=1,2,\cdots,q)$ be the poles (if there is any) of F(z) in E_2, and Γ_j: $|z-\zeta_j|<r_j(j=1,2,\cdots,q)$ be circles such that $\sum_{j=1}^{q}r_j<\varepsilon$ and that the closures $\bar{\Gamma}_j$ $(j=1,2,\cdots,q)$ are mutually disjoint and interior to D. The set $E=E_2-\bigcup_{j=1}^{q}\Gamma_j$ is a bounded closed set of points of D. Since

$$\Lambda(z\,|\,z\in E_2,\ |f_{n_k}(z)|\geqslant n_k)\geqslant\varepsilon,$$

there is a point $z_k\in E$ such that

$$|f_{n_k}(z_k)|\geqslant n_k. \tag{7.16}$$

On the other hand, since F(z) has no pole on E, it is bounded on E:

$$|F(z)|\leqslant A\ (0<A<+\infty)\ for\ z\in E.$$

Take a number $B>A$ and let $\mu=\mu(A,B)$ defined in Lemma 1.2. Since as $k\to+\infty$, $|f_{n_k}(z), F(z)|\to 0$ uniformly on E, there is k_0 such that when $k\geqslant k_0$, we have

$$|f_{n_k}(z), F(z)| < \mu \ for \ z \in E,$$

which implies

$$|f_{n_k}(z)| < B \ for \ z \in E.$$

This contradicts (7.16).

Note that in Corollary 7.6, the first inequality in (7.14) is equivalent to

$$\max_{z \in E_1} |f(z), \infty| \geqslant \frac{1}{\sqrt{2}}$$

and the second inequality is equivalent to

$$\Lambda(z \,|\, z \in E_2, \ |f(z), \infty| \leqslant \frac{1}{(1+b^2)^{\frac{1}{2}}}) \geqslant \varepsilon.$$

This suggests us to generalize the condition (7.4) by replacing the point ∞ by an arbitrary point $w_0 \in \hat{\mathrm{C}}$. In this way we get the following corollary:

Corollary 7.7. Let D, $E_i (i=1,2)$, $S_i (i=1,2,3)$ and ε have the same meaning as in Corollary 7.6. Let w_0 be a point of $\hat{\mathrm{C}}$ and δ $(0<\delta<1)$ a number. Then there exists a number η $(0<\eta<1)$ depending only on D, $E_i (i=1,2)$, $S_i (i=1,2,3)$, ε, w_0 and δ, having the following property: If $f(z)$ is a meromorphic function in D such that

$$\max_{z \in E_1} |f(z), w_0| \geqslant \delta, \ \Lambda(z \,|\, z \in E_2, \ |f(z), w_0| \leqslant \eta) \geqslant \varepsilon, \tag{7.17}$$

then the values of the function $f(z)$ cover in generalized sense at least one of the sets $S_i (i=1,2,3)$.

For the proof of this corollary, we need the following lemma:

Lemma 7.2. Let $f(z)$ be a meromorphic function in a domain D and w_0 a point of $\hat{\mathrm{C}}$. Then the function $|f(z), w_0|$ of z is continuous in D.

Proof. Consider a point $z_0 \in D$. We have

$$|f(z),w_0| \leqslant |f(z),f(z_0)| + |f(z_0),w_0|,$$
$$|f(z_0),w_0| \leqslant |f(z_0),f(z)| + |f(z),w_0|,$$

hence

$$\big||f(z),w_0| - |f(z_0),w_0|\big| \leqslant |f(z),f(z_0)|,$$

which imples that the function $|f(z),w_0|$ is continuous at z_0, by Lemma1. 4.

Now let us come back to the proof of Corollary 7. 7. Assume that the number $\eta(0<\eta<1)$ does not exist. Then for each positive integer n, there is a meromorphic function $f_n(z)$ in D such that

$$\max_{z\in E_1} |f_n(z),w_0| \geqslant \delta,\ \Lambda(z\,|\,z \in E_2, |f_n(z),w_0| \leqslant \frac{1}{n+1}) \geqslant \varepsilon \quad (7.18)$$

and that the values of the function $f_n(z)$ do not cover in generalized sense either of the sets $S_i(i=1,2,3)$. By Theorme 7. 4, the family $\{f_n(z)\ (n=1,2,\cdots)\}$ is normal in D. Accordingly we can find a subsequence $f_{n_k}(z)\ (k=1,2,\cdots)$ which is a C_0-sequence in D. Let $F(z)$ be the limit function of the sequence $f_{n_k}(z)\ (k=1, 2,\cdots)$, defined in D with respect to the spherical distance. In view of the first inequality in (7. 18), $F(z)$ can not be the constant w_0. Let $\zeta_j(j=1,2,\cdots,q)$ be the roots (if there is any) of the equation $F(z)=w_0$ in E_2, and $\Gamma_j: |z-\zeta_j|<r_j(j=1, 2,\cdots,q)$ be circles such that $\sum_{j=1}^{q} r_j<\varepsilon$ and the closures $\overline{\Gamma}_j(j=1,2,\cdots,q)$ are mutually disjoint and interior to D. The set $E=E_2-\bigcup_{j=1}^{q}\Gamma_j$ is a bounded closed set of points of D. Since

$$\Lambda(z\,|\,z \in E_2,\ |f_{n_k}(z),w_0| \leqslant \frac{1}{n_k+1}) \geqslant \varepsilon,$$

there is a point $z_k \in E$ such that

$$|f_{n_k}(z_k), w_0| \leqslant \frac{1}{n_k + 1}. \tag{7.19}$$

On the other hand, by Lemma 7.2, the function $|F(z), w_0|$ is continuous in D, hence

$$|F(z), w_0| \geqslant m \text{ for } z \in E,$$

where $m > 0$ is a constant. Next from the inequality

$$|F(z), w_0| \leqslant |F(z), f_{n_k}(z)| + |f_{n_k}(z), w_0|$$

and the fact that as $k \to +\infty$, $|F(z), f_{n_k}(z)| \to 0$ uniformly on E, we see that we can find k_0 such that for $k \geqslant k_0$, we have

$$|f_{n_k}(z), w_0| \geqslant \frac{m}{2} \text{ for } z \in E.$$

This contradicts (7.19).

Remark 7.3. In Corollaries 7.6, 7.7 and in other similar places, it is tacitly assumed that $\Lambda E_2 \geqslant \varepsilon$.

Instead of assuming the function f(z) satisfying a condition of variation of modulus of the form (7.14) in Corollary 7.6 or a condition of variation of spherical distance of the form (7.17) in Corollary 7.7, we can also assume the function f(z) to satisfy a condition involving the spherical derivative $\partial(z,f)$. (see (1.28))

Corollary 7.8. Let D be a domain, E a bounded closed set of points belonging to D and $S_i (i=1,2,3)$ three sets of meromorphic functions in D such that S_i and S_j are locally uniformly disjoint for $i \neq j$ $(i,j=1,2,3)$ and that $S_i (i=1,2)$ are normal families in D. Then there exists a positive number β depending only on D, E and $S_i (i=1,2,3)$ having the following property: If f(z) is a meromorphic function in D such that

$$\max_{z \in E} \partial(z, f) \geqslant \beta, \tag{7.19}$$

then the values of the funciton $f(z)$ cover in generalized sense at least one of the sets $S_i(i=1,2,3)$.

Proof. Assume that the positive number β does not exist. Then for each positive integer n, there is a meromorphic function $f_n(z)$ in D such that

$$\max_{z\in E} \partial(z, f_n) \geq n \tag{7.20}$$

and that the values of the function $f_n(z)$ do not cover in generalized sense either of the sets $S_i(i=1,2,3)$. By Theorem 7.4, the family $\{f_n(z)\ (n=1,2,\cdots)\}$ is normal in D. Consequently by Theorem 1.6, the sequence of functions $\partial(z,f_n)$ $(n=1,2,\cdots)$ is locally uniformly bounded in D, and hence it is uniformly bounded on E. This contradicts (7.20).

We can also assume the function $f(z)$ to satisfy a condition involving the spherical oscillation $O(z_0,r,f)$ defined as follows:

Definition 7.8. Let $f(z)$ be a meromorphic function in a domain D and c: $|z-z_0|<r$ a circle belonging to D. We define

$$O(z_0,r,f) = \sup_{z_1,z_2\in c} |f(z_1), f(z_2)|$$

which is called the spherical oscillation of the function $f(z)$ in c.

Definition 7.9. Consider a domain D and a bounded closed set E of points belonging to D. It is known that we can find a positive number ρ such that for each point $z_0\in E$, the circle $|z-z_0|<\rho$ belongs to D. We denote by $R(D,E)$ the least upper bound of the set of such positive nubmers ρ.

Corollary 7.9. Let D, E and $S_i(i=1,2,3)$ have the same meaning as in Corollary 7.8. Let $0<\varepsilon<1$ be a number. Then there exists a number $0<r<R(D,E)$ depending only on D, E, $S_i(i=1,2,3)$ and ε having the following property: If $f(z)$ is a meromorphic function in D such that

$$\sup_{z_0 \in E} O(z_0, r, f) \geqslant \varepsilon, \tag{7.21}$$

then the values of the function f(z) cover in generalized sense at least one of the sets $S_i (i=1,2,3)$.

Proof. Assume that the number r does not exist. Take a number $0<\rho<R(D,E)$. Then for each positive integer n, there is a meromorphic function $f_n(z)$ in D such that

$$\sup_{z_0 \in E} O(z_0, \frac{\rho}{n}, f_n) \geqslant \varepsilon \tag{7.22}$$

and that the values of the function $f_n(z)$ do not cover in generalized sense either of the sets $S_i (i=1,2,3)$. By Theorem 7.4, the family $\{f_n(z)\ (n=1,2,\cdots)\}$ is normal in D. By Theorem 1.5, the family $\{f_n(z)\ (n=1,2,\cdots)\}$ is equicontinuous in D with respect to the spherical distance. Since E is bounded closed, we can find a finite number of points $\zeta_j \in E\ (j=1,2,\cdots,m)$ such that

$$E \subset \bigcup_{j=1}^{m} (|z-\zeta_j| < \frac{\rho}{2}).$$

Consider the bounded closed set

$$\Gamma = \bigcup_{j=1}^{m} (|z-\zeta_j| \leqslant \rho) \subset D.$$

By the equicontinuity of the family $\{f_n(z)(n=1,2,\cdots)\}$, corresponding to the positive number $\varepsilon/2$, there is a number $\delta>0$ such that

$$|f_n(z_1), f_n(z_2)| < \frac{\varepsilon}{2} \text{ for } n \geqslant 1,\ z_1, z_2 \in \Gamma,\ |z_1 - z_2| < \delta.$$

Next take a positive integer $n_0>2$ such that

$$\frac{\rho}{n_0} < \frac{\delta}{2}.$$

Consider a point $z_0 \in E$. Then evidently

$$(|z - z_0| < \frac{\rho}{n_0}) \subset \Gamma$$

and hence for any two points z_1, z_2 of the circle $|z - z_0| < \rho/n_0$, we have

$$|f_{n_0}(z_1), f_{n_0}(z_2)| < \frac{\varepsilon}{2}$$

and therefore

$$O(z_0, \frac{\rho}{n_0}, f_{n_0}) \leqslant \frac{\varepsilon}{2}.$$

Since $z_0 \in E$ is arbitrary, we have

$$\sup_{z_0 \in E} O(z_0, \frac{\rho}{n_0}, f_{n_0}) \leqslant \frac{\varepsilon}{2}$$

which contradicts (7.22).

Theorem 7.5. Let D be a domain, $S_i (i=1,2,3)$ three sets of meromorphic functions in D, such that for $i \neq j$ $(i,j=1,2,3)$ S_i and S_j are locally uniformly disjoint and that $S_i (i=1,2)$ are normal families in D, and $p \geqslant 1$ an integer. Let $\mathscr{F}$ be the family of the functions $f(z)$ satisfying the following conditions:

1° $f(z)$ is meromorphic in D.

2° There are three functions $\varphi_i(z) \in S_i (i=1,2,3)$ such that each of the equations $f(z) = \varphi_i(z)$ $(i=1,2,3)$ has at most p distinct roots in D.

Then the family $\mathscr{F}$ is quasi-normal in D of order p at most.

For the proof of this theorem, we need some lemmas.

Lemma 7.3. Let $f_n(z)$ $(n=1,2,\cdots)$ be a sequence of meromorphic functions in a domain D, $\varphi_n(z)$, $\psi_n(z)$ $(n=1,2,\cdots)$ two sequences of holomorphic functions in D and $p \geqslant 1$ an integer. Suppose that the following conditions are satisfied:

1^0 For each n, each of the equations $f_n(z)=\infty$, $f_n(z)=\varphi_n(z)$, $f_n(z)=\psi_n(z)$ has at most p distinct roots in D.

2^0 For each point $z_0\in D$, we can find a circle $|z-z_0|<r$ belonging to D and two positive numbers M, δ such that

$$|\varphi_n(z)|<M,\ |\psi_n(z)|<M,\ |\varphi_n(z)-\psi_n(z)|>\delta\ (n=1,2,\cdots)$$

in the circle $|z-z_0|<r$.
Then the family $\{f_n(z)\ (n=1,2,\cdots)\}$ is quasi-normal in D of order p at most.

Proof. By the condition 2^0 in Lemma 7.3, the two sequences $\varphi_n(z)$, $\psi_n(z)$ $(n=1,2,\cdots)$ are locally uniformly bounded in D, hence by Corollary 1.3, we can find an increasing sequence of positive integers $n_k(k=1,2,\cdots)$ such that the two sequences $\varphi_{n_k}(z)$, $\psi_{n_k}(z)(k=1,2,\cdots)$ converge locally uniformly to holomorphic functions $\varphi(z)$, $\psi(z)$ respectively in D. Evidently by 2^0, we have

$$|\varphi(z)-\psi(z)|\geqslant\delta\ for\ |z-z_0|<r. \tag{7.23}$$

Now consider the sequence of functions

$$F_k(z)=\frac{f_{n_k}(z)-\varphi_{n_k}(z)}{\psi_{n_k}(z)-\varphi_{n_k}(z)}\ (k=1,2,\cdots)$$

meromorphic in D. By the condition 1^0 in Lemma 7.3, for $k\geqslant 1$, each of the three equations

$$F_k(z)=0,\ F_k(z)=1,\ F_k(z)=\infty$$

has at most p distinct roots in D, consequently by Corollary 5.9, the family $\{F_k(z)\ (k=1,2,\cdots)\}$ is quasi-normal in D of order p at most. Accordingly there is a subsequence $F_{k_s}(z)\ (s=1,2,\cdots)$ which is a C_0-sequence in a domain D-ω, where ω consists of at most p points of D. Evidently this is also true for the sequence

$$F_{k_s}(z)\{\psi_{m_s}(z)-\varphi_{m_s}(z)\}\ (s=1,2,\cdots;m_s=m_{k_s}),$$

by Theorem 1.1 and (7.23). Then by Lemma 7.1, we see that the identity

$$f_{m_s}(z) = F_{k_s}(z)\{\psi_{m_s}(z)-\varphi_{m_s}(z)\} + \varphi_{m_s}(z) \ (s = 1,2,\cdots)$$

implies that the sequence $f_{m_s}(z)$ $(s=1,2,\cdots)$ is a C_0-sequence in D-ω.

Evidently, in general, from every subsequence $f_{\lambda_k}(z)$ $(k=1,2,\cdots)$ of the sequence $f_n(z)$ $(n=1,2,\cdots)$, we can extract a subsequence $f_{\nu_s}(z)$ $(s=1,2,\cdots)$ which is a C_0-sequence in a domain D-σ, where σ consists of at most p points of D. Therefore the family $\{f_n(z)\ (n=1,2,\cdots)\}$ is quasi-normal in D of order p at most.

Lemma 7.4. Let $S: f_n(z)$ $(n=1,2,\cdots)$ be a sequence of meromorphic functions in a circle $|z-z_0|<r$, such that S is a C_0-sequence in the domain $0<|z-z_0|<r$ and that z_0 is a nonC_0-point of S which is irreducible with respect to the point z_0. (see Definition 5.6) Let $\sigma: a_n(z)$ $(n=1,2,\cdots)$ be a C_0-sequence of meromorphic functions in the circle $|z-z_0|<r$. Denote by f(z) and a(z) respectively the limit function of S in the domain $0<|z-z_0|<r$ and that of σ in the circle $|z-z_0|<r$, with respect to the spherical distance. If $f(z)\not\equiv a(z)$ in the domain $0<|z-z_0|<r$, then we can find a positive integer N such that when $n\geqslant N$, the equation

$$f_n(z) = a_n(z) \tag{7.24}$$

has at least one root in the circle $|z-z_0|<r$.

Proof. Distinguish several cases:

1) f(z) and a(z) meromorphic functions.

Assume first that $a(z_0)\neq\infty$. Since $f(z)-a(z)\not\equiv 0$, there is a circle $c: |z-z_0|=\rho$ $(0<\rho<r)$ on which $f(z)\neq\infty$, $a(z)\neq\infty$ and $f(z)-a(z)\neq 0$. Then by Lemma 5.13, we have

$$\lim_{n\to+\infty} f_n(z) = f(z), \ \lim_{n\to+\infty} a_n(z) = a(z) \ uniformly \ on \ c$$

and

$$|f(z) - a(z)| \geqslant m > 0 \text{ on } c.$$

Now suppose that there does not exist a positive integer N having the required property in Lemma 7. 4. Then we can find an increasing sequence of positive integers $n_k (k=1,2,\cdots)$ such that for $k \geqslant 1$, the equation

$$f_{n_k}(z) = a_{n_k}(z)$$

has no root in the circle $|z-z_0|<r$. Accordingly the sequence of functions

$$F_k(z) = \frac{1}{f_{n_k}(z) - a_{n_k}(z)} \quad (k = 1,2,\cdots)$$

are holomorphic in the circle $|z-z_0|<r$ and since it converges uniformly on c to $1/\{f(z)-a(z)\}$, it is uniformly convergent in the circle $|z-z_0|<\rho$. This implies that the point z_0 is a C_0-point of the sequence

$$\frac{1}{F_k(z)} = f_{n_k}(z) - a_{n_k}(z) \quad (k = 1,2,\cdots).$$

Now since $a(z_0) \neq \infty$, by Theorem 1. 1, we can find a circle $\Gamma: |z-z_0|<\rho_1 (0<\rho_1<\rho)$ and a positive integer k_0 such that in Γ, $a_{n_k}(z)$ $(k \geqslant k_0)$ and $a(z)$ are holomorphic and the sequence $a_{n_k}(z) (k \geqslant k_0)$ converges uniformly to $a(z)$. Then by Lemma 7. 1, the identity

$$f_{n_k}(z) = \frac{1}{F_k(z)} + a_{n_k}(z)$$

implies that the point z_0 is a C_0-point of the sequence $f_{n_k}(z)$ $(k=1,2,\cdots)$. This contradicts the hypothesis that the sequence S is irreducible with respect to z_0.

Next assume that $a(z_0)=\infty$. Then two cases are possible:

A) $f(z)\not\equiv 0$. In this case it is sufficient to apply the above result to $1/f_n(z)$ and $1/a_n(z)$ $(n \geqslant n_0)$.

B) $f(z)\equiv 0$. In this case we take a constant $\beta \neq 0$ and apply the result in case A) to $f_n(z)+\beta$, $a_n(z)+\beta$ $(n \geqslant 1)$.

2) $f(z)$ is the constant ∞ and $a(z)$ is a meromorphic function. Then two

cases are possible:

A_1) $a(z)\not\equiv 0$. In this case we apply the result in case 1) to $1/f_n(z)$ and $1/a_n(z)$ $(n\geqslant n_0)$.

B_1) $a(z)\equiv 0$ In this case we take a constant $\beta\neq 0$ and apply the result in case A_1) to $f_n(z)+\beta$, $a_n(z)+\beta$ $(n\geqslant 1)$.

3) $f(z)$ is a meromorphic function and $a(z)$ is the constant ∞. Then again we distinguish two cases:

A_2) $f(z)\not\equiv 0$. In this case we apply the result in case 1) to $1/f_n(z)$ and $1/a_n(z)$ $(n\geqslant n_0)$.

B_2) $f(z)\equiv 0$. In this case we take a constant $\beta\neq 0$ and apply the result in case A_2) to $f_n(z)+\beta$, $a_n(z)+\beta$ $(n\geqslant 1)$.

The proof of Lemma 7.4 is now complete.

Now let us prove Theorem 7.5. Let $f_n(z)$ $(n=1,2,\cdots)$ be a sequence of functions of the family $\mathscr{F}$. By hypothesis, to each $f_n(z)$ correspond three functions $a_n(z)\in S_1$, $b_n(z)\in S_2$, $c_n(z)\in S_3$ such that each of the three equations

$$f_n(z)=a_n(z),\ f_n(z)=b_n(z),\ f_n(z)=c_n(z)$$

has at most p distinct roots in D. Since S_i $(i=1,2)$ are normal families in D, there is an increasing sequence of positive integers n_k $(k=1,2,\cdots)$ such that the two sequences $a_{n_k}(z)$, $b_{n_k}(z)$ $(k=1,2,\cdots)$ are C_0-sequences in D. Let $a(z)$, $b(z)$ be respectively the limit functions defined in D with respect to the spherical distance, of $a_{n_k}(z)$, $b_{n_k}(z)$ $(k=1,2,\cdots)$.

Consider a point $z_0\in D$. Then one at least of the values $a(z_0)$ and $b(z_0)$ is finite, say $a(z_0)\neq\infty$. By Theorem 1.1, we can find a circle Γ_0: $|z-z_0|<r_0$ belonging to D and a positive integer k_0 such that the functions $a_{n_k}(z)$ $(k\geqslant k_0)$ and $a(z)$ are holomorphic in Γ_0 and the sequence $a_{n_k}(z)$ $(k\geqslant k_0)$ converges uniformly to $a(z)$ in Γ_0.

Then as in the proof of Theorem 7.4, we introduce the sequence of meromorphic functions in Γ_0:

$$F_k(z)=\frac{1}{f_{n_k}(z)-a_{n_k}(z)}\ (k\geqslant k_0)$$

and the two sequences of holomorphic functions in Γ_0:

$$B_k(z) = \frac{1}{b_{n_k}(z) - a_{n_k}(z)},\ C_k(z) = \frac{1}{c_{n_k}(z) - a_{n_k}(z)}\ (k \geqslant k_0).$$

We may assume that r_0 is small enough, so that in Γ_0 we have

$$|B_k(z)| < \frac{1}{\delta},\ |C_k(z)| < \frac{1}{\delta},\ |B_k(z) - C_k(z)| > \eta,$$

where δ and η are positive numbers. Since each of the three equations

$$F_k(z) = \infty,\ F_k(z) = B_k(z),\ F_k(z) = C_k(z)$$

has at most p distinct roots in Γ_0, we can apply Lemma 7. 3 and conclude that the family $\{F_k(z)\ (k \geqslant k_0)\}$ is quasi-normal in Γ_0 of order p at most. This implies that the family

$$\frac{1}{F_k(z)} = f_{n_k}(z) - a_{n_k}(z)\ (k \geqslant k_0)$$

is also quasi-normal in Γ_0 of order p at most. Then by Lemma 7. 1, the identity

$$f_{n_k}(z) = \frac{1}{F_k(z)} + a_{n_k}(z)$$

implies that the family $\{f_{n_k}(z)\ (k \geqslant k_0)\}$ is quasi-normal in Γ_0 of order p at most. Of course the same is true for the family $\{f_{n_k}(z)\ (k=1,2,\cdots)\}$. Since $z_0 \in D$ is arbitrary, we conclude, by Theorem 5. 1, that the family $\{f_{n_k}(z)\ (k=1,2,\cdots)\}$ is quasi-normal in D.

Accordingly we can extract from $f_{n_k}(z)\ (k=1,2,\cdots)$ a subsequence $f_{m_h}(z)\ (h=1,2,\cdots)$ which is a C_1-sequence in D. Then by Lemma 5. 5, we can extract from $f_{m_h}(z)\ (h=1,2,\cdots)$ a subsequence $f_{q_j}(z)\ (j=1,2,\cdots)$ which is either a C_0-sequence in D or an irreducible C_1-sequence in D. Consider the latter case and let E be the set of the nonC_0-points of the sequence $f_{q_j}(z)\ (j=1,2,\cdots)$ in D. Let f (z) be the limit function defined in D-E with respect to the spherical distance, of the sequence $f_{q_j}(z)\ (j=1,2,\cdots)$. Consider the seqeunces $a_{q_j}(z)$, $b_{q_j}(z)\ (j=1,2,\cdots)$ and their limit functions $a(z)$, $b(z)$ defined above. One at least of the two functions $a(z)$, $b(z)$ say $a(z)$ is such that $a(z)\not\equiv f(z)$ in D-E. Assume that there are ν points $z_i(i=1,2,\cdots,\nu)$ belonging to E. Let Γ_i: $|z-z_i|<r_i(i=1,2,\cdots,\nu)$ be ν circles belonging to D, without common point each other and Γ_i containing no point of E except z_i. Consider a circle Γ_i. Since $f(z)\not\equiv a(z)$ in the domain $0<|z-z_i|<r_i$ and the sequence $f_{q_j}(z)$ $(j=1,2,\cdots)$ is irreducible with respect to z_i, we can apply Lemma 7.4 and conclude that there is a positive integer J_i such that when $j\geqslant J_i$, the equation

$$f_{q_j}(z)=a_{q_j}(z) \tag{7.25}$$

has at least one root in the circle Γ_i. Let $J=\max_{1\leqslant i\leqslant\nu} J_i$, then when $j\geqslant J$, the equation (7.25) has at least one root in each of the circles $\Gamma_i(i=1,2,\cdots,\nu)$. It follows that $\nu\leqslant p$ and E consists of at most p points. Since $f_{q_j}(z)$ $(j=1,2,\cdots)$ is a subsequence of the sequence $f_n(z)$ $(n=1,2,\cdots)$, we have proved that the family $\mathscr{F}$ is quasi-normal in D of order p at most.

Definition 7.10. Let $f(z)$ be a meromorphic function in a domain D and S a set of meromorphic functions in D and $p\geqslant 2$ an integer. We say that the values of the funciton $f(z)$ cover in generalized sense the set S strictly at least p times, if for each function $\varphi(z)\in S$, the equation $f(z)=\varphi(z)$ has at least p distinct roots in D.

Corollary 7.10. Let D be a domain, $E_i(i=1,2)$ two bounded closed sets of points belonging to D, $S_i(i=1,2,3)$ three sets of meromorphic functions in D such that S_i and S_j are locally uniformly disjoint for $i\neq j$ $(i,j=1,2,3)$ and that S_i $(i=1,2)$ are normal families in D, ε and d positive numbers and $p\geqslant 2$ an integer. Then there exists a positive number b depending only on D, $E_i(i=1,2)$, $S_i(i=1,2,3)$, ε, d and p, having the following property: If $f(z)$ is a meromorphic function in D satisfying the following conditions:

1^0 $|f(z_j)|\leqslant 1$ for p points $z_j\in E_1(j=1,2,\cdots,p)$ with $|z_j-z_k|\geqslant d$ $(j,k=1,2,\cdots,p;j\neq k)$;

2^0 $\Lambda(z|z\in E_2,\ |f(z)|\geqslant b)\geqslant\varepsilon$;

then the values of the function $f(z)$ cover in generalized sense at least one of the sets $S_i(i=1,2,3)$ strictly at least p times.

Proof. Assume that the positive number b does not exist. Then for each positive integer n, there is a meromorphic function $f_n(z)$ in D satisfying the following conditions:

1) $|f_n(z_j^{(n)})| \leqslant 1$ for p points $z_j^{(n)} \in E_1 (j=1,2,\cdots,p)$ with $|z_j^{(n)} - z_k^{(n)}| \geqslant d$ $(j,k = 1,2,\cdots,p; j \neq k)$.

2) $\Lambda(z | z \in E_2, |f_n(z)| \geqslant n) \geqslant \varepsilon$.

3) The values of the function $f_n(z)$ do not cover in generalized sense either of the sets $S_i (i=1,2,3)$ strictly at least p times.

By Theorem 7.5, the family $\{f_n(z)\ (n=1,2,\cdots)\}$ is quasi-normal in D of order $p-1$ at most. Accordingly from the sequence $f_n(z)\ (n=1,2,\cdots)$ we can extract a subsequence $f_{n_k}(z)\ (k=1,2,\cdots)$ which is a C_0-sequence in a domain D-ω, where ω is a set consisting of at most $p-1$ points of D. Let f(z) be its limit function defined in D-ω, with respect to the spherical distance. f(z) can not be the constant ∞. In fact, we can find an increasing sequence of positive integers $k_s (s=1,2,\cdots)$ such that the p sequences

$$z_j^{(m_s)} (s = 1,2,\cdots)\ (j = 1,2,\cdots,p), m_s = n_{k_s}$$

converge respectively to limits $\zeta_j \in E_1 (j=1,2,\cdots,p)$ with

$$|\zeta_j - \zeta_l| \geqslant d\ \ (j,l = 1,2,\cdots,p; j \neq l).$$

(see Lemma 3.2) Among the points $\zeta_j (j=1,2,\cdots,p)$, there is one $\zeta_{j_0} \in D_1 - \omega$. If $f(z) \equiv \infty$, then in a circle $|z - \zeta_{j_0}| < \delta$, we have uniformly

$$\lim_{s \to +\infty} |f_{m_s}(z), \infty| = 0,$$

which is incompatible with the fact:

$$|f_{m_s}(z_{j_0}^{(m_s)})| \leqslant 1, \ \lim_{s \to +\infty} z_{j_0}^{(m_s)} = \zeta_{j_0}.$$

So f(z) is a meromorphic function in D-ω. Let $\alpha_j (j=1,2,\cdots,q)$ be the points of the set $E_2 \cap \omega$. Describe circles Γ_j: $|z - \alpha_j| < r_j (j=1,2,\cdots,q)$ such that $\sum_{j=1}^{q} r_j < \varepsilon/2$ and that the closures $\overline{\Gamma}_j \subset D$ $(j=1,2,\cdots,q)$ are mutually disjoint. Then the set $\sigma=E_2-\bigcup_{j=1}^{q}\Upsilon_j\subset D-\omega$ is bounded closed. Next let $\beta_j(j=1,2,\cdots,q')$ be the poles of the function f(z) on σ. Again describe circles Υ_j': $|z-\beta_j|<r_j'(j=1,2,\cdots,q')$ such that $\sum_{j=1}^{q'}r_j'<\frac{\varepsilon}{2}$ and that the closures $\overline{\Upsilon}_j'\subset D-\omega$ $(j=1,2,\cdots,q')$ are mutually disjoint. Then $\sigma_*=\sigma-\bigcup_{j=1}^{q'}\Upsilon_j'\subset D-\omega$ is a bounded closed set on which f(z) is finite.

By the condition 2), for each k there is a point $a_k\in\sigma_*$ such that

$$|f_{n_k}(a_k)|\geqslant n_k. \tag{7.26}$$

On the other hand, since f(z) is finite on σ_*, by Lemma 5.13, when $k\geqslant k_0$, we have

$$|f_{n_k}(z)|<B \text{ for } z\in\sigma_*,$$

where B is a positive constant independant of k. This contradicts (7.26).

Corollary 7.11. Let D, $E_i(i=1,2)$, $S_i(i=1,2,3)$, ε, d and p have the same meaning as in Corollary 7.10. Let w_0 be a point of $\hat{C}$ and $\delta(0<\delta<1)$ a number. Then there exists a number η $(0<\eta<1)$ depending only on D, $E_i(i=1,2)$, $S_i(i=1,2,3)$, ε, d, p, w_0 and δ, having the following property: If f(z) is a meromorphic function in D satisfying the following conditions:

1° $|f(z_j),w_0|\geqslant\delta$ for p points $z_j\in E_1(j=1,2,\cdots,p)$ with $|z_j-z_k|\geqslant d$ $(j,k=1,2,\cdots,p;j\neq k)$;

2° $\Lambda(z|z\in E_2,\ |f(z),w_0|\leqslant\eta)\geqslant\varepsilon$;

then the values of the function f(z) cover in generalized sense at least one of the sets $S_i(i=1,2,3)$ strictly at least p times.

Just as the proof of Corollary 7.7 is quite similar to that of Corollary 7.6, the proof of Corollary 7.11 is quite similar to that of Corallary 7.10. However for the sake of completeness, we give briefly the proof.

Proof. Assume that the number $\eta(0<\eta<1)$ does not exist. Then for each positive integer n, there is a meromorphic function $f_n(z)$ in D satisfying the following conditions:

1) $|f_n(z_j^{(n)}), w_0| \geqslant \delta$ for p points $z_j^{(n)} \in E_1 (j=1,2,\cdots,p)$ with $|z_j^{(n)}, z_k^{(n)}| \geqslant d$ $(j,k=1,2,\cdots,p; j\neq k)$.

2) $\Lambda(z | z \in E_2, \ |f_n(z), w_0| \leqslant \frac{1}{n+1}) \geqslant \varepsilon$.

3) The values of the function $f_n(z)$ do not cover in generalized sense either of the sets $S_i (i=1,2,3)$ strictly at least p times.

By Theoerm 7.5, the family $\{f_n(z) \ (n=1,2,\cdots)\}$ is quasi-normal in D of order $p-1$ at most. Accordingly from the sequence $f_n(z) \ (n=1,2,\cdots)$ we can extract a subsequence $f_{n_k}(z) \ (k=1,2,\cdots)$ which is a C_0-sequence in a domain D-ω, where ω is a set consisting of at most $p-1$ points of D. Let $f(z)$ be its limit function defined in D-ω, with respect to the spherical distance. By the condition 1), we see as in the proof of Corollary 7.10, that $f(z)$ is not the constant w_0. Having obtained this result, the proof can then be completed by a combination of the method used in the proof of Corollary 7.10 and that used in the proof of Corollary 7.7.

Corollary 7.12. Let D be a domain, E a bounded closed set of points belonging to D, $S_i (i=1,2,3)$ three sets of meromorphic functions in D such that S_i and S_j are locally uniformly disjoint for $i\neq j \ (i,j=1,2,3)$ and $S_i (i=1,2)$ are normal families in D, d a positivs number and $p\geqslant 2$ an integer. Then there exists a positive number β depending only on D, E, $S_i (i=1,2,3)$, d and p, having the following property: If $f(z)$ is a meromorphic function in D such that

$$\partial(z_j, f) \geqslant \beta \ (j = 1,2,\cdots,p)$$

for p points $z_j \in E \ (j=1,2,\cdots,p)$ with $|z_j - z_k| \geqslant d \ (j,k=1,2,\cdots,p; j\neq k)$, then the values of the function $f(z)$ cover in generalized sense at least one of the sets $S_i (i=1,2,3)$ strictly at least p times.

Proof. Assume that the positive number β does not exist. Then for each positive integer n, there is a meromorphic function $f_n(z)$ in D such that

$$\partial(z_j^{(n)}, f_n) \geqslant n \ (j = 1,2,\cdots,p)$$

for p points $z_j^{(n)} \in E \ (j=1,2,\cdots,p)$ with $|z_j^{(n)} - z_k^{(n)}| \geqslant d \ (j,k=1,2,\cdots,p; j\neq k)$, and that the values of the function $f_n(z)$ do not cover in generalized sense either of the sets $S_i(i=1,2,3)$ strictly at least p times. By Theorem 7.5, the family $\{f_n(z)\ (n=1,2,\cdots)\}$ is quasi-normal in D of order $p-1$ at most. From the sequence $f_n(z)\ (n=1,2,\cdots)$ we can extract a subsequence $f_{n_k}(z)\ (k=1,2,\cdots)$ which is a C_0-sequence in a domain D-ω, where ω is a set consisting of at most $p-1$ points of D. À fortiori, the family $\{f_{n_k}(z)\ (k=1,2,\cdots)\}$ is normal in D-ω, and by Theorem 1.6, the sequence of functions $\partial(z,f_{n_k})\ (k=1,2,\cdots)$ is locally uniformly bounded in D-ω. On the other hand, let $k_s(s=1,2,\cdots)$ be an increasing sequence of positive integers such that the p sequences $z_j^{(m_s)}(s=1,2,\cdots)\ (j=1,2,\cdots,p)$, $m_s=n_{k_s}$, converge respectively to limits $\zeta_j\in E\ (j=1,2,\cdots,p)$ with $|\zeta_j-\zeta_k|\geqslant d\ (j,k=1,2,\cdots,p;j\neq k)$. (see Lemma 3.2) One at least of the points $\zeta_j(j=1,2,\cdots,p)$, say $\zeta_{j_0}\in$ D-ω. Consequently the sequence $\partial(z,f_{n_k})\ (k=1,2,\cdots)$ is uniformly bounded in a circle $|z-\zeta_{j_0}|<\rho$. This is incompatible with the fact:

$$\partial(z_{j_0}^{(m_s)},f_{m_s})\geqslant m_s,\ \lim_{s\to+\infty} z_{j_0}^{(m_s)}=\zeta_{j_0}.$$

Corollary 7.13. Let D, E, $S_i(i=1,2,3)$, d and p have the same meaning as in Corollary 7.12. Let $0<\varepsilon<1$ be a number. Then there exists a number $0<r<R(D,E)$ depending only on D,E,$S_i(i=1,2,3)$, ε,d and p having the following property: If $f(z)$ is a meromorphic function in D such that

$$O(z_j,r,f)\geqslant\varepsilon\ (j=1,2,\cdots,p)$$

for p points $z_j\in E\ (j=1,2,\cdots,p)$ with $|z_j-z_k|\geqslant d\ (j,k=1,2,\cdots,p;j\neq k)$, then the values of the function $f(z)$ cover in generalized sense at least one of the sets $S_i(i=1,2,3)$ strictly at least p tiems.

Proof. Assume that the number r does not exist. Take a number $0<\rho<R(D,E)$. Then for each positive integer n, there is a meromorphic function $f_n(z)$ in D such that

$$O(z_j^{(n)},\frac{\rho}{n},f_n)\geqslant\varepsilon\ (j=1,2,\cdots,p) \qquad (7.27)$$

for p points $z_j^{(n)} \in E$ $(j=1,2,\cdots,p)$ with $|z_j^{(n)}-z_k^{(n)}| \geqslant d$ $(j,k=1,2,\cdots,p;j\neq k)$ and that the values of the function $f_n(z)$ do not cover in generalized sense either of the sets $S_i(i=1,2,3)$ strictly at least p times. By Theorem 7.5, the family $\{f_n(z)\ (n=1,2,\cdots)\}$ is quasi-normal in D of order $p-1$ at most. From the sequence $f_n(z)$ $(n=1,2,\cdots)$ we can extract a subsequence $f_{n_k}(z)$ $(k=1,2,\cdots)$ which is a C_0-sequence in a domain D-ω, where ω is a set consisting of at most $p-1$ points of D. À fortiori, the family $\{f_{n_k}(z)\ (k=1,2,\cdots)\}$ is normal in D-ω, and hence, by Theorem 1.5, the family $\{f_{n_k}(z)\ (k=1,2,\cdots)\}$ is equicontinuous in D-ω with respect to the spherical distance. Let $k_s(s=1,2,\cdots)$ be an increasing sequence of positive integers such that the p sequences $z_j^{(m_s)}(s=1,2,\cdots)$ $(j=1,2,\cdots,p)$, $m_s = n_{k_s}$ converge vespectively to limits $\zeta_j \in E$ $(j=1,2,\cdots,p)$ with $|\zeta_j-\zeta_k| \geqslant d$ $(j,k=1,2,\cdots,p;j\neq k)$. One at least of the points $\zeta_j(j=1,2,\cdots,p)$, say $\zeta_{j_0} \in$ D-ω. Since the family $\{f_{n_k}(z)\ (k=1,2,\cdots)\}$ is equicontinuous at the point ζ_{j_0}, we can find a circle c: $|z-\zeta_{j_0}|<h$ belonging to D such that

$$|f_{n_k}(z_1),f_{n_k}(z_2)| < \frac{\varepsilon}{2} \text{ for } k \geqslant 1,\ z_1,z_2 \in c.$$

Take a sufficiently large integer s such that

$$|z_{j_0}^{(m_s)} - \zeta_{j_0}| < \frac{h}{2},\ \frac{\rho}{m_s} < \frac{h}{2}.$$

Then the circle $|z-z_{j_0}^{(m_s)}|<\frac{\rho}{m_s}$ belongs to c, and hence

$$O(z_{j_0}^{(m_s)},\frac{\rho}{m_s},f_{m_s}) \leqslant \frac{\varepsilon}{2}$$

which is incompatible with (7.27).

7.3. CASE OF UNIVALENT AND MULTIVALENT FUNCTIONS

In this part the following lemma plays an important role:

Lemma 7.5. Let $\mathscr{F}$ be a normal family of holomorphic functions f(z) in a domain D satisfying the following conditions:

1^0 The functions $f(z)$ of the family $\mathscr{F}$ have no zero in D.

2^0 There does not exist a sequence of functions $f_n(z)$ $(n=1,2,\cdots)$ of the family $\mathscr{F}$, which converges locally uniformly to zero in D.

Let E be a bounded closed set of points belonging to D. Then there exists a number $\lambda>0$ such that for each function $f(z)\in\mathscr{F}$ we have

$$|f(z)| \geqslant \lambda \text{ for } z \in E. \tag{7.28}$$

Proof. Assume that such a number λ does not exist. Then to each positive integer n corresponds a function $f_n(z)\in\mathscr{F}$ such that

$$\min_{z\in E} |f_n(z)| < \frac{1}{n}. \tag{7.29}$$

From the sequence $f_n(z)(n=1,2,\cdots)$ we can extract a subsequence $f_{n_k}(z)$ $(k=1, 2,\cdots)$ converging locally uniformly to a holomorphic function $g(z)$ in D. By hypothesis $g(z)\not\equiv 0$. From (7.29) evidently $g(z)$ has at least one zero $z_0\in E$. Then by a well known theorem, when k is sufficiently large, $f_{n_k}(z)$ has zero in a certain circle with center z_0. So we get a contradiction.

Theorem 7.6. Let D be a domain and E_1, E_2 two bounded closed sets of points belonging to D. Then there exists a number $\rho>0$ depending only on D, E_1, E_2 having the following property: If $f(z)$ is a holomorphic and univalent function in D satisfying the following conditions:

1^0 $f(z)$ has a zero $z_0\in E_1$;

2^0 $|f'(z_0')|\geqslant 1$ at a point $z_0'\in E_2$;

then the values of $f(z)$ cover in generalized sense the set $H_D(|w|<\rho)$. (see Definition 7.3)

Proof. Denote by $\mathscr{F}$ the family of the functions $f(z)$ satisfying the conditions in Theorem 7.6. Assume that there does not exist a number $\rho>0$ having the required property. Then to each positive integer n corresponds a function $f_n(z)\in\mathscr{F}$ such that the values of $f_n(z)$ do not cover in generalized sense the set H_D $(|w|<\frac{1}{n})$. Let $z_n\in E_1$ be the zero of $f_n(z)$. We can find a subsequence $\zeta_k=z_{n_k}$ (k

$=1,2,\cdots)$ converging to a point $\zeta_0 \in E_1$. Consider a circle $|z-\zeta_0| \leqslant r$ belonging to D and the domain

$$D_0 = D - (|z - \zeta_0| \leqslant r/2).$$

Let k_0 be a positivs integer such that, for $k \geqslant k_0$, ζ_k lies in the circle $|z-\zeta_0| < r/2$. Then the family $\{g_k(z) = f_{n_k}(z)\ (k \geqslant k_0)\}$ is normal in D_0. (see the proof of Theorem 6. 12) By the conditions 1^0 and 2^0 in Theorem 7. 6, it is easy to see that the family $\{g_k(z)\ (k \geqslant k_0)\}$ satisfies the conditions in Lemma 7. 5 with respect to D_0. Hence there is a number $\rho > 0$ such that for each $k \geqslant k_0$ we have

$$|g_k(z)| \geqslant \rho \text{ for } |z - \zeta_0| = r.$$

Now consider an integer $k \geqslant k_0$ and a function $\varphi(z) \in H_D(|w| < \rho)$. By Definition 7. 3, $|\varphi(z)| < \rho$ in D. Consequently on the circle $|z-\zeta_0| = r$, we have $|\varphi(z)| < |g_k(z)|$ and by Rouché's theorem, the function $g_k(z) - \varphi(z)$ has a zero in the circle $|z-\zeta_0| < r$. It follows that the values of the function $g_k(z)$ cover in generalized sense the set $H_D(|w| < \rho)$, according to Definition 7. 2. This contradicts the fact that the values of the function $g_k(z)$ do not cover in generalized sense the set $H_D(|w| < \frac{1}{n_k})$, if $1/n_k < \rho$.

Theorem 7. 7. Let D be a domain, E_1, E_2 two bounded closed sets of points belonging to D and $p \geqslant 2$ an integer. Then there exists a number $\rho > 0$ depending only on D, E_1, E_2 and p, having the following property: If $f(z)$ is a holomorphic and p-valent function in D satisfying the following conditions:

1^0 $f(z)$ has a zero $z_0 \in E_1$ of order p;

2^0 $|f^{(p)}(z_0')| \geqslant 1$ at a point $z_0' \in E_2$;

then the values of $f(z)$ cover in generalized sense the set $H_D(|w| < \rho)$ at least p times. (see Definition 7. 4)

This theorem is proved in the same way as for Theorem 7. 6.

In the above two theorems, the functions are assumed to be holomorphic. However the same method is also applicable to the case of meromorphic univalent and multivalent functions as shown by the following two theorems.

Theorem 7.8. Let D be a domain and $E_i(i=1,2,3)$ three bounded closed sets of points belonging to D, such that

$$E_1 \cap E_3 = \varnothing, \ E_2 \cap E_3 = \varnothing. \tag{7.30}$$

Then there exists a number $\rho>0$ depending only on D and $E_i(i=1,2,3)$ having the following property: If f(z) is a meromorphic and univalent function in D satisfying the following conditions:

1^0 f(z) has a zero $z_0 \in E_1$;

2^0 $|f'(z_0')|\geqslant 1$ at a point $z_0' \in E_2$;

3^0 f(z) has a pole $z_0'' \in E_3$;

then the values of f(z) cover in generalized sense the set $H_D(|w|<\rho)$.

Proof. Denote by $\mathscr{F}$ the family of the functions f(z) satisfying the conditions in Theorem 7.8. Assume that there does not exist a number $\rho>0$ having the required property. Then to each positive integer n corresponds a function $f_n(z)\in\mathscr{F}$ such that the values of $f_n(z)$ do not cover in generalized sense the set $H_D(|w|<\frac{1}{n})$. Let $z_n\in E_3$ be the pole of $f_n(z)$. We can find a subsequence $\zeta_k=z_{n_k}(k=1,2,\cdots)$ converging to a point $\zeta_0\in E_3$. Consider a circle $\overline{\Gamma}: |z-\zeta_0|\leqslant r$ belonging to D and such that E_1 and E_2 are exterior to $\overline{\Gamma}$. Such a circle $\overline{\Gamma}$ can be found by (7.30). Let k_0 be a positive integer such that for $k\geqslant k_0$, ζ_k lies in the circle Γ: $|z-\zeta_0|<r$. Then the functions $g_k(z)=f_{n_k}(z)$ $(k\geqslant k_0)$ are holomorphic and univalent in the domain $D_\cdot=D-\overline{\Gamma}$. Since E_1 and E_2 belong to $D_\cdot$, we can apply Theorem 7.6 to $D_\cdot$, E_1, E_2 and conclude that there is a number $\rho_\cdot>0$ such that, for $k\geqslant k_0$, the values of $g_k(z)$ cover in generalized sense the set $H_{D_\cdot}(|w|<\rho_\cdot)$. This contradicts the fact that the values of $g_k(z)$ do not cover in generalized sense the set $H_D(|w|<\frac{1}{n_k})$, if $1/n_k<\rho_\cdot$.

Theorem 7.9. Let D be a domain, $E_i(i=1,2,3)$ three bounded closed sets of points belonging to D satisfying the condition (7.30) and $p\geqslant 2$ an integer. Then there exists a number $\rho>0$ depending only on D, $E_i(i=1,2,3)$ and p, having the following property: If f(z) is a meromorphic and p-valent function in D satisfying the following conditions:

1^0 $f(z)$ has a zero $z_0 \in E_1$ of order p;

2^0 $|f^{(p)}(z_0')| \geqslant 1$ at a point $z_0' \in E_2$;

3^0 $f(z)$ has a pole $z_0'' \in E_3$ of order p;

then the values of $f(z)$ cover in generalized sense the set $H_D(|w| < \rho)$ at least p times.

This theorem is proved in the same way as for Theorem 7.8, except that here we apply Theorem 7.7.

Remark 7.4. In Theorems 7.6 and 7.7, the condition 2^0 ensures that there does not exist a sequence of functions of the corresponding family $\mathscr{F}$ converging locally uniformly to zero in D. Any other condition C having the same effect can be used instead of the condition 2^0. Of course, then the number ρ will depend also on C. For example, we may use the condition: $|f^{(m)}(z_0')| \geqslant 1$ at a point $z_0' \in E_2$, where $m \geqslant 0$ is an integer. Then ρ will depend also on m. The same remark is also applicable to Theorems 7.8 and 7.9 in the sense that any condition C which can be used instead of the condition 2^0 in Theorems 7.6 and 7.7, can also be used instead of the condition 2^0 in Theorems 7.8 and 7.9.

Remark 7.5. In Theorems 7.7 and 7.9, the condition 1^0 may be replaced by "$f(z)$ has k $(k \geqslant 2)$ zeros $a_j \in E_1 (j=1,2,\cdots,k)$ of orders $\alpha_j (j=1,2,\cdots,k)$ respectively with $\sum_{j=1}^{k} \alpha_j = p$ and $|a_j - a_{j'}| \geqslant \delta$ $(1 \leqslant j, j' \leqslant k, j \neq j'; \delta > 0)$." Also in Theorem 7.9, the condition 3^0 may be replaced by "$f(z)$ has h $(h \geqslant 2)$ poles $b_j \in E_3 (j=1,2,\cdots,h)$ of orders $\beta_j (j=1,2,\cdots,h)$ respectively with $\sum_{j=1}^{h} \beta_j = p$ and $|b_j - b_{j'}| \geqslant \eta$ $(1 \leqslant j, j' \leqslant h, j \neq j'; \eta > 0)$." Here k, $\alpha_j (j=1,2,\cdots,k)$, δ and h, $\beta_j (j=1,2,\cdots,h)$, η are given in advance; the number ρ will also depend on them.

8

Q_m-NORMAL FAMILIES OF MEROMORPHIC FUNCTIONS

We have defined in Chapter 1 the notions of C_0-point, C_0-sequence and normal family of meromorphic functions, and in Chapter 5 the notions of C_1-point, C_1-sequence and quasi-normal family of meromorphic functions. Let $S: f_n(z)$ $(n=1,2,\cdots)$ be a sequence of meromorphic functions in a domain D. Denote by E the set of nonC_0-points of S in D and by E' the set of the points of accumulation in D of E. Then $E=\varnothing$, if S is a C_0-sequence in D, and $E'=\varnothing$, if S is a C_1-sequence in D. It is natural to go still further and introduce notions corresponding respectively to the conditions $E''=\varnothing$, $E'''=\varnothing$, $\cdots$. In what follows we shall define for each integer $m\geqslant 0$, the notions C_m-point, C_m-sequence and Q_m-normal family of meromorphic functions, corresponding to the condition $E^{(m)}=\varnothing$. In particular a Q_0-normal family of meromorphic functions is a normal family of meromorphic funtions and a Q_1-normal family of meromorphic functions is a quasi-normal family of meromorphic functions, defined respectively in Chapter 1 and Chapter 5. Here Q is the first letter of the prefix quasi—.

8. 1. SOME NOTIONS AND THEOREMS

Definition 8. 1. Let S: $f_n(z)$ $(n=1,2,\cdots)$ be a sequence of meromorphic functions in a domain D and z_0 a point of D. We say that z_0 is a C_0-point of S, if there is a circle $\Gamma: |z-z_0|<r$ belonging to D such that S is uniformly convergent in Γ with respect to the spherical distance. (see Definition 1. 4) Next we say that z_0 is a C_1-point of S, if there is a circle $|z-z_0|<r$ belonging to D such that each point of the domain $0<|z-z_0|<r$ is a C_0-point of S. (see Definition 5. 1) Similarly we say that z_0 is a C_2-point of S, if there is a circle $|z-z_0|<r$ belonging to D such that each point of the domain $0<|z-z_0|<r$ is a C_1-point of S. In gener-

al, for an integer $m \geqslant 1$, after having defined the notion of C_{m-1}-point of S, we say that z_0 is a C_m-point of S, if there is a circle $|z-z_0|<r$ belonging to D such that each point of the domain $0<|z-z_0|<r$ is a C_{m-1}-point of S. In this way, the notion of C_m-point of S is defined for each integer $m \geqslant 0$.

Lemma 8.1. If, for an integer $m \geqslant 0$, z_0 is a C_m-point of S, then the following assertions hold:

1^0 z_0 is a C_{m+1}-point of S.

2^0 There is a circle $|z-z_0|<r$ belonging to D such that each point of this circle is a C_m-point of S.

3^0 z_0 is a C_m-point of every subsequence of S.

This lemma is easily proved by mathematical induction.

Definition 8.2. $m \geqslant 0$ being an integer, we say that a sequence S of meromorphic functions in a domain D is a C_m-sequence in D, if each point of D is a C_m-point of S.

Lemma 8.2. If, for an integer $m \geqslant 0$, S is a C_m-sequence in D, then S is a C_{m+1}-sequence in D, and every subsequence of S is a C_m-sequence in D.

This lemma follows immediately from Lemma 8.1.

Definition 8.3. Let S be a sequence of meromorphic functions in a domain D and $m \geqslant 0$ an integer. A point z_0 of D is said to be a nonC_m-point of S, if z_0 is not a C_m-point of S.

Theorem 8.1. Let S be a sequence of meromorphic functions in a domain D. If, for an integer $m \geqslant 0$, S is a C_m-sequence in D, then the set E of nonC_0-points of S in D is at most enumerable.

Proof. For the case $m=0$, the set E is empty. For the case $m=1$, the set E has no point of accumulation in D, hence by Lemma 5.6, the set E is at most enumerable. To complete the proof, suppose that Theorem 8.1 is true for an integer $m \geqslant 1$ and consider a C_{m+1}-sequence S of meromorphic functions in a domain D. Let σ be the set of nonC_m-points of S in D. Then σ has no point of accumulation in D, and hence is at most enumerable by Lemma 5.6. On the other hand, S is a C_m-sequence in the domain $D_1=D\text{-}\sigma$. Consequently, by assumption, the set E_1 of the nonC_0-points of S in D_1 is at most enumerable. Evidently the set $E\subset\sigma\cup E_1$, hence E is at most enumeralbe. Thus Theorem 8.1 is also true for the integer $m+1$.

Definition 8.4. let E be a set of points in a domain D. Denote by E' the set of the points of accumulation of E in D, by E'' the set of the points of accumulation of E' in D, in general, by $E^{(m)}$ the set of the points of accumulation of $E^{(m-1)}$ in D. If, for an integer $m\geqslant 0$, the sets $E^{(j)}(j=0,1,\cdots,m;E^{(0)}=E)$ are all nonempty, we say that the set E has the property W_m with respect to D.

Theorem 8.1 can be precised as follows:

Theorem 8.2. Let S be a sequence of meromorphic functions in a domain D, and $m\geqslant 0$ an integer. In order that S is a C_m-sequence in D, it is necessary and sufficient that the set of nonC_0-points of S in D does not have the property W_m with respect to D.

Proof. The necessity of the condition can be proved as follows: For $m=0,1$, the necessity is evident. Now assume that the condition is necessary for an integer $m\geqslant 1$. Consider a C_{m+1}-sequence S of meromorphic functions in a domain D, and suppose that the set E of the nonC_0-points of S in D has the property W_{m+1} with respect to D. Let z_0 be a point of $E^{(m+1)}$ and $|z-z_0|<r$ a circle belonging to D, such that in the domain $d:0<|z-z_0|<r$ the sequence S is a C_m-sequence. Then, since the sets $E^{(m)}\cap d$, $E^{(m-1)}\cap d,\cdots,E\cap d$ are all nonempty, the set $E\cap d$ has the property W_m with respect to d. We are thus led to a contradition with the assumption.

The sufficiency of the condition is also proved by mathematical induction as follows: For $m=0,1$, the sufficiency is evident. Assume that the condition is sufficient for an integer $m\geqslant 1$. Suppose that a sequence S of meromorphic functions in a domain D is not a C_{m+1}-sequence in D. Then there is a point z_0 of D, which is not a C_{m+1}-point of S. Consider a circle $|z-z_0|<r$ belonging to D. Then S is not a C_m-sequence in the domain $d:0<|z-z_0|<r$. Hence, by assumption, the set of nonC_0-points of S in d has the property W_m with respect to d.

Consequently the set E of the nonC_0-points of S in D has the property W_m with respect to D and the set $E^{(m)} \cap d$ is nonempty. Since r may be arbitrarily small, $z_0 \in E^{(m+1)}$. Hence E has the property W_{m+1} with respect to D.

Definition 8.5. Let $\mathscr{F}$ be a family of meromorphic functions in a domain D and $m \geqslant 0$ an integer. We say that the family $\mathscr{F}$ is Q_m-normal in D, if from every sequence of functions of the family $\mathscr{F}$, we can extract a subsequence which is a C_m-sequence in D. $\mathscr{F}$ is said to be Q_m-normal at a point z_0 of D, if there is a circle $\Gamma: |z-z_0|<r$ belonging to D such that $\mathscr{F}$ is Q_m-normal in Γ.

In particular, a Q_0-normal family is a normal family and a Q_1-normal family is a quasi-normal family, according respectively to Definition 1.5 and Definition 5.2.

Evidently according to Definition 8.5, if $\mathscr{F}$ is Q_m-normal in D, then $\mathscr{F}$ is Q_m-normal at each point of D. Conversely we have the following theorem:

Theorem 8.3. Let $\mathscr{F}$ be a family of meromorphic functions in a domain D and $m \geqslant 0$ an integer. If $\mathscr{F}$ is Q_m-normal at each point of D, then $\mathscr{F}$ is Q_m-normal in D.

Proof. By Theorems 1.4 and 5.1, we know already that Theorem 8.3 is true in the cases $m=0,1$. In general, Theorem 8.3 is proved in the same way as for Theorem 1.4. In order to get the proof of Theorem 8.3, it is sufficient to replace in the proof of Theorem 1.4, the word "normal" by "Q_m-normal" the term "C_0-sequence" by "C_m-sequence" and the term "C_0-point" by "C_m-point".

Lemma 8.3. If, for an integer $m \geqslant 0$, a family $\mathscr{F}$ of meromorphic functions in a domain D is Q_m-normal in D, then $\mathscr{F}$ is Q_{m+1}-normal in D.

This lemma is an immediate consequence of Lemma 8.2. Conversely we have the following theorem:

Theorem 8.4. Let $\mathscr{F}$ be a family of meromorphic functions in a domain D and $m \geqslant 1$ an integer. Assume that the following conditions are satisfied:

1^0 $\mathscr{F}$ is Q_m-normal in D.

2^0 There is a number δ $(0<\delta<1)$ such that for each function $f(z) \in \mathscr{F}$,

there exist two values $a_j(f)\in\hat{C}$ $(j=1,2)$ such that $|a_1(f),a_2(f)|>\delta$ and that the function $f(z)$ does not take the values $a_j(z)$ $(j=1,2)$ in D.
Then the family $\mathscr{F}$ is Q_0-normal in D.

Proof. By Theorem 5. 2, Theorem 8. 4 is true for the case $m=1$. Now assume that Theorem 8. 4 holds for an integer $m\geqslant 1$. Consider a Q_{m+1}-normal family $\mathscr{F}$ of meromorphic functions in a domain D, satisfying the condition 2^0 in Theorem 8. 4. Let S be a sequence of functions of $\mathscr{F}$. Then from S we can extract a subsequence S_1 which is a C_{m+1}-sequence in D. Denote by σ the set of the nonC_m-points of S_1 in D. By Definition 8. 1, σ has no point of accumulation in D. In the domain $\Delta=D-\sigma$, S_1 is a C_m-sequence. À fortiori, S_1 constitutes a Q_m-normal family in Δ. Since S_1 satisfies the condition 2^0 in Theorem 8. 4, so by assumption, S_1 is Q_0-normal in Δ. From the sequence S_1 we can then extract a subsequence S_2 which is a C_0-sequence in Δ. Then S_2 is a C_1-sequence in D and, à fortiori, constitutes a Q_1-normal family in D. Again since S_2 satisfies the condition 2^0 in Theorem 8. 4, so by Theorem 5. 2, S_2 is Q_0-normal in D. Finally from S_2 we can extract a subsequence S_3 which is a C_0-sequence in D. This proves that $\mathscr{F}$ is Q_0-normal in D and hence Theorem 8. 4 also holds for the integer $m+1$.

From Theorem 8. 4, we deduce the following corollary:

Corollary 8. 1. Let $S: f_n(z)$ $(n=1,2,\cdots)$ be a sequence of meromorphic functions in a domain D and $m\geqslant 1$ an integer. Assume that the following conditions are satisfied:

1^0 S is a C_m-sequence in D.

2^0 There is a number $\delta(0<\delta<1)$ such that for each integer $n\geqslant 1$, there exist two values $a_{nj}\in\hat{C}$ $(j=1,2)$ such that $|a_{n_1},a_{n_2}|>\delta$ and that the function $f_n(z)$ does not take the values $a_{nj}(j=1,2)$ in D.

Then the sequence S is a C_0-sequence in D.

Proof. By the condition 1^0 in Corollary 8. 1, S considered as a family of functions is, à fortiori, Q_m-normal in D. Next by the condition 2^0 in Corollary 8. 1, S is Q_0-normal in D. On the other hand, by the condition 1^0 in Corollary 8. 1 and Theorem 8. 1, S has a C_0-point z_0 in D. Consequently we can find a circle $|z-z_0|<r$ belonging to D, in which S is a C_0-sequence. Hence we can apply Theo-

rem 6. 2 and conclude that S is a C_0-sequence in D.

8. 2. AN EXISTENCE THEOREM

In view of Lemma 8. 3, it is natural to ask the following question: Given arbitrarily an integer $m \geqslant 1$, can we always find a family of meromorphic functions in a domain D, which is Q_m-normal in D, without being Q_{m-1}-normal in D? We are going to give an affirmative answer to this question. First of all, we point out that, for this purpose, it is sufficient to prove the following theorem:

Theorem 8. 5. Given any integer $m \geqslant 1$, we can find a sequence S of meromorphic functions in a domain D, such that S is a C_m-sequence in D, without being a C_{m-1}-sequence in D, and that the same is true for every subsequence of S.

In fact, such a sequence S of meromorphic functions in a domain, considered as a family of meromorphic functions in the same domain, evidently satisfies the required conditions in the above question.

To prove Theorem 8. 5, consider first the case $m=1$. In this case, the construction of S is known and easy. In fact, the sequence of functions

$$f_n(z) = nz \ (n = 1, 2, \cdots)$$

are holomorphic in **C** and has only one nonC_0-point, namely the point $z=0$. Hence this is a C_1-sequence in **C**, without being a C_0-seqence in **C**. This is also true for each of its subsequences.

In order to prove Theorem 8. 5 for the case $m \geqslant 2$, we need the following two lemmas:

Lemma 8. 4. Let E be a set of points of **C** satisfying the following conditions:

1^0 E is infinite and enumerable.

2^0 $E \cap E'$ is empty.

Then we can construct a sequence S of holomorphic functions in **C** such that the set $E \cup E'$ is precisely the set of nonC_0-points of S in **C**, and that the same is true for every subsequence of S.

Proof. Since E is enumerable, it can be represented by a sequence of points $a_n (n=1,2,\cdots)$. By condition 2^0 in Lemma 8. 4, to each point a_n corresponds a number $0<\delta_n<1$ such that the domain $0<|z-a_n|<2\delta_n$ contains no point of E, consequently

$$|a_m-a_n| \geqslant 2\delta_n \ for \ m \neq n. \tag{8.1}$$

We may suppose that $\delta_{n+1}<\delta_n (n=1,2,\cdots)$. Next take two sequences of numbers $p_n (n=1,2,\cdots)$ and $d_n (n=1,2,\cdots)$ such that

$$p_n > 0 \quad (n = 1,2,\cdots), \ \lim_{n\to+\infty} p_n = +\infty \tag{8.2}$$

$$0 < d_n < 1 \quad (n = 1,2,\cdots), \ \lim_{n\to+\infty} d_n = 0 \tag{8.3}$$

and construct a sequence of functions $S: f_n(z) (n=1,2,\cdots)$ as follows:

$$f_1(z) = \frac{p_1}{d_1\delta_1}(z-a_1)$$

$$f_2(z) = \frac{p_2}{d_2^2\delta_2^2}(z-a_1)(z-a_2)$$

$$\cdots\cdots\cdots$$

$$f_n(z) = \frac{p_n}{d_n^n\delta_n^n}(z-a_1)(z-a_2)\cdots(z-a_n)$$

$$\cdots\cdots\cdots\cdots\cdots\cdots$$

Consider the point a_1 and a circle $c_1: |z-a_1|<\delta$ with $\delta<\delta_1$. By (8. 1), if $m\neq 1$, then in c_1 we have

$$|z-a_m| \geqslant |a_1-a_m|-|z-a_1| > 2\delta_1-\delta > 2\delta_1-\delta_1 = \delta_1.$$

It follows that in c_1 we have

$$|f_n(z)| \geqslant \frac{p_n}{\delta_1^n}|(z-a_1)(z-a_2)\cdots(z-a_n)| \geqslant \frac{p_n}{\delta_1}|z-a_1| \geqslant p_n|z-a_1|.$$

By (8. 2), this inequality shows that a_1 is a nonC_0-point of S. In general, consider a point a_n and a circle c_n: $|z-a_n|<\delta$ with $\delta<\delta_n$. By (8.1), if $m\neq n$, then in c_n we have

$$|z-a_m| \geqslant |a_n-a_m|-|z-a_n| > 2\delta_n-\delta > 2\delta_n-\delta_n = \delta_n.$$

It follows that, for any integer $k\geqslant 0$, we have in c_n,

$$|f_{n+k}(z)| \geqslant p_{n+k}|z-a_n|, \tag{8.4}$$

and hence a_n is a nonC_0-point of S.

Thus we have shown that each point of E is a nonC_0-point of S. Of course each point of E' (if E' is nonempty) is also a nonC_0-point of S.

Now consider a point $z_0\notin E\cup E'$. Then we can find a circle $|z-z_0|<r$ which contains no point of E. Consider the circle c: $|z-z_0|<r/2$. For $m\geqslant 1$, we have in c,

$$|z-a_m| \geqslant |z_0-a_m|-|z-z_0| > r-\frac{r}{2} = \frac{r}{2}.$$

Let N be a positive integer such that $d_n<r/2$ for $n\geqslant N$. Then, for $n\geqslant N$, we have in c,

$$|f_n(z)| > p_n.$$

Hence z_0 is a C_0-point of S.

It is then proved that the set $E\cup E'$ is precisely the set of nonC_0-points of S. This is also true for any subsequence $f_{n_j}(z)(j=1,2,\cdots)$ of S, because, from (8.4), we have in c_n,

$$|f_{n_j}(z)| \geqslant p_{n_j}|z-a_n|,$$

provided that j is sufficiently large.

Lemma 8.5. Let $p\geqslant 1$ be an integer and $a<b$ be two real numbers. Then we can find a set of points E of the closed interval $[a,b]$ satisfying the following

conditions:

1^0 E is infinite and enumerable.

2^0 $E \cap E'$ is empty.

3^0 The set $E \cup E'$ has the property W_p with respect to **C** and $(E \cup E')^{(p+1)}$ is empty.

Proof. When $p=1$, it is sufficient to take the sequence of points

$$E: x_n = a + \frac{1}{n}(b-a) \ (n = 1,2,\cdots). \tag{8.5}$$

Evidently E satisfies the conditions 1^0 and 2^0. Moreover, we have

$$E' = (a), \ (E \cup E')' = (E \cup (a))' = a,$$

hence the condition 3^0 is also satisfied.

Now suppose that Lemma 8.5 is proved for an integer $p \geqslant 1$. Consider the integer $p+1$ and a closed interval $[a,b]$. Let $x_n (n=1,2,\cdots)$ be defined by (8.5) and consider the sequence of closed intervals: $I_j: [x_{2j}, x_{2j-1}] (j=1,2,\cdots)$. By hypothesis, to each j corresponds a set of points E_j of the closed interval I_j satisfying the conditions $1^0, 2^0, 3^0$ for the integer p. Consider the set of points $E = \bigcup_{j=1}^{\infty} E_j$ of the closed interval $[a,b]$.

It is easy to see that E satisfies the conditions 1^0 and 2^0. On the other hand, it is also easy to see that

$$E' = (\bigcup_{j=1}^{\infty} E_j') \cup (a)$$

which implies that

$$E \cup E' = (\bigcup_{j=1}^{\infty} H_j) \cup (a),$$

where $H_j = E_j \cup E_j'$. Since $H_j \subset I_j$, we see that

$$(E \cup E')' = (\bigcup_{j=1}^{\infty} H_j)' \cup (a) = (\bigcup_{j=1}^{\infty} H_j') \cup (a).$$

Then we have successively

$$(E\cup E')''=(\bigcup_{j=1}^{\infty}H_j'')\cup(a),$$
$$\cdots\cdots\cdots\cdots\cdots\cdots\cdots\cdots\cdots\cdots$$
$$(E\cup E')^{(p)}=(\bigcup_{j=1}^{\infty}H_j^{(p)})\cup(a).$$

Since $H_j^{(p+1)}$ is empty $(j=1,2,\cdots)$, we see that

$$(E\cup E')^{(p+1)}=(a).$$

Consequently the set $E\cup E'$ has the property W_{p+1} with respect to $\mathbf{C}$ and $(E\cup E')^{(p+2)}$ is empty. Therefore Lemma 8. 5 is also true for the integer $p+1$.

Now let us return to the proof of Theorem 8. 5 for the case $m\geqslant 2$. Given an integer $m\geqslant 2$, consider the integer $p=m-1\geqslant 1$. By Lemma 8. 5, we can find a set of points E of $\mathbf{C}$, satisfying the conditions $1^0,2^0,3^0$ in Lemma 8. 5. From the first two conditions, by Lemma 8. 4, we can construct a sequence S of holomorphic functions in $\mathbf{C}$ such that the set $E\cup E'$ is precisely the set of nonC_0-points of S in $\mathbf{C}$ and that the same is true for every subsequence of S. Since the set $E\cup E'$ satisfies the condition 3^0 in Lemma 8. 5, then by Theorem 8. 2, S is a C_{p+1}-sequence in $\mathbf{C}$, without being a C_p-sequence in $\mathbf{C}$, and the same is true for every subsequence of S. Since $p+1=m$, $p=m-1$, S satisfies the conditions required by Theorem 8. 5 with $D=\mathbf{C}$.

8.3. NECESSARY AND SUFFICIENT CONDITIONS FOR Q_m-NORMALITY

Definition 8. 6. Let $S: f_n(z)\,(n=1,2,\cdots)$ be a sequence of meromorphic functions in a domain D and z_0 a point of D. We say that z_0 is a μ_1-point of S, if for each circle $\overline{\gamma}: |z-z_0|\leqslant r$ belonging to D, we have

$$\lim_{n\to+\infty}\max_{z\in\overline{\gamma}}\partial(z,f_n)=+\infty.$$

z_0 is called a μ_2-point of S, if for each circle $\gamma: |z-z_0|<r$ belonging to D, S has a

μ_1-point z' in the domain $0<|z-z_0|<r$. In general, $m\geqslant 2$ being an integer, z_0 is called a μ_m-point of S, if for each circle $\Gamma: |z-z_0|<r$ belonging to D, S has a μ_{m-1}-point in the domain $0<|z-z_0|<r$.

Let S' be a subsequence of S. Obviously if z_0 is a μ_1-point of S, then z_0 is a μ_1-point of S'. By mathematical induction, we see that in general, $m\geqslant 1$ being an integer, if z_0 is a μ_m-point of S, then z_0 is a μ_m-point of S'.

Theorem 8. 6. Let $\mathscr{F}$ be a family of meromorphic functions in a domain D. In order that $\mathscr{F}$ be Q_0-normal (or normal) in D, it is necessary and sufficient that every sequence S: $f_n(z)(n=1,2,\cdots)$ of functions of $\mathscr{F}$ has no μ_1-point in D.

Proof. By Theorem 1. 6, the condition is evidently necessary. To prove that the condition is also sufficient, assume that it is satisfied and consider a sequence S: $f_n(z)(n=1,2,\cdots)$ of functions of $\mathscr{F}$. Then the sequence of functions $\partial(z,f_n)(n=1,2,\cdots)$ must be locally uniformly bounded in D. In fact, otherwise there is a point $z_0\in D$ such that we can not find a circle $\Gamma: |z-z_0|<r$ belonging to D and a number $M>0$ such that for $n\geqslant 1$, we have

$$\partial(z,f_n)\leqslant M$$

in Γ. Take two sequences of numbers $r_k>0(k=1,2,\cdots)$ and $M_k>0(k=1,2,\cdots)$ such that

$$\lim_{k\to+\infty} r_k=0,\quad \lim_{k\to+\infty} M_k=+\infty$$

and that the circles $\overline{\Gamma}_k: |z-z_0|\leqslant r_k(k=1,2,\cdots)$ belong to D. Then there is an integer $n_1\geqslant 1$ such that

$$\max_{z\in\overline{\Gamma}_1}\partial(z,f_{n_1})>M_1.$$

Next there is an integer $n_2>n_1$ such that

$$\max_{z\in\overline{\Gamma}_2}\partial(z,f_{n_2})>M_2.$$

Continuing in this way, we get successively a sequence of integers $n_k(k=1,2,\cdots;n_{k+1}>n_k)$ such that for $k\geqslant 1$, we have

$$\max_{z\in\bar{\gamma}_k}\partial(z,f_{n_k})>M_k.$$

Consider a circle $\bar{\Gamma}:|z-z_0|\leqslant r$ belonging to D. Let $k_0\geqslant 1$ be an integer such that $r_k<r$ for $k\geqslant k_0$. Then for $k\geqslant k_0$ we have

$$M_k<\max_{z\in\bar{\gamma}_k}\partial(z,f_{n_k})\leqslant\max_{z\in\bar{\Gamma}}\partial(z,f_{n_k}),$$

hence

$$\lim_{k\to+\infty}\max_{z\in\bar{\Gamma}}\partial(z,f_{n_k})=+\infty.$$

z_0 is then a μ_1-point of the sequence of functions $f_{n_k}(z)(k=1,2,\cdots)$, incompatible with the hypothesis.

So we have shown that the sequence of functions $\partial(z,f_n)(n=1,2,\cdots)$ is locally uniformly bounded in **D**. Consequently by Theorem 1.6, the sequence **S**: $f_n(z)$ $(n=1,2,\ldots)$ constitutes a $\mathbf{Q}_0$-normal family in **D** and hence we can extract from S a subsequence S′ which is a C_0-sequence in D.

Definition 8. 7. $m\geqslant 1$ being an integer, let $\mathscr{F}$ be a family of meromorphic functions in a domain D and $\nu\geqslant 0$ an integer. We say that $\mathscr{F}$ is Q_m-normal of order ν at most in D, if from every sequence of functions of the family $\mathscr{F}$, we can extract a subsequence which is a C_m-sequence in D and has at most ν non-C_{m-1}-points in D.

In particular when $\nu=0$, $\mathscr{F}$ is Q_{m-1}-normal in D.

Theorem 8. 7. Let $\mathscr{F}$ be a family of meromorphic functions in a domain D and $\nu\geqslant 0$ an integer. In order that $\mathscr{F}$ be Q_1-normal (or quasi-normal) of order ν at most in D, it is necessary and sufficient that every sequence $S:f_n(z)$ $(n=1,2,\cdots)$ of functions of $\mathscr{F}$ has at most ν μ_1-points in D.

Proof. In the case $\nu=0$, Theorem 8.7 reduces to Theorem 8.6. Consider the case $\nu \geqslant 1$. Assume that the family $\mathscr{F}$ is Q_1-normal of order ν at most in D. Let $S: f_n(z)(n=1,2,\cdots)$ be a sequence of functions of $\mathscr{F}$. Suppose, on the contrary, that S has $\nu+1$ μ_1-points $z_j(j=1,2,\cdots,\nu+1)$ in D. By hypothesis, from S we can extract a subsequence S' which is a C_1-sequence in D and has at most ν nonC_0-points in D. Denote by E the set of nonC_0-points of S' in D. Then in the domain $D'=D-E$, S' is a C_0-sequence and hence constitutes a Q_0-normal family. Among the points $z_j(j=1,2,\cdots,\nu+1)$, there is at least one, for instance $z_1\in D'$. By Theorem 8.6, S' has no μ_1-point in D'. But on the other hand, z_1 is a μ_1-point of S'. So we get a contradiction.

Next assume that the condition of Theorem 8.7 is satisfied. Let S be a sequence of functions of $\mathscr{F}$. If every subsequence of S has no μ_1-point in D, then by Theorem 8.6, S constitutes a Q_0-normal family in D, hence we can extract from S a subsequence S_0 which is a C_0-sequence in D. Otherwise S has a subsequence S' which has a μ_1-point z_1 in D. If every subsequence of S' has no μ_1-point in the domain $D-(z_1)$, then we can extract from S' a subsequence S_1 which is a C_0-sequence in $D-(z_1)$. Otherwise S' has a subsequence S'' which has a μ_1-point z_2 in $D-(z_1)$. Continuing in this way, after at most ν steps, we get a subsequence $S^{(\nu)}$ of S, such that $S^{(\nu)}$ has ν μ_1-points $z_j(j=1,2,\cdots,\nu)$ in D. In the domain $D-(z_1,z_2,\cdots,z_\nu)$, every subsequence of $S^{(\nu)}$ has no μ_1-point, hence we can extract from $S^{(\nu)}$ a subsequence S_ν which is a C_0-sequence in $D-(z_1,z_2,\cdots,z_\nu)$.

Theorem 8.8. Let $\mathscr{F}$ be a family of meromorphic functions in a domain D. In order that $\mathscr{F}$ be Q_1-normal (or quasi-normal) in D, it is necessary and sufficient that every sequence $S: f_n(z)$ $(n=1,2,\cdots)$ of functions of $\mathscr{F}$ has no μ_2-point in D.

Proof. Assume that the family $\mathscr{F}$ is Q_1-normal in D. Let S be a sequence of functions of $\mathscr{F}$. Then from S we can extract a subsequence S' which is a C_1-sequence in D. Suppose, on the contrary, that S has a μ_2-point z_0 in D. Since z_0 is a C_1-point of S', we can find a circle $|z-z_0|<r$ belonging to D such that each point of the domain $d: 0<|z-z_0|<r$ is a C_0-point of S'. S' is then a C_0-sequence in d, and hence constitutes a Q_0-normal family in d. By Theorem 8.6, S' has no μ_1-point in d. But by Definition 8.6, S has a μ_1-point z' in d. Since z' is also a μ_1-point of S', we arrive at a contradiction.

Now assume that the condition of Theorem 8.8 is satisfied. Consider a circle $\overline{\Gamma}: |z-z_0| \leq r$ belonging to D. By Theorem 8.3, it is sufficient to show that the family $\mathscr{F}$ is Q_1-normal in the circle $\Gamma: |z-z_0| < r$. Let S be a sequence of functions of $\mathscr{F}$. Suppose, on the contrary, that we can not extract from S a subsequence which is a C_1-sequence in Γ. Then S must have a subsequence $S^{(1)}$ which has a μ_1-point z_1 in Γ, because otherwise, by Theorem 8.6, S constitutes a Q_0-normal family in Γ and we can extract from S a subsequence which is a C_0-sequence in Γ. For the same reason, $S^{(1)}$ must have a subsequence $S^{(2)}$ which has a μ_1-point z_2 in the domain $\Gamma - (z_1)$. Continuing in this way, we get a sequence of sequences $S^{(k)}(k=1,2,\cdots)$ satisfying the following conditions:

1) For each $k \geq 1$, $S^{(k+1)}$ is a subsequence of $S^{(k)}$.

2) For each $k \geq 1$, $S^{(k)}$ has a μ_1-point z_k in Γ.

3) The sequence of points $z_k(k=1,2,\cdots)$ are distinct.

Let S' be the diagonal sequence of the sequences $S^{(k)}(k=1,2,\cdots)$. Then each point z_k is a μ_1-point of S', because beginning from the k-th term of S', all the following terms of S' belong to $S^{(k)}$. Let z_0 be a limiting point of the sequence of points $z_k(k=1,2,\cdots)$. Then $z_0 \in \overline{\Gamma}$ and z_0 is a μ_2-point of S', incompatible with the hypothesis that the condition of Theorem 8.8 is satisfied.

In general we have the following two theorems:

Theorem 8.9. $m \geq 1$ being an integer, let $\mathscr{F}$ be a family of meromorphic functions in a domain D and $\nu \geq 0$ an integer. In order that $\mathscr{F}$ be Q_m-normal of order ν at most in D, it is necessary and sufficient that every sequence $S: f_n(z)$ $(n=1,2,\cdots)$ of functions of $\mathscr{F}$ has at msot ν μ_m-points in D.

Theorem 8.10. $m \geq 1$ being an integer, let $\mathscr{F}$ be a family of meromorphic functions in a domain D. In order that $\mathscr{F}$ be Q_m-normal in D, it is necessary and sufficient that every sequence $S: f_n(z)$ $(n=1,2,\cdots)$ of functions of $\mathscr{F}$ has no μ_{m+1}-point in D.

Proof. From Theorems 8.7 and 8.8, we know that Theorems 8.9 and 8.10 are true for $m=1$. Assume that Theorems 8.9 and 8.10 hold for an integer $m \geq 1$. Let us first prove the following theorem:

Let $\mathscr{F}$ be a family of meromorphic functions in a domain D and $\nu \geq 0$ an integer. In order that $\mathscr{F}$ be Q_{m+1}-normal of order ν at most in D, it is necessary and sufficient that every sequence of functions of $\mathscr{F}$ has at most ν μ_{m+1}-points in D.

For the case $\nu=0$, this theorem reduces to Theorem 8.10. Consider the case $\nu\geqslant 1$. Assume that $\mathscr{F}$ is Q_{m+1}-normal of order ν at most in D. Let S be a sequence of functions of $\mathscr{F}$. Suppose, on the contrary, that S has $\nu+1$ μ_{m+1}-points z_j $(j=1,2,\cdots,\nu+1)$ in D. From S we can extract a subsequence S′ which is a C_{m+1}-sequence in D and has at most ν nonC_m-points in D. Denote by E the set of nonC_m-points of S′ in D. Then in the domain $D'=D-E$, S′ is a C_m-sequence and hence constitutes a Q_m-normal family. By Theorem 8.10, S′ has no μ_{m+1}-point in D′. But among the points $z_j(j=1,2,\cdots,\nu+1)$ there is at least one, for instance $z_1\in D'$. Since z_1 is also a μ_{m+1}-point of S′, we get a contradiction.

Conversely assume that the condition of the above theorem is satisfied. Let S be a sequence of functions of $\mathscr{F}$. If every subsequence of S has no μ_{m+1}-point in D, then by Theorem 8.10, S constitutes a Q_m-normal family in D, hence we can extract from S a subsequence S_0 which is a C_m-sequence in D. Otherwise S has a subsequence S′ which has a μ_{m+1}-point z_1 in D. If every subsequence of S′ has no μ_{m+1}-point in $D-(z_1)$, then we can extract from S′ a subsequence S_1 which is a C_m-sequence in $D-(z_1)$. As in the proof of Theorem 8.7, continuing in this way, after at most ν steps, we get a subsequence $S^{(\nu)}$ of S, such that $S^{(\nu)}$ has ν μ_{m+1}-points $z_j(j=1,2,\cdots,\nu)$ in D. In the domain $D-(z_1,z_2,\cdots,z_\nu)$, every subsequence of $S^{(\nu)}$ has no μ_{m+1}-point, hence we can extract from $S^{(\nu)}$ a subsequence S_ν which is a C_m-sequence in $D-(z_1,z_2,\cdots,z_\nu)$.

Next let us prove the following theorem:

Let $\mathscr{F}$ be a family of meromorphic functions in a domain D. In order that $\mathscr{F}$ be Q_{m+1}-normal in D, it is necessary and sufficient that every sequence of functions of $\mathscr{F}$ has no μ_{m+2}-point in D.

Assume that $\mathscr{F}$ is Q_{m+1}-normal in D. Let S be a sequence of functions of $\mathscr{F}$. Then from S we can extract a subsequence S′ which is a C_{m+1}-sequence in D. Suppose, on the contrary, that S has a μ_{m+2}-point z_0 in D. Since z_0 is a C_{m+1}-point of S′, we can find a circle $|z-z_0|<r$ belonging to D such that each point of the domain $d: 0<|z-z_0|<r$ is a C_m-point of S′. S′ is then a C_m-sequence in d and hence constitutes a Q_m-normal family in d. By Theorem 8.10, S′ has no μ_{m+1}-point in d. But by Definition 8.6, S has a μ_{m+1}-point z' in d. Since z' is also a μ_{m+1}-point of S′, we get a contradiction.

Conversely assume that every sequence of functions of $\mathscr{F}$ has no μ_{m+2}-point in D. Consider a circle $\overline{\Upsilon}: |z-z_0| \leqslant r$ belonging to D. By Theorem 8.3, it is sufficient to show that the family $\mathscr{F}$ is Q_{m+1}-normal in the circle $\Upsilon: |z-z_0| < r$. Let S be a sequence of functions of $\mathscr{F}$. Suppose, on the contrary, that we can not extract from S a subsequence which is a C_{m+1}-sequence in Υ. Then S must have a subsequence $S^{(1)}$ which has a μ_{m+1}-point z_1 in Υ, because otherwise, by Theorem 8.10, S constitutes a Q_m-normal family in Υ, and we can extract from S a subsequence which is a C_m-sequence in Υ. For the same reason, $S^{(1)}$ must have a subsequence $S^{(2)}$ which has a μ_{m+1}-point z_2 in the domain $\Upsilon-(z_1)$. As in the proof of Theorem 8.8, continuing in this way, we get a sequence of sequences $S^{(k)}(k=1, 2,\cdots)$ satisfying the following conditions:

1) For each $k \geqslant 1$, $S^{(k+1)}$ is a subsequence of $S^{(k)}$.

2) For each $k \geqslant 1$, $S^{(k)}$ has a μ_{m+1}-point z_k in Υ.

3) The sequence of points $z_k(k=1,2,\cdots)$ are distinct.

Let S' be the diagonal sequence of the sequences $S^{(k)}(k=1,2,\cdots)$. Then each point z_k is a μ_{m+1}-point of S'. Let z_0 be a limiting point of the sequence of points $z_k(k=1,2,\cdots)$. Then $z_0 \in \overline{\Upsilon}$ and z_0 is a μ_{m+2}-point of S', incompatible with the hypothesis.

In what follows, we deduce from Theorems 8.7—8.10, some generalizations of Marty's theorem. For this we need to give the following definitions:

Definition 8.8. Consider a domain D. Let $\sigma^{(1)}: z_j(1 \leqslant j \leqslant n)$ be a system of points of D. We say that $\sigma^{(1)}$ satisfies the condition (Γ_1) with respect to D, if either $n=1$ or $n>1$ and the n points $z_j(1 \leqslant j \leqslant n)$ are distinct. Let $\sigma^{(1)}: z_j(j \geqslant 1)$ be a sequence of points of D, we say that $\sigma^{(1)}$ satisfies the condition (Γ_1') with respect to D, if the points $z_j(j \geqslant 1)$ are distinct and the limit $\lim\limits_{j \to +\infty} z_j = \lambda$ exists, with $\lambda = \lambda(\sigma^{(1)}) \in D$.

Let $\sigma^{(2)}: z_{ij}(1 \leqslant i \leqslant n, j \geqslant 1)$ be a double sequence of points of D. We say that $\sigma^{(2)}$ satisfies the condition (Γ_2) with respect to D, if for each $i(1 \leqslant i \leqslant n)$, the sequence $\sigma_i^{(1)}: z_{ij}(j \geqslant 1)$ satisfies the condition (Γ_1') with respect to D and the system $\sigma^{(1)}: \lambda_i = \lambda(\sigma_i^{(1)})\ (1 \leqslant i \leqslant n)$ satisfies the condition (Γ_1) with respect to D. Let $\sigma^{(2)}: z_{ij}(i \geqslant 1, j \geqslant 1)$ be a double sequence of points of D. We say that $\sigma^{(2)}$ satisfies the condition (Γ_2') with respect to D, if for each i $(i \geqslant 1)$, the sequence $\sigma_i^{(1)}: z_{ij}(j \geqslant 1)$ satisfies the condition (Γ_1') with respect to D, and the sequence $\sigma^{(1)}: \lambda_i = \lambda(\sigma_i^{(1)})\ (i \geqslant 1)$ satisfies the condition (Γ_1') with respect to D.

In general, let $\sigma^{(m)}: z_{j_1 j_2 \cdots j_m}(1 \leqslant j_1 \leqslant n, j_2 \geqslant 1, \cdots, j_m \geqslant 1)$ be a m-ple ($m \geqslant 3$) sequence of points of D. We say that $\sigma^{(m)}$ satisfies the condition (Γ_m) with respect to D, if for each system $j_1, j_2, \cdots, j_{m-1}(1 \leqslant j_1 \leqslant n, j_2 \geqslant 1, \cdots, j_{m-1} \geqslant 1)$, the sequence $\sigma^{(1)}_{j_1 j_2 \cdots j_{m-1}}: z_{j_1 j_2 \cdots j_{m-1} j_m}(j_m \geqslant 1)$ satisfies the condition (Γ'_1) with respect to D, and the $(m-1)$-ple sequence $\lambda_{j_1 j_2 \cdots j_{m-1}} = \lambda(\sigma^{(1)}_{j_1 j_2 \cdots j_{m-1}})(1 \leqslant j_1 \leqslant n, j_2 \geqslant 1, \cdots, j_{m-1} \geqslant 1)$ satisfies the condition (Γ_{m-1}) with respect to D. Let $\sigma^{(m)}$: $z_{j_1 j_2 \cdots j_m}(j_1 \geqslant 1, j_2 \geqslant 1, \cdots, j_m \geqslant 1)$ be a m-ple ($m \geqslant 3$) sequence of points of D. We say that $\sigma^{(m)}$ satisfies the condition (Γ'_m) with respect to D, if for each system $j_1, j_2, \cdots, j_{m-1}(j_1 \geqslant 1, j_2 \geqslant 1, \cdots, j_{m-1} \geqslant 1)$, the sequence $\sigma^{(1)}_{j_1 j_2 \cdots j_{m-1}}: z_{j_1 j_2 \cdots j_{m-1} j_m}(j_m \geqslant 1)$ satisfies the condition (Γ'_1) with respect to D, and the $(m-1)$-ple sequence $\lambda_{j_1 j_2 \cdots j_{m-1}} = \lambda(\sigma^{(1)}_{j_1 j_2 \cdots j_{m-1}})(j_1 \geqslant 1, j_2 \geqslant 1, \cdots, j_{m-1} \geqslant 1)$ satisfies the condition (Γ'_{m-1}) with respect to D.

Definition 8. 9. Let $\mathscr{F}$ be a family of meromorphic functions in a domain D and $z_j(j=1,2,\cdots,n)$ a system of points of D. We say that the family $\mathscr{F}$ satisfies the condition (M) with respect to the system $z_j(j=1,2,\cdots,n)$, if there exist circles $\overline{\Gamma}_j: |z-z_j| \leqslant r_j(j=1,2,\cdots,n)$ belonging to D and a number $A>0$ such that for each function $f(z) \in \mathscr{F}$, we have

$$\min_{1 \leqslant j \leqslant n} (\max_{z \in \overline{\Gamma}_j} \partial(z, f)) \leqslant A.$$

Theorem 8. 11. Let $\mathscr{F}$ be a family of meromorphic functions in a domain D and $\nu \geqslant 0$ an integer. In order that $\mathscr{F}$ be Q_1-normal (or quasi-normal) of order ν at most in D, it is necessary and sufficient that $\mathscr{F}$ satisfies the condition (M) with respect to every system of points $z_j(1 \leqslant j \leqslant \nu+1)$ of D satisfying the conditon (Γ_1) with respect to D.

Proof. For the case $\nu=0$, Theorem 8. 11 reduces to Theorem 1. 6. Consider the case $\nu \geqslant 1$. First assume that $\mathscr{F}$ is Q_1-normal of order ν at most in D. Let $z_j(1 \leqslant j \leqslant \nu+1)$ be a system of distinct points of D. Suppose, on the contrary, that $\mathscr{F}$ does not satisfy the conditon (M) with respect to the system $z_j(1 \leqslant j \leqslant \nu+1)$. Take a number $\rho>0$ such that the circles $|z-z_j| \leqslant \rho$ $(1 \leqslant j \leqslant \nu+1)$ belong to D. Then to each integer $n \geqslant 1$, corresponds a function $f_n(z) \in \mathscr{F}$ such that

$$\min_{1\leqslant j\leqslant \nu+1}\{\max_{z\in\overline{\Gamma}_j^{(n)}}\partial(z,f_n)\}>n,$$

where $\overline{\Gamma}_j^{(n)}$ denotes the circle $|z-z_j|\leqslant\frac{\rho}{n}$. Hence

$$\max_{z\in\overline{\Gamma}_j^{(n)}}\partial(z,f_n)>n\ (1\leqslant j\leqslant \nu+1).$$

Consider the point z_1 and a circle $\overline{\Gamma}:|z-z_1|\leqslant r$ belonging to D. Let $n_0\geqslant 1$ be an integer such that for $n\geqslant n_0$ we have $\rho/n<r$. Then for $n\geqslant n_0$, we have

$$\max_{z\in\overline{\Gamma}}\partial(z,f_n)\geqslant\max_{z\in\overline{\Gamma}_1^{(n)}}\partial(z,f_n)>n$$

and hence

$$\lim_{n\to+\infty}\max_{z\in\overline{\Gamma}}\partial(z,f_n)=+\infty.$$

This shows that z_1 is a μ_1-point of the sequence of functions $S:f_n(z)$ $(n=1,2,\cdots)$. Similarly $z_j(j=2,\cdots,\nu+1)$ are also μ_1-points of S. Consequently S has $\nu+1$ μ_1-points in D. But by Theorem 8.12 and the hypothesis, S has at most ν μ_1-points in D. So we get a contradicition.

Next assume that the condition of Theorem 8.11 is satisfied. Let $S:f_n(z)$ $(n=1,2,\cdots)$ be a sequence of functions of $\mathscr{F}$. By Theorem 8.7, it is sufficient to show that S has at most ν μ_1-points in D. Suppose, on the contrary, that S has $\nu+1$ μ_1-points $z_j(1\leqslant j\leqslant\nu+1)$ in D. Take arbitrarily a system of circles $\overline{\Gamma}_j:|z-z_j|\leqslant r_j(1\leqslant j\leqslant\nu+1)$ belonging to D and a number $A>0$, then by Definition 8.6, we have

$$\lim_{n\to+\infty}\max_{z\in\overline{\Gamma}_j}\partial(z,f_n)=+\infty\ (1\leqslant j\leqslant \nu+1).$$

Consequently we can find an integer $N\geqslant 1$ such that

$$\max_{z\in\overline{\Gamma}_j}\partial(z,f_N)>A\ (1\leqslant j\leqslant \nu+1).$$

This contradicts the hypothesis that the condition of Theorem 8. 11 is satisfied.

Theorem 8. 12. Let $\mathscr{F}$ be a family of meromorphic functions in a domain D. In order that $\mathscr{F}$ be Q_1-normal (or quasi-normal) in D, it is necessary and sufficient that for every sequence $z_j(j\geqslant 1)$ of points of D satisfying the condition (Γ_1') with respect to D, we can find an integer $p\geqslant 1$ such that $\mathscr{F}$ satisfies the condition (M) with respect to the system $z_j(1\leqslant j\leqslant p)$.

Proof. First assume that $\mathscr{F}$ is Q_1-normal in D. Let $z_j(j\geqslant 1)$ be a sequence of points of D satisfying the condition (Γ_1') with respect to D. Suppose, on the contrary, that we can not find an integer $p\geqslant 1$ having the required property. Take a circle $\overline{\gamma}_{11}: |z-z_1|\leqslant r_1\leqslant 1$ belonging to D, then there is a function $f_1(z)\in \mathscr{F}$ such that

$$\max_{z\in\overline{\gamma}_{11}} \partial(z,f_1) > 1.$$

Next take two circles $\overline{\gamma}_{2j}: |z-z_j|\leqslant r_2\leqslant 1/2\ (j=1,2)$ belonging to D, then there is a function $f_2(z)\in\mathscr{F}$ such that

$$\min_{1\leqslant j\leqslant 2}\left(\max_{z\in\overline{\gamma}_{2j}} \partial(z,f_2)\right) > 2.$$

In general, take p circles $\overline{\gamma}_{pj}: |z-z_j|\leqslant r_p\leqslant 1/p\ (j=1,2,\cdots,p)$ belonging to D, then there is a function $f_p(z)\in\mathscr{F}$ such that

$$\min_{1\leqslant j\leqslant p}\left(\max_{z\in\overline{\gamma}_{pj}} \partial(z,f_p)\right) > p.$$

Obviously for any point z_j and any circle $\overline{\gamma}: |z-z_j|\leqslant r$ belonging to D, we have

$$\lim_{p\to+\infty}\max_{z\in\overline{\gamma}} \partial(z,f_p) = +\infty.$$

So by Definition 8. 6, the points $z_j(j\geqslant 1)$ are all μ_1-points of the sequence $S: f_p(z)\ (p=1,2,\cdots)$ of functions of $\mathscr{F}$. Since the sequence of points $z_j(j\geqslant 1)$ satisfies the condition (Γ_1') with respect to D, z_j converges to a point $\lambda\in D$. λ is then a μ_2-point of S. But by the hypothesis and Theorem 8.8, S has no μ_2-point in D. Thus we get a contradiction.

Now conversely assume that the condition of Theorem 8.12 is satisfied. Let $S: f_n(z)$ $(n=1,2,\cdots)$ be a sequence of functions of $\mathscr{F}$. By Theorem 8.8, it is sufficient to show that S has no μ_2-point in D. In fact, suppose, on the contrary, that S has a μ_2-point ζ in D. Then there is a sequence of distinct μ_1-points $z_j(j\geqslant 1)$ of S, such that $\lim\limits_{j\to+\infty} z_j=\zeta$. The sequence of points $z_j(j\geqslant 1)$ satisfies then the condition (Γ_1') with respect to D. It is easy to see that for this sequence of points $z_j(j\geqslant 1)$, we can not find an integer $p\geqslant 1$ having the required property in Theorem 8.12, incompatible with the hypothesis.

Theorem 8.13. Let $\mathscr{F}$ be a family of meromorphic functions in a domain D and $\nu\geqslant 0$ an integer. In order that $\mathscr{F}$ be Q_2-normal of order ν at most in D, it is necessary and sufficient that for every double sequence $z_{ij}(1\leqslant i\leqslant \nu+1, j\geqslant 1)$ of points of D satisfying the condition (Γ_2) with respect to D, we can find an integer $p\geqslant 1$ such that $\mathscr{F}$ satisfies the condition (M) with respect to the system z_{ij} $(1\leqslant i\leqslant \nu+1, 1\leqslant j\leqslant p)$.

Proof. For the case $\nu=0$, Theorem 8.13 reduces to Theorem 8.12. Consider the case $\nu\geqslant 1$. First assume that $\mathscr{F}$ is Q_2-normal of order ν at most in D. Let $\sigma^{(2)}: z_{ij}(1\leqslant i\leqslant \nu+1, j\geqslant 1)$ be a double sequence of points of D satisfying the condition (Γ_2) with respect to D. Suppose, on the contrary, that we can not find an integer $p\geqslant 1$ having the required property. $p\geqslant 1$ being an integer, take a number $r_p(0<r_p<1/p)$ such that the circles $\overline{\gamma}_{ij}: |z-z_{ij}|\leqslant r_p(1\leqslant i\leqslant \nu+1,\ 1\leqslant j\leqslant p)$ belong to D, then there is a function $f_p(z)\in\mathscr{F}$ such that

$$\min_{\substack{1\leqslant i\leqslant \nu+1\\ 1\leqslant j\leqslant p}} (\max_{z\in\overline{\gamma}_{ij}} \partial(z, f_p)) > p.$$

It is easy to see that each point z_{ij} of $\sigma^{(2)}$ is a μ_1-point of the sequence of functions $S: f_p(z)$ $(p=1,2,\cdots)$. Since $\sigma^{(2)}$ satisfies the condition (Γ_2) with respect to D, for each $i(1\leqslant i\leqslant \nu+1)$, the limit $\lim\limits_{j\to+\infty} z_{ij}=\zeta_i$ exists and the points $\zeta_i(1\leqslant i\leqslant \nu+1)$ are distinct and belong to D. The points $\zeta_i(1\leqslant i\leqslant \nu+1)$ are then μ_2-points of S, incompatible with the hypothesis, by Theorem 8.9.

Now assume that the condition of Theorem 8.13 is satisfied and let us show that $\mathscr{F}$ is Q_2-normal of order ν at most in D. Let $S: f_n(z)(n=1,2,\cdots)$ be a sequence of functions of $\mathscr{F}$. By Theorem 8.9, it is sufficient to show that S has at most ν μ_2-points in D. In fact, suppose, on the contrary, that S has $\nu+1$ μ_2-points $\zeta_i(1\leqslant i\leqslant \nu+1)$ in D. Then for each i $(1\leqslant i\leqslant \nu+1)$, we can find a sequence $z_{ij}(j\geqslant 1)$ of distinct μ_1-points of S, such that

$$\lim_{j\to+\infty} z_{ij} = \zeta_i.$$

In this way we get a double sequence $\sigma^{(2)}: z_{ij}(1\leqslant i\leqslant \nu+1, j\geqslant 1)$ satisfying the condition (Γ_2) with respect to D. Since the points $z_{ij}(1\leqslant i\leqslant \nu+1, j\geqslant 1)$ are all μ_1-points of S, it is easy to see that we can not find an integer $p\geqslant 1$ such that $\mathscr{F}$ satisfies the condition (M) with respect to the system $z_{ij}(1\leqslant i\leqslant \nu+1, 1\leqslant j\leqslant p)$, incompatible with the hypothesis.

Theorem 8. 14. Let $\mathscr{F}$ be a family of meromorphic functions in a domain D. In order that $\mathscr{F}$ be Q_2-normal in D, it is necessary and sufficient that for every double sequence $z_{ij}(i\geqslant 1, j\geqslant 1)$ of points of D satisfying the condition (Γ_2') with respect to D, we can find an integer $p\geqslant 1$ such that $\mathscr{F}$ satisfies the condition (M) with respect to the system $z_{ij}(1\leqslant i\leqslant p, 1\leqslant j\leqslant p)$.

Proof. First assume that $\mathscr{F}$ is Q_2-normal in D. Let $\sigma^{(2)}: z_{ij}(i\geqslant 1, j\geqslant 1)$ be a double sequence of points of D satisfying the condition (Γ_2') with respect to D. Suppose, on the contrary, that we can not find an integer $p\geqslant 1$ having the required property. Then as in the proof of Theorem 8. 13, we can find a sequence of functions $S: f_p(z)$ $(p=1,2,\cdots)$ of $\mathscr{F}$ such that each point z_{ij} of $\sigma^{(2)}$ is a μ_1-point of S. Since $\sigma^{(2)}$ satisfies the condition (Γ_2') with respect to D, by Definition 8. 8, for each i $(i\geqslant 1)$, the limit $\lim\limits_{j\to+\infty} z_{ij}=\zeta_i$ exists, $\zeta_i\in D$ and $\lim\limits_{i\to+\infty}\zeta_i=Z$ exists with $Z\in D$. The points $\zeta_i(i\geqslant 1)$ are then μ_2-points of S and Z is a μ_3-point of S. This is incompatible with the hypothesis, by Theorem 8. 10.

Now conversely assume that the condition of Theorem 8. 14 is satisfied and let us show that $\mathscr{F}$ is Q_2-normal in D. Let $S: f_n(z)$ $(n=1,2,\cdots)$ be a sequence of functions of $\mathscr{F}$. By Theorem 8. 10, it is sufficient to show that S has no μ_3-point in D. In fact, suppose, on the contrary, that S has a μ_3-point Z in D. Then S has a sequence of distinct μ_2-points $\zeta_i(i\geqslant 1)$ in D such that

$$\lim_{i\to+\infty} \zeta_i = Z.$$

Next for each i ($i\geqslant 1$), there is a sequence of distinct μ_1-points $z_{ij}(j\geqslant 1)$ in D of S, such that

$$\lim_{j\to+\infty} z_{ij} = \zeta_i.$$

In this way we get a double sequence $\sigma^{(2)}: z_{ij}(i\geqslant 1, j\geqslant 1)$ satisfying the condition (Γ'_2) with respect to D. Since the points $z_{ij}(i\geqslant 1, j\geqslant 1)$ are all μ_1-points of S, it is easy to see that for the double sequence $\sigma^{(2)}$ we can not find an integer $p\geqslant 1$ having the required property in Theorem 8. 14. So we get a contradiction with the hypothesis.

By the same method we can prove the following two general theorems:

Theorem 8. 15. $m\geqslant 2$ being an integer, let $\mathscr{F}$ be a family of meromorphic functions in a domain D and $\nu\geqslant 0$ an integer. In order that $\mathscr{F}$ be Q_m-normal of order ν at most in D, it is necessary and sufficient that for every m-ple sequence $z_{j_1 j_2 \cdots j_m}(1\leqslant j_1\leqslant \nu+1, j_2\geqslant 1, \cdots, j_m\geqslant 1)$ of points of D satisfying the condition (Γ_m) with respect to D, we can find an integer $p\geqslant 1$ such that $\mathscr{F}$ satisfies the condition (M) with respect to the system $z_{j_1 j_2 \cdots j_m}(1\leqslant j_1\leqslant \nu+1, 1\leqslant j_2\leqslant p, \cdots, 1\leqslant j_m \leqslant p)$.

Theorem 8. 16. $m\geqslant 2$ being an integer, let $\mathscr{F}$ be a family of meromorphic functions in a domain D. In order that $\mathscr{F}$ be Q_m-normal in D, it is necessary and sufficient that for every m-ple sequence $z_{j_1 j_2 \cdots j_m}(j_1\geqslant 1, j_2\geqslant 1, \cdots, j_m\geqslant 1)$ of points of D satisfying the condition (Γ'_m) with respect to D, we can find an integer $p\geqslant 1$ such that $\mathscr{F}$ satisfies the condition (M) with respect to the system $z_{j_1 j_2 \cdots j_m}(1\leqslant j_1\leqslant p, 1\leqslant j_2\leqslant p, \cdots, 1\leqslant j_m\leqslant p)$.

8. 4. CONDITIONS AND CRITERIONS OF Q_m-NORMALITY

Definition 8. 10. $m\geqslant 0$ being an integer, a condition K is said to be a condition of Q_m-normality, if it has the following properties:

1^0 If a meromorphic function f(z) in a domain D satisfies the condition K in

D, then f(z) satisfies the condition K in every subdomain of D.

2^0 If each function of a family $\mathscr{F}$ of meromorphic functions in a domain D satisfies the condition K in D, then the family $\mathscr{F}$ is Q_m-normal in D.

Definition 8.11. $m \geqslant 1$ and $\nu \geqslant 0$ being two integers, a condition K is said to be a condition of Q_m-normality of order ν at most, if K has the property 1^0 in Definition 8.10 and has the following property:

2^0 If each function of a family $\mathscr{F}$ of meromorphic functions in a domain D satisfies the condition K in D, then the family $\mathscr{F}$ is Q_m-normal of order ν at most in D.

Consider a domain D and a finite number of circles $\Gamma_j |z-z_j|<r_j (j=1,2,\cdots,p)$ such that their closures $\overline{\Gamma}_j: |z-z_j| \leqslant r_j (j=1,2,\cdots,p)$ belong to D. We say that the circles $\Gamma_j (j=1,2,\cdots,p)$ are distinct, if $\overline{\Gamma}_j (j=1,2,\cdots,p)$ are mutually disjoint.

Definition 8.12. $m \geqslant 0$ being an integer, let K be a condition of Q_m-normality and $\varepsilon>0$ a number. Let f(z) be a meromorphic function in a domain D. A circle Γ: $|z-z_0|<r$ is called a (K,ε)-circle of the function f(z) in D, if it has the following properties:

$1^0 \overline{\Gamma}: |z-z_0| \leqslant r$ belongs to D.

2^0 $r<\varepsilon$.

3^0 The function f(z) does not satisfy the condition K in Γ.

Theorem 8.17. $m \geqslant 0$ being an integer, let K be a condition of Q_m-normality and $\varepsilon>0$ a number. Let $\mathscr{F}$ be a family of meromorphic functions in a domain D. If each function f(z) of the family $\mathscr{F}$ has no (K,ε)-circle in D, then the family $\mathscr{F}$ is Q_m-normal in D.

Proof. Consider a point z_0 of D and let $\Upsilon: |z-z_0|<r$ having the first two properties 1^0 and 2^0 in Definition 8.12. Then, by hypothesis, each function f(z) of the family $\mathscr{F}$ satisfies the condition K in Υ. Hence the family $\mathscr{F}$ is Q_m-normal in Υ. Since z_0 is arbitrary, $\mathscr{F}$ is Q_m-normal in D by Theroem 8.3.

Theorem 8.18. Let $\mathscr{F}$ be a family of meromorphic functions in a domain

D. $m \geqslant 0$ being an integer, let K be a condition of Q_m-normality and $\varepsilon > 0$ a number. Let $\nu \geqslant 0$ be an integer. If each function $f(z)$ of $\mathscr{F}$ has at most ν distinct (K, ε)-circles in D, then $\mathscr{F}$ is Q_{m+1}-normal of order ν at most in D.

In the case $\nu = 0$, by Theorem 8.17 the family $\mathscr{F}$ is Q_m-normal in D, hence the conclusion of Thoerm 8.18 holds.

In the case $\nu > 0$, Theorem 8.18 is proved in the same way as for Theorem 5.3 in the case $\nu > 0$. It is sufficient to replace in the proof of Theorem 5.3, the word "normal" by "Q_m-normal", the term "C_0-sequence" by "C_m-sequence", the term "C_1-sequence" by "C_{m+1}-sequence" and the term "nonC_0-point" by "nonC_m-point", in order to get the proof of Theoerm 8.18.

Denoting by K′ the condition of Theorem 8.27, namely "f(z) has at most ν distinct (K, ε)-circles in D", then by Theorem 8.27, K′ is a condition of Q_{m+1}-normality of order ν at most. The condition K′ depends on m, K, ε and ν. We express this dependence in writing

$$K' = \Omega(m, K, \varepsilon, \nu). \tag{8.6}$$

Now let K_0 be a condition of Q_0-normality. Let $\varepsilon_m > 0$ $(m = 0, 1, 2, \cdots)$ be a sequence of numbers and $\nu_m \geqslant 0$ $(m = 0, 1, 2, \cdots)$ a sequence of integers. Then

$$K_1 = \Omega(0, K_0, \varepsilon_0, \nu_0)$$

is a condition of Q_1-normality of order ν_0 at most;

$$K_2 = \Omega(1, K_1, \varepsilon_1, \nu_1)$$

is a condition of Q_2-normality of order ν_1 at most; in general, for $m \geqslant 1$,

$$K_m = \Omega(m-1, K_{m-1}, \varepsilon_{m-1}, \nu_{m-1})$$

is a condition of Q_m-normality of order ν_{m-1} at most. We may say that the sequence of conditions K_m $(m = 1, 2, \cdots)$ is generated by the condition K_0 and the two sequences ε_m $(m = 0, 1, 2, \cdots)$ and ν_m $(m = 0, 1, 2, \cdots)$.

Next we give some generalizations of Theorems 3.9, 5.4, 3.17 and Corollary 5.7. For this we need the following definition:

Definition 8. 13. Given a sequence of integers $I_1=1, I_m\geqslant 2$ $(m=2,3,\cdots)$ and a sequence of numbers $\eta_m>0$ $(m=1,2,\cdots)$, consider a meromorphic function $f(z)$ in a domain D, a circle $\Upsilon: |z-z_0|<r$ and a value $a\in\hat{C}$. We say that Υ is a $(a)_1$-circle of the function $f(z)$ in D, if Υ has the following properties:

A_1) $\overline{\Upsilon}: |z-z_0|\leqslant r$ belongs to D.

B_1) $r<\eta_1$.

C_1) The function $f(z)$ takes the value a in Υ.

We say that Υ is a $(a)_2$-circle of the function $f(z)$ in D, if Υ has the property A_1) and the following properties:

B_2) $r<\eta_2$.

C_2) The function $f(z)$ has at least I_2 distinct $(a)_1$-circles in Υ.

In general, $m\geqslant 2$ being an integer, we say that Υ is a $(a)_m$-circle of the function $f(z)$ in D, if Υ has the property A_1) and the following properties:

B_m) $r<\eta_m$.

C_m) The function $f(z)$ has at least I_m distinct $(a)_{m-1}$-circles in Υ.

Lemma 8. 6. Let $m\geqslant 1$ be an integer, $\delta(0<\delta<1)$ a number and $N\geqslant 1$ an integer. Let $\mathscr{F}$ be a family of meromorphic functions in a domain D. Assume that for each function $f(z)\in\mathscr{F}$, there are three values $a_j=a_j(f)\in\hat{C}$ $(j=1,2,3)$ such that

$$|a_j,a_{j'}|>\delta\ (j,j'=1,2,3;j\neq j') \tag{8.7}$$

and that for each j $(1\leqslant j\leqslant 3)$ $f(z)$ has at most N distinct $(a_j)_m$-circles in D. Then the family $\mathscr{F}$ is Q_m-normal of order 3N at most in D.

Proof. In the case $m=1$, Lemma 8. 6 follows from Corollary 5. 3. Now assume that Lemma 8. 6 is true for an integer $m\geqslant 1$. Then given a number $\delta(0<\delta<1)$ and an integer $N\geqslant 1$, we deduce from Lemma 8. 6 a corresponding condition $K_m(\delta,N)$ which is a condition of Q_m-normality of order 3N at most. In order to complete the proof of Lemma 8. 6, it is sufficient to show that the condition $K_{m+1}(\delta,N)$ is a condition of Q_{m+1}-normality of order 3N at most. To see this, let $f(z)$ be a meromorphic function in a domain D and assume that there are three values $a_j\in\hat{C}$ $(j=1,2,3)$ such that (8. 7) holds and that for each $j(1\leqslant j\leqslant 3)$ $f(z)$

has at most N distinct $(a_j)_{m+1}$-circles in D. Consider the condition $K'=K_m(\delta, I_{m+1})$, where I_{m+1} is given in Definition 8.13. We know already that K' is a condition of Q_m-normality of order $3I_{m+1}$ at most. Suppose that the function f(z) has a (K', η_{m+1})-circle Γ in D, where η_{m+1} is given in Definition 8.13. Then by Definitions 8.12 and 8.13, we see that among the three values $a_j(j=1,2,3)$, there is at least one value a_j such that f(z) has $I_{m+1}+1$ distinct $(a_j)_m$-circles in Γ, hence Γ is a $(a_j)_{m+1}$-circle of f(z) in D, by Definition 8.13. It follows that the function f(z) has at most 3N distinct (K', η_{m+1})-circles in D. Consequently the condition $K_{m+1}(\delta, N)$ is a condition of Q_{m+1}-normality of order 3N at most, by Theorem 8.18.

Theorem 8.19. Let $m\geqslant 0$ be an integer and $\delta(0<\delta<1)$ a number. Let $\mathscr{F}$ be a family of meromorphic functions in a domain D. Assume that for each function $f(z)\in\mathscr{F}$, there are three values $a_j=a_j(f)\in\hat{\mathbf{C}}$ $(j=1,2,3)$ such that (8.7) holds and that for each $j(1\leqslant j\leqslant 3)$ f(z) has no $(a_j)_{m+1}$-circle in D. Then the family $\mathscr{F}$ is Q_m-normal in D.

This theorem is a generalization of Theorem 3.9 which corresponds to the case $m=0$.

Proof. Assume $m\geqslant 1$. Consider a point $z_0\in D$ and a circle $\gamma: |z-z_0|<r$ such that $\overline{\gamma}: |z-z_0|\leqslant r$ belongs to D and $r<\eta_{m+1}$, where η_{m+1} is given in Definition 8.13. Let $f(z)\in\mathscr{F}$. Then for each $j(1\leqslant j\leqslant 3)$ f(z) has at most I_{m+1} distinct $(a_j)_m$-circles in γ, where I_{m+1} is given in Definition 8.13 and $a_j=a_j(f)$, because otherwise, γ would be a $(a_j)_{m+1}$-circle of f(z) in D, incompatible with the hypothesis. Consequently by Lemma 8.6, the family $\mathscr{F}$ is Q_m-normal of order $3I_{m+1}$ at most in γ. Since the point z_0 is arbitrary, hence by Theorem 8.3, the family $\mathscr{F}$ is Q_m-normal in D.

In what follows we shall try to obtain a generalization of Theorem 5.4. For this we need first to give some definitions and prove some lemmas.

Definition 8.14. $m\geqslant 1$ being an integer, let $S: f_n(z)$ $(n=1,2,\cdots)$ be a C_m-sequence of meromorphic functions in a domain D. Assume that S has a non-C_{m-1}-point z_0 in D. Then two cases are possible:

1^0 We can extract from the sequence S a subsequence S' of which z_0 is a

C_{m-1}-point. In this case, we say that S is reducible with respect to the point z_0.

2^0 z_0 is a nonC_{m-1}-point of every subsequence of S. In this case, we say that S is irreducible with respect to the point z_0.

The sequence S is said to be an irreducible C_m-sequence in D, if S has nonC_{m-1}-points in D with respect to each of which S is irreducible.

Lemma 8. 7. $m \geqslant 1$ being an integer, let S: $f_n(z)$ $(n=1,2,\cdots)$ be a C_m-sequence of meromorphic functions in a domain D. Assume that S has nonC_{m-1}-points in D. Then we can extract from S a subsequence S′ which is either a C_{m-1}-sequence in D or an irreducible C_m-sequence in D.

This lemma is proved in the same way as for Lemma 5. 5. In order to get the proof of Lemma 8. 7, it is sufficient to replace in the proof of Lemma 5. 5, the term "C_1-sequence" by "C_m-sequence", the term "C_0-point" by "C_{m-1}-point" and the term "nonC_0-point" by "nonC_{m-1}-point".

Definition 8. 15. Let S be a sequence of meromorphic functions in a domain D. S is called a completely irreducible C_1-sequence in D, if S is an irreducible C_1-sequence in D. S is called a completely irreducible C_2—sequence in D, if S satisfies the following conditions:

1) S is an irreducible C_2-sequence in D.

2) E being the set of nonC_1-points of S in D, S is a completely irreducible C_1-sequence in the domain D−E.

In general, $m \geqslant 2$ being an integer, S is called a completely irreducible C_m-sequence in D, if S satisfies the following conditions:

1′) S is an irreducible C_m-sequence in D.

2′) E being the set of nonC_{m-1}-points of S in D, S is a completely irreducible C_{m-1}-sequence in the domain D−E.

Lemma 8. 8. $m \geqslant 1$ being an integer, let S be an irreducible C_m-sequence of meromorphic functions in a domain D. Then we can find a subsequence S′ of S, such that S′ is a completely irreducible C_m-sequence in D.

Proof. By Definition 8. 15, Lemma 8. 8 is trivial in the case $m=1$. Consider the case $m=2$. Let S be an irreducible C_2-sequence of meromorphic functions in a domain D. E being the set of nonC_1-points of S in D, S is a C_1-sequence in the domain $G=D-E$. S must have nonC_0-points in G, because otherwise, S would be a C_0-sequence in G and each point of E would be a C_1-point of S. So by Lemma 8.7, we can extract from S a subsequence S′ which is either a C_0-sequence in G or an irreducible C_1-sequence in G. S′ can not be a C_0-sequence in G, because S is irreducible with respect to each point of E. Hence S′ is an irreducible C_1-sequence in G. Moreover evidently S′ is also an irreducible C_2-sequence in D. Thus Lemma 8.8 is proved in the case $m=2$.

Now suppose that Lemma 8.8 is true for an integer $m\geqslant 2$ and let us show that it is also true for $m+1$. In fact, let S be an irreducible C_{m+1}-sequence of meromorphic functions in a domain D. E being the set of the nonC_m-points of S in D, S is a C_m-sequence in the domain $G=D-E$. In G, S has necessarily non-C_{m-1}-points. By Lemma 8.7, we can extract for S a subsequence S_1 which is either a C_{m-1}-sequence in G or an irreducible C_m-sequence in G. The first alternative can not occur, because each point of E is a nonC_m-point of S_1. So S_1 is an irreducible C_m-sequence in G. Hence, by assumption, we can extract from S_1 a subsequence S′ which is a completely irreducible C_m-sequence in G. Since S′ is a subsequence of S, S′ is an irreducible C_{m+1}-sequence in D and its set of nonC_m-points is E. Consequently S′ is a completely irreducible C_{m+1}-sequence in D.

Lemma 8.9. $m\geqslant 1$ being an integer, let S: $f_n(z)$ $(n=1,2,\cdots)$ be a completely irreducible C_m-sequence of meromorphic functions in a domain D and let E be the set of nonC_{m-1}-points of S in D. Then we can find a value $w_0\in\hat{C}$ having the following properties:

1^0 For any point $z_0\in E$, any circle $\Upsilon: |z-z_0|<r$ belonging to D and any sequence of values $w_n\in\hat{C}$ $(n=1,2,\cdots)$ such that

$$\lim_{n\to+\infty}|w_n, W|=0,\ W\neq w_0, \tag{8.8}$$

we can find a circle $\Upsilon': |z-z_0|<r'$ $(0<r'<r)$ and a positive integer N such that, when $n\geqslant N$, Υ' is a $(w_n)_m$-circle of the function $f_n(z)$ in Υ.

2^0 For any point $z_0\in E$, any circle $\Upsilon: |z-z_0|<r$ belonging to D and any value $w\in\hat{C}$, $w\neq w_0$, we can find a circle $\Upsilon': |z-z_0|<r'$ $(0<r'<r)$ and a positive integer N such that, when $n\geqslant N$, Υ' is a $(w)_m$-circle of the function $f_n(z)$ in Υ.

Proof. By Definition 8.15 and Lemma 5.7, Lemma 8.9 is true in the case $m=1$. Now assume that Lemma 8.9 is true for an integer $m\geqslant 1$ and let us show that it is also true for $m+1$. In fact, let $S: f_n(z)(n=1,2,\cdots)$ be a completely irreducible C_{m+1}-sequence of meromorphic functions in a domain D and let E be the set of nonC_m-points of S in D. By Definition 8.15, S is a completely irreducible C_m-sequence in the domain $G=D-E$. Let E_1 be the set of nonC_{m-1}-points of S in G. By assumption, we can find a value $w_0\in\hat{C}$ having the following property: For any point $z_0\in E_1$, any circle $\Upsilon: |z-z_0|<r$ belonging to G and any sequence of values $w_n\in\hat{C}(n=1,2,\cdots)$ such that (8.8) holds, we can find a circle $\Upsilon': |z-z_0|<r'$ $(0<r'<r)$ and a positive integer N such that, when $n\geqslant N$, Υ' is a $(w_n)_m$-circle of the function $f_n(z)$ in Υ.

Now consider a point $\zeta_0\in E$, a circle $C: |z-\zeta_0|<\rho$ belonging to D and a sequence of values $v_n\in\hat{C}(n=1,2,\cdots)$ such that,

$$\lim_{n\to+\infty} |v_n, V| = 0, \quad V\neq w_0.$$

Consider a circle $c': |z-\zeta_0|<\rho'$, $\rho'<\min(\rho,\eta_{m+1})$, such that the domain $g: 0<|z-\zeta_0|<\rho'$ belongs to the domain G, where η_{m+1} is given in Definiiton 8.13. ζ_0 is necessarily a point of accumulation of the set E_1. Let $z_j(j=1,2,\cdots,I_{m+1})$ be I_{m+1} points of $E_1\cap g$, where I_{m+1} is given in Definition 8.13. Consider I_{m+1} circles $\Upsilon_j: |z-z_j|<r_j(j=1,2,\cdots,I_{m+1})$ belonging to g and disjoint with each other. Then by the above property of the value w_0, to each $j(1\leqslant j\leqslant I_{m+1})$ corresponds a circle $\Upsilon_j': |z-z_j|<r_j'(0<r_j'<r_j)$ and a positive integer N_j such that, when $n\geqslant N_j$, Υ_j' is a $(v_n)_m$-circle of the function $f_n(z)$ in Υ_j. Set

$$N = \max(N_1,N_2,\cdots,N_{I_{m+1}}).$$

Then when $n\geqslant N$, the function $f_n(z)$ has I_{m+1} distinct $(v_n)_m$-circles $\Upsilon_j'(j=1,2,\cdots,I_{m+1})$ in c', hence c' is a $(v_n)_{m+1}$-circle of the function $f_n(z)$ in the circle c.

Next consider again a point $\zeta_0\in E$, a circle $c: |z-\zeta_0|<\rho$ belonging to D and a value $v\in\hat{C}$, $v\neq w_0$. Setting $v_n=v$ $(n=1,2,\cdots)$, by the result just obtained, we can find a circle $c': |z-z_0|<\rho'$ $(0<\rho'<\rho)$ and a positive integer N such that, when $n\geqslant N$, c' is a $(v)_{m+1}$-circle of the function $f_n(z)$ in the circle c.

This completes the proof that Lemma 8.9 is also true for the integer $m+1$.

Now we are in a position to prove the following theorem:

Theorem 8.20. Let $m \geqslant 1$ be an integer, $\delta(0<\delta<1)$ a number and $p_j(j=1,2,3)$ three nonnegative integers such that $p_1 \leqslant p_2 \leqslant p_3$. Let $\mathscr{F}$ be a family of meromorphic functions in a domain D. Assume that for each function $f(z) \in \mathscr{F}$, there are three values $a_j(f) \in \hat{C}$ $(j=1,2,3)$ such that

$$|a_j(f), a_{j'}(f)| > \delta \ (1 \leqslant j, j' \leqslant 3, j \neq j')$$

and that for each $j(1 \leqslant j \leqslant 3)$, $f(z)$ has at most p_j distinct $(a_j(f))_m$-circles in D. Then the family $\mathscr{F}$ is Q_m-normal of order p_2 at most in D.

This theorem is a generalization of Theorem 5.4 which corresponds to the case $m=1$.

Proof. Let $N \geqslant 1$ be an integer such that $p_3 \leqslant N$. Then by hypothesis and Lemma 8.6, the family $\mathscr{F}$ is Q_m-normal of order 3N at most in D. Consider a sequence S of functions of $\mathscr{F}$. We have to show that from S we can extract a subsequence S' which is a C_m-sequence in D and has at most p_2 nonC_{m-1}-points in D. For this we first extract from S a subsequence S_1 which is a C_m-sequence in D and has at most 3N nonC_{m-1}-points in D. Assume that S_1 has nonC_{m-1}-points in D. Then by Lemma 8.7, we can extract from S_1 a subsequence S_2 which is either a C_{m-1}-sequence in D or an irreducible C_m-sequence in D. Assume that S_2 is an irreducible C_m-sequence in D. Then by Lemma 8.8, we can extract from S_2 a subsequence S_3 which is a completely irreducible C_m-sequence in D. Let S_3 be the sequence of functions $f_n(z)$ $(n=1,2,\cdots)$ and set

$$a_j^{(n)} = a_j(f_n) \ (n = 1,2,\cdots)(j = 1,2,3).$$

By Lemma 3.2, we can find an increasing sequence of positive integers $n_q(q=1,2,\cdots)$ such that for each $j(1 \leqslant j \leqslant 3)$, there is a value $A_j \in \hat{C}$ such that

$$\lim_{q \to +\infty} |a_j^{(n_q)}, A_j| = 0$$

with

$$|A_j, A_{j'}| \geqslant \delta \ (1 \leqslant j, j' \leqslant 3, j \neq j').$$

Set $g_q(z)=f_{n_q}(z)$ $(q=1,2,\cdots)$ and

$$\beta_j^{(q)} = \alpha_j^{(n_q)}(q = 1,2,\cdots)\ (j = 1,2,3).$$

It is easy to see that the sequence of functions S' : $g_q(z)$ $(q=1,2,\cdots)$ is also a completely irreducible C_m-sequence in D. Moreover we have

$$\lim_{q\to+\infty} |\beta_j^{(q)}, A_j| = 0\ (j = 1,2,3)$$

and for each $j(1\leqslant j\leqslant 3)$, $g_q(z)$ has at most p_j distinct $(\beta_j^{(q)})_m$-circles in D.

Applying Lemma 8. 9 to the sequence of functions S', we conclude that there is a value $w_0\in\hat{C}$ having the property 1^0 in that lemma, with respect to S'. One at least of the two values $A_j(j=1,2)$ is different from w_0, for example $A_2\neq w_0$. Assume that S' has μ nonC_{m-1}-points $z_k(k=1,2,\cdots,\mu)$ in D. Let $\Gamma_k: |z-z_k|<r_k(k=1,2,\cdots,\mu)$ be μ circles belonging to D and disjoint with each other. Then for each $k(1\leqslant k\leqslant\mu)$ we can find a circle $\Gamma_k': |z-z_k|<r_k'(0<r_k'<r_k)$ and a positive integer Q_k such that, when $q\geqslant Q_k$, Γ_k' is a $(\beta_2^{(q)})_m$-circle of the function $g_q(z)$ in Γ_k. Let $Q=\max(Q_1,Q_2,\cdots Q_\mu)$. Then when $q\geqslant Q$, the function $g_q(z)$ has μ distinct $(\beta_2^{(q)})_m$-circles in D. Hence $\mu\leqslant p_2$, and S' is a C_m-sequence in D and has at most p_2 nonC_{m-1}-points in D.

In what follows, we give some generalizations of Theorem 3. 17 and Corollary 5. 7.

Lemma 8. 10. Let $m\geqslant 1$ be an integer, M, $\delta(0<\delta<M)$ two numbers and $N\geqslant 1$, $v\geqslant 1$ two integers. Let $\mathscr{F}$ be a family of meromorphic functions in a domain D. Assume that for each function $f(z)\in\mathscr{F}$, there are two values $a(f)$ and $b(f)$ such that

$$|a(f)| < M,\ \delta < |b(f)| < M \tag{8.9}$$

and that $f(z)$ has at most N distinct $(a(f))_m$-circles in D and $f^{(v)}(z)$ has at most N distinct $(b(f))_m$-circles in D. Then the family $\mathscr{F}$ is Q_m-normal of order 2N at most in D.

Proof. By Corollary 5.7 and Definition 8.13, Lemma 8.10 is true in the case $m=1$. Now assume that Lemma 8.10 is true for an integer $m\geqslant 1$. Then given two numbers M, δ $(0<\delta<M)$ and two integers $N\geqslant 1$, $\nu\geqslant 1$, we deduce from Lemma 8.10 a corresponding condition $K_m(M,\delta,N,\nu)$ which is a condition of Q_m-normality of order $2N$ at most. In order to complete the proof of Lemma 8.10, it is sufficient to show that the condition $K_{m+1}(M,\delta,N,\nu)$ is a condition of Q_{m+1}-normality of order $2N$ at most. To see this, let $f(z)$ be a meromorphic function in a domain D and assume that there are two values $a(f)$ and $b(f)$ such that (8.9) holds and that $f(z)$ has at most N distinct $(a(f))_{m+1}$-circles in D and $f^{(\nu)}(z)$ has at most N distinct $(b(f))_{m+1}$-circles in D. Consider the condition $K'=K_m(M,\delta,I_{m+1},\nu)$, where I_{m+1} is given in Definition 8.13. We know already that K' is a condition of Q_m-normality of order $2I_{m+1}$ at most. Suppose that the function $f(z)$ has a (K',η_{m+1})-circle Γ in D, where η_{m+1} is given in Definition 8.13. Then by Definition 8.12 and Definition 8.13, either $f(z)$ has $I_{m+1}+1$ distinct $(a(f))_m$-circles in Γ, hence Γ is a $(a(f))_{m+1}$-circle of $f(z)$ in D, or $f^{(\nu)}(z)$ has $I_{m+1}+1$ distinct $(b(f))_m$-circles in Γ, hence Γ is a $(b(f))_{m+1}$-circle of $f^{(\nu)}(z)$ in D. Consequently the function $f(z)$ has at most $2N$ distinct (K',η_{m+1})-circles in D. Hence the condition $K_{m+1}(M,\delta,N,\nu)$ is a condition of Q_{m+1}-normality of order $2N$ at most, by Theorem 8.18.

Theorem 8.21. Let $m\geqslant 0$ be an integer, M, $\delta(0<\delta<M)$ two numbers and $\nu\geqslant 1$ an integer. Let $\mathscr{F}$ be a family of meromorphic functions in a domain D. Assume that for each function $f(z)\in\mathscr{F}$, there are two values $a(f)$ and $b(f)$ such that (8.9) holds and that $f(z)$ has no $(a(f))_{m+1}$-circle in D and $f^{(\nu)}(z)$ has no $(b(f))_{m+1}$-circle in D. Then the family $\mathscr{F}$ is Q_m-normal in D.

This theorem is a generalization of Theorem 3.17 which corresponds to the case $m=0$.

Proof. Assume $m\geqslant 1$. Consider a point $z_0\in D$ and a circle $\gamma: |z-z_0|<r$ such that $\overline{\gamma}: |z-z_0|\leqslant r$ belongs to D and $r<\eta_{m+1}$, where η_{m+1} is given in Definition 8.13. Let $f(z)\in\mathscr{F}$. Then $f(z)$ has at most I_{m+1} distinct $(a(f))_m$-circles in γ, where I_{m+1} is given in Definition 8.13, because otherwise, γ would be a $(a(f))_{m+1}$-circle of $f(z)$ in D. Similarly $f^{(\nu)}(z)$ has at most I_{m+1} distinct $(b(f))_m$-circles in γ. Hence by Lemma 8.10, the family $\mathscr{F}$ is Q_m-normal of order $2I_{m+1}$ at most in γ. Since the point z_0 is arbitrary, hence by Theorem 8.3, the family $\mathscr{F}$

is Q_m-normal in D.

Theorem 8. 22. Let $m \geqslant 1$ be an integer, $M, \delta (0 < \delta < M)$ two numbers, $p \geqslant 0, q \geqslant 0$ two integers and $\nu \geqslant 1$ an integer. Let $\mathscr{F}$ be a family of meromorphic functions in a domain D. Assume that for each function $f(z) \in \mathscr{F}$, there are two values $a(f)$ and $b(f)$ such that (8. 9) holds and that $f(z)$ has at most p distinct $(a(f))_m$-circles in D and $f^{(\nu)}(z)$ has at most q distinct $(b(f))_m$-circles in D. Then the family $\mathscr{F}$ is Q_m-normal of order $p+q$ at most in D.

This theorem is a generalization of Corollary 5. 7 which corresponds to the case $m=1$.

Proof. Assume $m \geqslant 2$. Consider the condition $K' = K_{m-1}(M, \delta, I_m, \nu)$ deduced from Lemma 8. 10, which is a condition of Q_{m-1}-normality of order $2I_m$ at most, where I_m is given in Definition 8. 13. Let $f(z)$ be a function of the family $\mathscr{F}$, so that there are two values $a(f)$ and $b(f)$ such that (8. 9) holds and that $f(z)$ has at most p distinct $(a(f))_m$-circles in D and $f^{(\nu)}(z)$ has at most q distinct $(b(f))_m$-circles in D. Suppose that $f(z)$ has a (K', η_m)-circle Γ in D, where η_m is given in Definition 8. 13. Then either $f(z)$ has $I_m + 1$ distinct $(a(f))_{m-1}$-circles in Γ, hence Γ is a $(a(f))_m$-circle of $f(z)$ in D, or $f^{(\nu)}(z)$ has $I_m + 1$ distinct $(b(f))_{m-1}$-circles in Γ, hence Γ is a $(b(f))_m$-circle of $f^{(\nu)}(z)$ in D. Consequently $f(z)$ has at most $p+q$ distinct (K', η_m)-circles in D and, by Theorem 8. 18, the family $\mathscr{F}$ is Q_m-normal of order $p+q$ at most in D.

8. 5. OTHER CRITERIONS OF Q_m-NORMALITY

In what follows we give some criterions of Q_m-normality of a different kind. Consider a domain D and a nonempty set E of points of D. Let $m \geqslant 1$ be an integer and assume that E does not have the property W_m with respect to D, according to Definition 8. 4. Let us investigate the question: Under what condition the set $D-E$ is a domain? First if $m=1$, then it is easy to see that D-E is a domain. Assume $m \geqslant 2$ and $E' \neq \varnothing$. We are going to show that in order that $D-E$ is a domain, it is necessary and sufficient that $E' \subset E$.

In fact, assume that $D-E$ is a domain. Let $z_0 \in E'$ and suppose, on the contrary, that $z_0 \notin E$. Then since $E' \subset D$, we have $z_0 \in D-E$. Hence there is a circle $\Gamma: |z-z_0| < r$ such that $\Gamma \subset D-E$, incompatible with the condition $z_0 \in E'$. Hence $E' \subset E$.

Conversely assume that $E' \subset E$. Consider first the case $m=2$, so that by hypothesis, $E''=\emptyset$. Then the set $G=D-E'$ is a domain. Consequently, if $E=E'$, then $D-E$ is a domain. Assume $E\neq E'$. Define $E_1=E-E'$, then

$$D - E = G - E_1.$$

Consider a point $z_0\in G$. Then we can find a circle $\gamma: |z-z_0|<r$ belonging to D such that the domain $0<|z-z_0|<r$ contains no point of E, hence z_0 is not a point of accumulation of E_1. Consequently $G-E_1$ and hence $D-E$ is a domain.

Now assume that the condition $E' \subset E$ is sufficient for the case of an integer $m\geqslant 2$, and let us show that it is also sufficient for the case of the integer $m+1$. In fact, assume that E does not have the property W_{m+1} with respect to D, and distinguish two cases:

1)E does not have the property W_m with respect to D, then by assumption, $D-E$ is a domain.

2) E has the property W_m with respect to D. Then $E^{(m)}\neq\emptyset$, $E^{(m+1)}=\emptyset$. Then $(E')^{(m-1)}=E^{(m)}\neq\emptyset$, $(E')^{(m)}=\emptyset$. Consequently the set E' does not have the property W_m with respect to D. On the other hand, from the condition $E' \subset E$, we deduce $(E')' \subset E'$. Hence, by assumption, the set $G=D-E'$ is a domain. Then by the same procedure used in the case $m=2$, we conclude that $D-E$ is a domain.

Theorem 8. 23. Let $\mathscr{F}$ be a family of meromorphic functions in a domain D. Let E be a nonempty set of points of D and $m\geqslant 1$ an integer. Assume that the following conditions are satisfied:

1^0 E does not have the property W_m with respect to D and the set $D-E$ is a domain.

2^0 The family $\mathscr{F}$ is Q_0-normal in the domain $D-E$.

Then the family $\mathscr{F}$ is Q_m-normal in D.

Proof. Let S be a sequence of functions of the family $\mathscr{F}$. Then by the condition 2^0 in Theorem 8. 23, we can extract from S a subsequence S' which is a C_0-sequence in $D-E$. Denote by σ the set of nonC_0-points of S' in D. Then $\sigma\subset E$ and hence by the condition 1^0 in Theorem 8. 23, σ does not have the property W_m with respect to D. Consequently by Theorem 8. 2, S' is a C_m-sequence in D.

More generally we have the following theroem:

Theorem 8.24. Let $\mathscr{F}$ be a family of meromorphic functions in a domain D. Let E be a nonempty set of points of D and $m\geqslant 1, k\geqslant 0$ two integers. Assume that the following conditions are satisfied:

1^0 E does not have the property W_m with respect to D and D$-$E is a domain.

2^0 The family $\mathscr{F}$ is Q_k-normal in the domain D$-$E.

Then the family $\mathscr{F}$ is Q_{m+k}-normal in D.

For the proof of this theorem, we need the following lemmas:

Lemma 8.11. Let D be a domain and $m\geqslant 1, k\geqslant 0$ two integers. Let E be a nonempty set of points of D, such that E does not have the property W_m with respect to D and that D$-$E is a domain. Let σ be a set of points of the domain D$-$E, such that σ does not have the property W_k with respect to D$-$E. Then σ does not have the property W_{m+k} with respect to D.

Proof. Denote by $\sigma^{(j)}(D)(j=1,2,\cdots)$ the successive derived sets (not necessarily nonempty) of σ with respect to D, and by $\sigma^{(j)}(D-E)(j=1,2,\cdots)$ the successive derived sets of σ with respect to D$-$E. Suppose, on the contrary, that σ has the property W_{m+k} with respect to D, so that the sets $\sigma^{(j)}(D)(j=0,1,\cdots,m+k;\sigma^{(0)}(D)=\sigma)$ are all nonempty. It is easy to see successively that

$$\sigma'(D)\cap(D-E)\subset\sigma'(D-E),$$
$$\sigma''(D)\cap(D-E)\subset\sigma''(D-E),$$
$$\cdots\cdots\cdots\cdots\cdots\cdots\cdots\cdots$$
$$\sigma^{(k)}(D)\cap(D-E)\subset\sigma^{(k)}(D-E).$$

By hypothesis, $\sigma^{(k)}(D-E)$ is empty, hence $\sigma^{(k)}(D)\cap(D-E)$ is empty and $\sigma^{(k)}(D)\subset E$. Since $\sigma^{(k+i)}(D)(i=0,1,\cdots,m)$ are all nonempty, $\sigma^{(k)}(D)$ has the property W_m with respect to D. On the other hand, since $\sigma^{(k)}(D)\subset E$ and E does not have the property W_m with respect to D, $\sigma^{(k)}(D)$ can not have the property W_m with respect to D. Thus we get a contradiction.

Lemma 8.12. Let D be a domain, $E_j(j=1,2)$ two sets of points of D and

$m_j \geqslant 0$ $(j=1,2)$ two integers. Assume that E_j does not have the property W_{m_j} with respect to D, for $j=1,2$. Then the set $E=E_1 \cup E_2$ does not have the property W_p with respect to D, where $p=\max(m_1, m_2)$.

Proof. Denote by $E', E'', \cdots$ and $E_j', E_j'', \cdots$ respectively the successive derived sets of E and E_j in D. It is easy to see that the relation $E=E_1 \cup E_2$ always implies the relation

$$E' = E_1' \cup E_2',$$

whether E_j' is empty or not. Then we have successively

$$E'' = E_1'' \cup E_2'', \cdots, E^{(p)} = E_1^{(p)} \cup E_2^{(p)}.$$

Now suppose, on the contrary, that the set E has the property W_p with respect to D. Then, by definition, the sets $E^{(i)}$ $(i=0,1,\cdots,p)$ are all nonempty. In particular $E^{(p)}$ is nonempty. Hence one at least of the two sets $E_j^{(p)}$ $(j=1,2)$, say $E_1^{(p)}$, is nonempty. This implies that the sets $E_1^{(i)}$ $(i=0,1,\cdots,p)$ are all nonempty. Since $m_1 \leqslant p$, E_1 has the property W_{m_1} with respect to D, incompatible with the hypothesis.

Now let us prove Theorem 8.24. Let S be a sequence of functions of $\mathscr{F}$. By the condition 2^0 in Theorem 8.24, we can extract from S a subsequence S′ which is a C_k-sequence in the domain D−E. Denote by σ the set of nonC_0-points of S′ in D−E, then by Theorem 8.2, σ does not have the property W_k with respect to D−E. Hence by Lemma 8.11, σ does not have the property W_{m+k} with respect to D. Next by Lemma 8.12, the set $E \cup \sigma$ does not have the property W_{m+k} with respect to D. Denote by σ_1 the set of the nonC_0-points of S′ in D, evidently $\sigma_1 \subset E \cup \sigma$ and hence σ_1 does not have the property W_{m+k} with respect to D. Finally by Theorem 8.2, S′ is a C_{m+k}-sequence in D.

In Theorems 8.23 and 8.24, the set E is fixed and independent of the function $f(z) \in \mathscr{F}$. In what follows, we extend Theorem 8.24 to the case that the set E is variable with the function $f(z) \in \mathscr{F}$.

Theorem 8.25. Let $\mathscr{F}$ be a family of meromorphic functions in a domain D. Let $\{\sigma\}$ be a collection of nonempty sets σ of D and $m \geqslant 1, k \geqslant 0$ two integers.

Let K be a condition of Q_k-normality. Assume that the following conditions are satisfied:

1^0 For each set $\sigma \in \{\sigma\}$, the set $D-\sigma$ is a domain.

2^0 For every sequence of sets $\sigma_n (n=1,2,\cdots)$ of $\{\sigma\}$, we can find a subsequence $\sigma_{n_p} (p=1,2,\cdots)$ and a nonempty set E of points of D, having the following properties:

a) E does not have the property W_m with respect to D and the set $D-E$ is a domain.

b) For each point $z_0 \in D-E$, there exist a circle $\Upsilon: |z-z_0|<r$ belonging to $D-E$ and an integer $p_0 \geqslant 1$ such that when $p \geqslant p_0$, the set $\sigma_{n_p} \cap \Upsilon$ is empty.

3^0 To each function $f(z) \in \mathscr{F}$ corresponds a set $\sigma=\sigma(f) \in \{\sigma\}$ such that the function $f(z)$ satisfies the condition K in the domain $D-\sigma(f)$.

Then the family $\mathscr{F}$ is Q_{m+k}-normal in D.

Proof. Let $S: f_n(z)$ $(n=1,2,\cdots)$ be a sequence of functions of the family $\mathscr{F}$. Then by the condition 2^0 in Theorem 8. 25, we can find a subsequence $\sigma_{n_p}= \sigma(f_{n_p})(p=1,2,\cdots)$ of the sequence of sets $\sigma_n=\sigma(f_n)(n=1,2,\cdots)$ and a nonempty set E of points of D having the properties a) and b). Consider the subsequence $S': f_{n_p}(z)(p=1,2,\cdots)$ of S and a point $z_0 \in D-E$. Then by the property b), there exist a circle $\Upsilon: |z-z_0|<r$ and an integer p_0 satisfying the conditions in b). It follows that when $p \geqslant p_0$, the circle Υ belongs to $D-\sigma_{n_p}$ and hence by the condition 3^0, the function $f_{n_p}(z)$ satisfies the condition K in the circle Υ. This implies that the family of functions $f_{n_p}(z)$ $(p \geqslant p_0)$ and hence the family of functions $S': f_{n_p}(z)$ $(p=1,2,\cdots)$ is Q_k-normal in Υ. Since the point $z_0 \in D-E$ is arbitrary, we conclude that the family S' is Q_k-normal in the domain $D-E$. Consequently, by Theorem 8. 24 the family S' is Q_{m+k}-normal in D and hence we can extract from S' a subsequence S'' which is a C_{m+k}-sequence in D.

9

APPLICATIONS OF THE THEORY OF Q_m-NORMAL FAMILIES OF MEROMORPHIC FUNCTIONS

9. 1. UNIFORM CONVERGENCE OF SEQUENCES OF MEROMORPHIC FUNCTIONS

Theorem 9. 1. Let $S: f_n(z)(n=1,2,\cdots)$ be a sequence of holomorphic functions in a domain D satisfying the following conditions:

1^0 For an integer $m \geqslant 1$, S belongs to a Q_m-normal family of holomorphic functions in D.

2^0 There exists a set E of points of D, such that for the same integer m, E has the property W_m with respect to D (see Definition 8. 4), and that at each point of E, S converges to a finite limit.

Then S converges locally uniformly to a holomorphic function in D.

This theorem is a generalization of Theorem 6. 1 which corresponds to the case $m=1$. For the proof of this theorem, we need the following lemma:

Lemma 9. 1. Under the conditions of Theorem 9. 1, we can extract from the sequence S a subsequence S′ which converges locally uniformly to a holomorphic function in D.

Proof. For the case $m=1$, this lemma has already been proved in the proof of Theorem 6. 1. To complete the proof of Lemma 9. 1, suppose that it is true for an integer $m \geqslant 1$, and consider the case corresponding to the integer $m+1$. In this case, we can extract from the sequence S a subsequence S′ which is a

C_{m+1}-sequence in D. Let σ be the set of nonC_m-points of S' in D. Then S' is a C_m-sequence in the domain $D_1 = D-\sigma$ and , à fortiori, S' considered as a family of functions is Q_m-normal in D_1. By considering a point $z_0 \in E^{(m+1)}$, we see as in the proof of Theorem 6. 1, that there is a point $z_1 \in E^{(m)} \cap D_1$. Then it is easy to see that the set $E_1 = E \cap D_1$ has the property W_m with respect to D_1. Moreover S' converges to a finite limit at each point of E_1. Hence, by assumption, we can extract from S' a subsequence S'' which is locally uniformly convergent in D_1. Then, as in the proof of Theorem 6. 1, we see that S'' is locally uniformly convergent in D. The proof of Lemma 9. 1 is now complete.

Now let us prove Theorem 9. 1. First by Lemma 9. 1, we can extract from the sequence S a subsequence S' which converges locally uniformly to a holomorphic function $F(z)$ in D. Then, as in the proof of Theorem 6. 1, we can show that the sequence S converges locally uniformly to the function $F(z)$ in D.

Definition 9. 1. Let E be a set of points in a domain D, and $m \geqslant 0, \nu > 0$ two integers. We say that E has the property $W_{m,\nu}$ with respect ot D, if E has the property W_m with respect to D and the set $E^{(m)}$ consists of at least ν points.

Corollary 9. 1. Let $m \geqslant 2$ and $\nu \geqslant 0$ be two integers. Let $S: f_n(z)(n=1,2,\cdots)$ be a sequence of holomorphic functions in a domain D satisfying the following conditions:

1^0 S belongs to a family $\mathscr{F}$ of holomorphic functions in D, which is Q_m-normal of order ν at most in D.

2^0 There exists a set E of points of D, such that E has the property $W_{m-1,\nu+1}$ with respect to D, and that at each point of E, S converges to a finite limit. Then S converges locally uniformly to a holomorphic function in D.

Proof. By the condition 1^0, we can extract from the sequence S, a subsequence S' which is a C_m-sequence in D and has at most ν nonC_{m-1}-points in D. Denoting by e the set of nonC_{m-1}-points of S' in D. Then in the domain $D_1 = D-e$, S' is a C_{m-1}-sequence. S' considered as a family is then Q_{m-1}-normal in D_1. By the condition 2^0, the set $E^{(m-1)}$ consists of at least $\nu+1$ points. Hence there is a point $z_0 \in D_1 \cap E^{(m-1)}$. Consider a circle $\Gamma: |z-z_0| < r$ such that $\bar{\Gamma}: |z-z_0| \leqslant r$ belongs to D_1. Then the set $E_1 = E \cap \Gamma$ has the property W_{m-1} with respect to Γ, à fortiori E_1 has the property W_{m-1} with respect to D_1. Moreover at each point of E_1, S' converges to a finite limit. Consequently by Theorem 9. 1, S' converges locally uniformly to a holomorphic function in D_1. Next we see easily that S' converges locally uniformly to a holomorphic function $F(z)$ in D. Finally we see that the sequence S converges locally uniformly to the function $F(z)$ in D, by the usual method.

Theorem 9. 2. Let $S: f_n(z)(n=1,2,\cdots)$ be a sequence of meromorphic functions in a domain D satisfying the following conditions:

1^0 For an integer $m\geqslant 1$, S belongs to a Q_m-normal family $\mathscr{F}$ of meromorphic functions in D.

2^0 S has no exceptional point in D.

3^0 There exists a set E of points of D, such that for the same integer m, E has the property W_m with respect to D, and that at each point $z\in E$, S converges (with respect to the spherical distance) to a limit $\lambda(z)$ such that for every value $a\in\hat{C}$, the set $\{z|z\in E, \lambda(z)=a\}$ does not have the property W_m with respect to D. Then S is a C_0-sequence in D and the limit function of S (with respect to the spherical distance) is a non-constant meromorphic function in D.

This theorem is a generalization of Theorem 6. 3 which corresponds to the case $m=1$. For the proof of this theorem, we need the following lemma:

Lemma 9. 2. If for an integer $m\geqslant 0$, a sequence of meromorphic functions $f_n(z)(n=1,2,\cdots)$ in a domain D is a C_m-sequence in D, then for any complex number c, the sequence $f_n(z)+c(n=1,2,\cdots)$ is a C_m-sequence in D. Moreover if $f_n(z)\not\equiv 0$ in $D(n=1,2,\cdots)$, the sequence $1/f_n(z)(n=1,2,\cdots)$ is also a C_m-sequence in D.

Proof. By Lemmas 5. 2 and 5. 3, Lemma 9. 2 is true in the cases $m=0$ and $m=1$. By mathematical induction, we can prove Lemma 9. 2 in general, by basing upon Definitions 8. 2 and 8. 1.

Now let us return to the proof of Theorem 9. 2. Denote as usual by $E^{(j)}(j=1,2,\cdots)$ the derived sets of E in the domain D. Then by the condition 3^0, the sets $E^{(j)}(j=0,1,2,\cdots,m)$ are nonempty. Consider a point $z_0\in E^{(m)}$ and let us show that z_0 is a C_0-point of the sequence $S: f_n(z)(n=1,2,\cdots)$. By the condition 2^0, there exist a circle $\Gamma: |z-z_0|<r$ belonging to D, a value a and a positive integer N, such that for each $n\geqslant N$, the function $f_n(z)$ does not take the value a in

Υ. Define for $n \geqslant N$,

$$F_n(z) = \begin{cases} f_n(z) \ if \ a = \infty \\ \dfrac{1}{f_n(z)-a} \ if \ a \ is \ finite. \end{cases} \tag{9.1}$$

Then by Lemma 9. 2, the sequence $F_n(z)(n \geqslant N)$ belongs to a Q_m-normal family of holomorphic functions in Υ. (This is because that, by the condition 1^0, the sequence $\{f_n(z)(n \geqslant N)\}$ constitutes a Q_m-normal family in Υ. Then by Lemma 9. 2, this is also true for the sequence $\{F_n(z)(n \geqslant N)\}$.) On the other hand, the set $E_1 = E \cap \Upsilon$ has the property W_m with respect to Υ, and by the condition 3^0, the set

$$e_1 = \{z \,|\, z \in E_1, \lambda(z) = a\} \tag{9.2}$$

does not have the property W_m with respect to Υ. Since

$$E_1 = (E_1 - e_1) \cup e_1,$$

the set E_1-e_1 has the property W_m with respect to Υ, by Lemma 8. 12. Moreover at each point of E_1-e_1, the sequence $F_n(z)(n \geqslant N)$ converges to a finite limit. It follows that we can apply Theorem 9. 1 to the sequence of holomorphic functions $F_n(z)(n \geqslant N)$ in Υ and the set $E_1 - e_1 \subset \Upsilon$, and conclude that the sequence $F_n(z)(n \geqslant N)$ converges locally uniformly to a holomorphic function in Υ. À fortiori, $F_n(z)(n \geqslant N)$ is a C_0-sequence in Υ. Then by Lemma 9. 2, the sequence $S: f_n(z)(n = 1, 2, \cdots)$ is a C_0-sequence in Υ. Hence z_0 is a C_0-point of S.

Now we are going to show that each point of D is a C_0-point of S. Consider a point $z_1 \in D(z_1 \neq z_0)$. Join z_0 and z_1 by a polygonal line $Z(t)(0 \leqslant t \leqslant 1)$ lying in D with $Z(0) = z_0$ and $Z(1) = z_1$. Since z_0 is a C_0-point of S, there is a number $\tau(0 < \tau < 1)$ such that each point of the polygonal line $Z(t)(0 \leqslant t \leqslant \tau)$ is a C_0-point of S. Let τ_* be the least upper bound of the set of such numbers τ. We have $0 < \tau_* \leqslant 1$. Then each point of the polygonal line $Z(t)(0 \leqslant t < \tau_*)$ is a C_0-point of S. To each $t(0 \leqslant t < \tau_*)$ corresponds a circle $\Upsilon_t: |z - Z(t)| < r_t$ belonging to D, such that S is a C_0-sequence in Υ_t, where $\Upsilon_0 = \Upsilon$. Consider the set

$$d = \bigcup_{0 \leqslant t < \tau_*} \Gamma_t .$$

It is easy to see that d is a subdomain of D and S is a C_0-sequence in d. Let f(z) be the limit function (with respect to the spherical distance) of S in d. On the set E_1, we have $f(z) = \lambda(z)$. By the condition 3^0, $\lambda(z)$ is nonconstant in E_1. Consequently f(z) is a nonconstant meromorphic function in d.

Consider the point $z_* = Z(\tau_*)$. Again by the condition 2^0, there exist a circle $\Gamma_* : |z - z_*| < r_*$ belonging to D, a valuc a_* and a positive integer N_*, such that for each $n \geqslant N_*$, the function $f_n(z)$ does not take the value a_* in Γ_*. Evidently we can find a line segment $L: Z(t)\,(t_1 \leqslant t \leqslant t_2; 0 < t_1 < t_2 < \tau_*)$ belonging to Γ_*. L considered as a set of points has the property W_m with respect to Γ_*, and the set

$$\{z \mid z \in L, f(z) = a_*\}$$

consists of at most a finite number of points. Consequently, as for the point z_0, we conclude that the point z_* is a C_0-point of S. It follows that we can not have $\tau_* < 1$. Therefore $\tau_* = 1$ and the point $z_1 = Z(1)$ is a C_0-point of S.

Thus we have proved that S is a C_0-sequence in D. Since, by hypothesis, the limit function (with respect to the spherical distance) of S is nonconstant on the set E, it must be a nonconstant meromorphic function in D.

In what follows, we shall give an application of Theorem 9. 1. For this we need the following definition:

Definition 9. 2. Let f(z) be a meromorphic function in a domain D. Let $\Gamma : |z - z_0| < r$ be a circle and $a \in \hat{C}$ be a value. We say that Γ is a generalized $(a)_1$-circle of the function f(z) in D, if Γ has the following properties:

$P_1)$ $\bar{\Gamma} : |z - z_0| \leqslant r$ belongs to D.

$Q_1)$ The function f(z) takes the value a in Γ.

We say that Γ is a generalized $(a)_2$-circle of the function f(z) in D, if Γ has the property $P_1)$ and the following property:

$Q_2)$ The function f(z) has at least I_2 distinct generalized $(a)_1$-circles in Γ, where I_2 is given in Definition 8. 13. In general, $m \geqslant 2$ being an integer, we say that Γ is a generalized $(a)_m$-circle of the function f(z) in D, if Γ has the property $P_1)$ and the following property:

Q_m) The function f(z) has at least I_m distinct generalized $(a)_{m-1}$-circles in Υ, where I_m is given in Definition 8. 13.

Lemma 9. 3. Let $w=cz$ be a linear function which maps a domain D onto a domain G, where $c\neq 0$ is a constant. Let f(w) be a meromorphic function in G and f(cz) the corresponding meromorphic function in D. Let $\Upsilon: |z-z_0|<r$ be a circle, $a\in\hat{C}$ a value and $m\geqslant 1$ an integer. If the circle Υ is a $(a)_m$-circle of the function f(cz) in D, then the image circle $\Gamma: |w-w_0|<\rho(w_0=cz_0, \rho=|c|r)$ of Υ by the mapping $w=cz$ is a generalized $(a)_m$-circle of the function f(w) in G.

Proof. Consider first the case $m=1$. Since, by Definition 8. 12, the circle $\bar{\Upsilon}: |z-z_0|\leqslant r$ belongs to D, evidently the circle $\bar{\Gamma}: |w-w_0|\leqslant\rho$ belongs to G. Next since the function f(cz) takes the value a in Υ, the function f(w) takes the value a in Γ. Hence by Definition 9. 2, Γ is a generalized $(a)_1$-circle of the function f (w) in G. So Lemma 9. 3 is true in the case $m=1$.

Now assume that Lemma 9. 3 is true for an integer $m\geqslant 1$, we are going to show that it also holds for the integer $m+1$. In fact, suppose that the circle Υ: $|z-z_0|<r$ is a $(a)_{m+1}$-circle of the function f(cz) in D. Then by Definition 8. 13, the circle $\bar{\Upsilon}: |z-z_0|\leqslant r$ belongs to D, and hence the circle $\bar{\Gamma}: |w-w_0|\leqslant\rho$ belongs to G. Consider a circle $\lambda: |z-z_*|<r_*$ which is a $(a)_m$-circle of the function f(cz) in Υ. Since the function $w=cz$ maps the circle $\Upsilon: |z-z_0|<r$ onto the circle $\Gamma: |w-w_0|<\rho$, hence, by assumption, the image circle $\Lambda: |w-w_*|<\rho_* (w_*=cz_*, \rho_*=|c| r_*)$ of λ by the mapping $w=cz$ is a generalized $(a)_m$-circle of the function f(w) in Γ. So by Definition 8. 13, the function f(w) has at least I_{m+1} distinct generalized $(a)_m$-circles in Γ. Consequently Γ is a generalized $(a)_{m+1}$-circle of f(w) in G.

Theorem 9. 3. Let f(z) be a holomorphic function in a domain Ω defined by

$\Omega: |z|>r_0, \alpha<\arg z<\beta\ (r_0>0, 0\leqslant\alpha<\beta\leqslant 2\pi)$, and let $m\geqslant 1$ be an integer. Let ω and ω_1 be two bounded domains defined respectively by

$$\omega: r_0<|z|<R,\ \alpha<\arg z<\beta$$

and

$$\omega_* : r_1 < |z| < R_1, \ \alpha_1 < \arg z < \beta_1$$
$$(r_0 < r_1 < R_1 < R, \alpha < \alpha_1 < \beta_1 < \beta).$$

Let E be a set of points of ω_*, such that E has the property W_m with respect to ω. Let $k = R_1/r_1$ and E_j($j \geq 1$ integer) be the set of the values of the function $k^j z$ on E. Define

$$\sigma = \bigcup_{j=0}^{\infty} E_j \quad (E_0 = E).$$

Assume that the following conditions are satisfied:

1^0 There are a number $\rho > r_0$ and two finite values a and b($a \neq b$) such that in the domain Ω_1 defined by

$$\Omega_1 : |z| > \rho, \ \alpha < \arg z < \beta,$$

the function f(z) has no generalized $(a)_{m+1}$-circle and no generalized $(b)_{m+1}$-circle.

2^0 As $|z| \to +\infty$, $z \in \sigma$, f(z) tends to a finite limit L.

Then for any domain Ω' defined by

$$\Omega' : |z| > r_0, \alpha' < \arg z < \beta' \ (\alpha < \alpha' < \beta' < \beta),$$

as $|z| \to +\infty$, $z \in \Omega'$, f(z) tends to L.

This Theorem is a generalization of a theorem of Montel. (see[36])

Proof. Consider an integer $n \geq 1$ and the linear function $k^n z$. Evidently in the domain ω, the set of the values of this function is the domain

$$\omega_n : k^n r_0 < |z| < k^n R, \ \alpha < \arg z < \beta.$$

Taking an integer $N \geq 1$ such that for $n \geq N$, we have $k^n r_0 > \rho$. Then for $n \geq N$, ω_n belongs to the domain Ω_1. Consequently by Lemma 9. 3 and the condition 1^0, for $n \geq N$, the function $f_n(z) = f(k^n z)$ has no $(a)_{m+1}$-circle and no $(b)_{m+1}$-circle in ω. Of course $f_n(z)$ has no $(\infty)_{m+1}$-circle in ω. Hence by Theorem 8. 19, the family of functions $f_n(z)$($n \geq N$) is Q_m-normal in ω. The family $f_n(z)$($n = 1, 2, \cdots$) is also

Q_m-normal in ω. On the other hand, by the condition 2^0, at each point of the set E, the sequence $f_n(z)(n=1,2,\cdots)$ converges to the limit L. Hence by Theorem 9.1, the sequence $f_n(z)(n=1,2,\cdots)$ converges locally uniformly to a holomorphic function F(z) in ω. Evidently $F(z)\equiv L$ in ω.

Consider a domain ω' defined by

$$\omega' : r_1 \leqslant |z| \leqslant R_1, \alpha' < \arg z < \beta' \ (\alpha < \alpha' < \beta' < \beta).$$

Then the sequence $f_n(z)(n=1,2,\cdots)$ converges uniformly in ω' to the constant L. Evidently this implies that as $|z|\to+\infty$, $z\in\Omega'$, f(z) tends to L.

9. 2. DISTRIBUTION OF THE VALUES OF MEROMORPHIC FUNCTIONS IN THE NEIGHBORHOOD OF AN ESSENTIAL SINGULARITY.

Lemma 9. 4. Let $S: f_n(z)(n=1,2,\cdots)$ be a sequence of meromorphic functions in a domain D, z_0 a point of D and $m\geqslant 1$ an integer. Let $D_1\subset D$ be a subdomain of D, such that $z_0\in D_1$. Assume that z_0 is a μ_m-point of S. Then S considered as a sequence of meromorphic functions in D_1, z_0 is also a μ_m-point of S.

Proof. By Definition 8. 6, evidently Lemma 9. 4 is true in the case $m=1$. Then by mathematical induction, we see that Lemma 9. 4 holds in general, in basing upon Definition 8. 6.

For the sake of convenience we introduce several notations which will be used in what follows. Consider a linear function $cz(c\neq 0)$ and a circle $|z-z_0|<r$. We shall denote by $\Gamma(z_0,r;c)$ the circle $|z-cz_0|<|c|r$ which is the set of the values taken by the function cz in the circle $|z-z_0|<r$. Similarly we shall denote by $\overline{\Gamma}(z_0,r;c)$ the circle $|z-cz_0|\leqslant|c|r$ which is the set of the values taken by the function cz in the circle $|z-z_0|\leqslant r$.

Consider a meromorphic function f(z) in the finite part of a domain $|z|>r_0$ $(r_0>0)$. Let D be a domain in a bounded domain $r_0<|z|<R$. Let $c_n(n=1,2,\cdots)$ be a sequence of complex numbers such that

$$|c_1| > 1, \quad k|c_n| \leqslant |c_{n+1}| \ (n = 1,2,\cdots), \ k = \frac{R}{r_0} \tag{9.3}$$

and consider the sequence of functions

$$S: f_n(z) = f(c_n z)(n = 1,2,\cdots) \tag{9.4}$$

meromorphic in the domain D. We are going to find necessary and sufficient conditions in order that the sequence S considered as a family of functions, is not Q_0-normal in D, or not Q_1-normal of order ν at most in D, or not Q_1-normal in D, etc., and then deduce consequences on the distribution of the values of the function f(z) in the neighborhood of the point ∞.

1) Assume that the family S is not Q_0-normal in D. Then by Theorem 8. 6, we can find a subsequence:

$$S': f_{n_p}(z) = f(c_{n_p} z)(p = 1,2,\cdots) \tag{9. 5}$$

of the sequence S, such that S′ has a μ_1-point $z_0 \in D$. According to Definition 8. 6, for each circle $|z-z_0| \leqslant \varepsilon$ belonging to D, we have

$$\lim_{p\to+\infty} \max_{|z-z_0|\leqslant\varepsilon} \partial(z, f_{n_p}) = +\infty. \tag{9. 6}$$

Thus we have shown that in order that the family S is not Q_0-normal in D, it is necessary that we can find a subsequence S′ defined by (9. 5) of the sequence S and a point $z_0 \in D$ such that (9. 6) holds for any circle $|z-z_0| \leqslant \varepsilon$ belonging to D. Evidently this condition is also sufficient in order that the family S is not Q_0-normal in D.

From this result we deduce the following theorem:

Theorem 9. 4. In order that the family S is not Q_0-normal in D, it is necessary and sufficient that we can find an increasing sequence of positive integers $n_p (p=1,2,\cdots)$ and a point $z_0 \in D$, such that for any circle $|z-z_0| \leqslant \varepsilon$ belonging to D, we have

$$\lim_{p\to+\infty} \max_{z\in\overline{T}(z_0,\varepsilon,c_{n_p})} |z|\partial(z,f) = +\infty. \tag{9. 7}$$

Proof. First of all, in the domain D we have

$$\partial(z,f_{n_p}) = |c_{n_p}|\partial(c_{n_p}z,f).$$

Consider a point z of the circle $|z-z_0|\leqslant\varepsilon$. Then $c_{n_p}z$ is a point of $\overline{\Gamma}(z_0,\varepsilon;c_{n_p})$ and

$$\partial(z,f_{n_p}) \leqslant \frac{1}{r_0}|c_{n_p}z|\partial(c_{n_p}z,f),$$

hence

$$\max_{|z-z_0|\leqslant\varepsilon}\partial(z,f_{n_p}) \leqslant \frac{1}{r_0}\max_{z\in\overline{\Gamma}(z_0,\varepsilon;c_{n_p})}|z|\partial(z,f). \tag{9.8}$$

Conversely consider a point z of $\overline{\Gamma}(z_0,\varepsilon;c_{n_p})$. Then $z=c_{n_p}z'$, where z' is a point of the circle $|z-z_0|\leqslant\varepsilon$, and

$$\begin{aligned}|z|\partial(z,f) &= |c_{n_p}z'|\partial(c_{n_p}z',f) \leqslant R|c_{n_p}|\partial(c_{n_p}z'f)\\ &= R\partial(z',f_{n_p}),\end{aligned}$$

hence

$$\max_{z\in\overline{\Gamma}(z_0,\varepsilon;c_{n_p})}|z|\partial(z,f) \leqslant R\max_{|z-z_0|\leqslant\varepsilon}\partial(z,f_{n_p}). \tag{9.9}$$

(9.8) shows that (9.6) implise (9.7). Conversely (9.9) shows that (9.7) implies (9.6). Consequently Theorem 9.4 follows from the above result.

Now assume that the condition (9.7) is satisfied. Then (9.6) holds for any circle $|z-z_0|\leqslant\varepsilon$ belonging to D. This implies that the sequence S′ defined by (9.5) considered as a family is not Q_0-normal at the point z_0. Moreover every subsequence S″ of S′ considered as a family is also not Q_0-normal at the point z_0.

Consider a circle $|z-z_0|<\varepsilon$ belonging to D and a number $\delta(0<\delta<1)$. Then by Theorem 3.9, we can find an integer $p_0\geqslant 1$ such that for each $p\geqslant p_0$, in the circle $|z-z_0|<\varepsilon$, the function $f_{n_p}(z)$ takes every value $w\in\hat{\mathbf{C}}$ except at most the values w which belong to two spherical circles $|w, w_j| \leqslant \delta (j=1,2)$. Consequently for each $p \geqslant p_0$, in the circle $\Gamma(z_0, \varepsilon; c_{n_p})$, the function $f(z)$ takes every value $w \in \hat{C}$ except at most the values w which belong to two spherical circles $|w, w_j| \leqslant \delta (j=1,2)$.

The circles $\Gamma(z_0, \varepsilon; c_{n_p})(p \geqslant p_0)$ are therefore filling circles of the function $f(z)$.

2) Assume that for an integer $\nu \geqslant 0$, the family S defined by (9. 2) is not Q_1-normal of order ν at most in the domain D. Then by Theorem 8. 7, we can find a subsequence S' defined by (9. 5) of S, such that S' has $\nu+1$ μ_1-points $z_j (j=1,2,\cdots,\nu+1)$ in D. Hence for any one z_0 of the points $z_j (j=1,2,\cdots,\nu+1)$ and any circle $|z-z_0| \leqslant \varepsilon$ belonging to D, (9. 6) holds.

Thus in order that the family S is not Q_1-normal of order ν at most in D, it is necessary that we can find a subsequence S' defined by (9. 5) of S and $\nu+1$ points $z_j (j=1,2,\cdots,\nu+1)$ in D, such that for any one z_0 of the points $z_j (j=1,2,\cdots,\nu+1)$ and any circle $|z-z_0| \leqslant \varepsilon$ belonging to D, (9. 6) holds. Evidently by Theorem 8. 7, this condition is also sufficient in order that the family S is not Q_1-normal of order ν at most in D.

By means of (9. 8) and (9. 9), we deduce from this result the following theorem:

Theorem 9. 5. In order that the family S is not Q_1-normal of order ν at most in the domain D, it is necessary and sufficient that we can find an increasing sequence of positive integers $n_p (p=1,2,\cdots)$ and $\nu+1$ points $z_j (j=1,2,\cdots,\nu+1)$ in D, such that for any one z_0 of the points $z_j (j=1,2,\cdots,\nu+1)$ and any circle $|z-z_0| \leqslant \varepsilon$ belonging to D, we have (9. 7).

Now assume that the condition of Theorem 9. 5 is satisfied. Then by Theorem 8. 7, the sequence S' defined by (9. 5), considered as a family is not Q_1-normal of order ν at most in D, and this is also true for every subsequence S'' of S'.

Denoting by $\Delta(D, c_{n_p})$ the domain which is the set of the values taken by the function $c_{n_p} z$ in the domain D, then by Theorem 5. 4, we see that given arbitrarily a number $\delta (0 < \delta < 1)$, we can find an integer $p_0 \geqslant 1$ such that when $p \geqslant p_0$, in the domain $\Delta(D, c_{n_p})$, the equation

$$f(z) = w$$

has at least $\nu+1$ distinct roots for every value $w\in\hat{C}$, except the values w which belong to two spherical circles $|w,w_j|\leqslant\delta(j=1,2)$.

The domains $\Delta(D,c_{n_p})(p\geqslant p_0)$ are filling domains of the function $f(z)$.

3) Assume that the family S defined by (9.4) is not Q_1-normal in the domain D. Then from the proof of Theorem 8. 12, we see that we can find a subsequence S' defined by (9. 5) of the sequence S and a sequence of points $z_j(j\geqslant 1)$ of the domain D, satisfying the condition (Γ_1') with respect to D, such that the points $z_j(j\geqslant 1)$ are all μ_1-points of the sequence S'. By Definition 8. 8, the points $z_j(j\geqslant 1)$ are distinct and the limit $\lim_{j\to+\infty} z_j=\lambda$ exists with $\lambda\in D$. Hence by Definition 8. 6, λ is a μ_2-point of S'.

Thus in order that the family S is not Q_1-normal in D, it is necessary that we can find a subsequence S' defined by (9. 5) of S and a sequence of points $z_j(j\geqslant 1)$ of D, satisfying the condition (Γ_1') with respect to D, such that the points z_j $(j\geqslant 1)$ are all μ_1-points of S'. Evidently by Theorem 8. 12 this condition is also sufficient in order that the family S is not Q_1-normal in D.

By means of (9. 8) and (9. 9), we deduce from this result the following theorem:

Theorem 9. 6. In order that the family S is not Q_1-normal in the domain D, it is necessary and sufficient that we can find an increasing sequence of positive integers $n_p(p=1,2,\cdots)$ and a sequence of points $z_j(j\geqslant 1)$ of D, satisfying the condition (Γ_1') with respect to D, such that for any one z_0 of the points $z_j(j\geqslant 1)$ and any circle $|z-z_0|\leqslant\varepsilon$ belonging to D, we have (9. 7).

Now assume that the condition of Theorem 9. 6 is satisfied. Then the $\lim_{j\to+\infty} z_j=\lambda$ exists and $\lambda\in D$ is a μ_2-point of the sequence S' defined by (9. 5). Consider a circle $|z-\lambda|<\varepsilon$ belonging to the domain D. By Lemma 9. 4, S' considered as a sequence of meromorphic functions in the circle $|z-\lambda|<\varepsilon$, λ is also a μ_2-point of S'. Consequently by Theorem 8. 8, S' considered as a family, is not Q_1-normal in the circle $|z-\lambda|<\varepsilon$. This is also true for any subsequence S'' of S'.

Consider a circle $|z-\lambda|<\varepsilon$ belonging to the domain D and a number $\delta(0<\delta<1)$. Then by Theorem 8. 19 we can find an integer $p_0\geqslant 1$ such that when $p\geqslant p_0$, the function $f_{n_p}(z)=f(c_{n_p}z)$ has a $(w)_2$-circle in the circle $|z-\lambda|<\varepsilon$, for every value $w\in\hat{C}$, except at most the values w which belong to two spherical circles $|w,w_j|\leqslant\delta(j=1,2)$. Consequently by Lemma 9. 3, when $p\geqslant p_0$, the function $f(z)$ has a generalized $(w)_2$-circle in the circle $\Gamma(\lambda,\varepsilon;c_{n_p})$, for every value $w\in\hat{C}$, except at most the values w which belong to two spherical circles $|w, w_j| \leqslant \delta (j=1, 2,)$.

The circles $\Gamma(\lambda, \varepsilon; c_{n_p})(p \geqslant p_0)$ are filling circles of the function $f(z)$, in extended sense.

By the same procedure, we can also obtain similar results from Theorems 8. 13, 8. 14, 8. 15 and 8. 16. On the other hand, besides the filling circles and filling domains (in usual sense or in extended sense) of the function $f(z)$, we can also get Julia directions and Julia curves (in usual sense or in extended sense) of the function $f(z)$, provided that we choose suitablely the sequence of constants $c_n (n=1, 2, \cdots)$ satisfying (9.3). In fact, it is easy to see that for Julia directions, it is sufficient that $c_n (n=1, 2, \cdots)$ is a sequence of positive numbers satisfying the condition (9.3). On the other hand, consider a curve $z=\varphi(t)(0 \leqslant t < +\infty)$, where $\varphi(t)$ is a continuous complex valued function for $0 \leqslant t < +\infty$, such that when t increases from 0 to $+\infty$, $|\varphi(t)|$ increases from 0 to $+\infty$. Then it is easy to see that for Julia curves, it is sufficient to choose $c_n = \varphi(t_n)(n=1, 2, \cdots)$ such that the condition (9.3) is satisfied (see Chapter 2, section 2. 1).

9. 3. LIMITATION OF THE MODULUS OF MEROMORPHIC FUNCTIONS

We are going to give some generalizations of Schottky's theorem.

A) Case of holomorphic functions

Lemma 9. 5. $m \geqslant 0$ being an integer, let $\mathscr{F}$ be a Q_m-normal family of holomorphic functions in a domain D. Suppose that there is a set $E \subset D$ satisfying the following conditions:

1^0 E has the property W_m with respect to D.

2^0 At each point of E the family $\mathscr{F}$ is uniformly bounded. Then the family $\mathscr{F}$ is uniformly bounded on each bounded closed set $\sigma \subset D$.

Proof. Consider a bounded closed set $\sigma \subset D$. Assume, on the contrary, that the family $\mathscr{F}$ is not uniformly bounded on σ. Then to each positive integer n corresponds a function $f_n(z)$ of the family $\mathscr{F}$, such that

$$\max_{z \in \sigma} |f_n(z)| > n. \tag{9. 10}$$

From the sequence S: $f_n(z)(n=1,2,\cdots)$ we can extract a subsequence S' : $f_{n_k}(z)$ $(k=1,2,\cdots)$ which is a C_m-sequence in D. Let e be the set of nonC_0-points of the sequence S' in D. By Theorem 8. 2, e does not have the property W_m with respect to D. Since $e'\subset e$, the set $D_0=D\text{-}e$ is a domain , by a result obtained in section 8. 5. In D_0, the sequence S' is a C_0-sequence. By the condition 1^0, there is a point $z_0\in E\cap D_0$. By the condition 2^0, the sequence S' is bounded at z_0. So the limit function f(z) of this sequence in D_0 is a holomorphic function . So we can apply Theorem 9. 1 to the sequence S' and conclude that it converges locally uniformly to a holomorphic function in D. But this is in contradiction with (9. 10), and the Lemma 9. 5 is proved.

Theorem 9. 7. Given a domain D, an integer $m\geqslant 0$, a condition K of Q^m-normality, a set $E\subset D$ having the property W_m with respect to D, a positive function M(z) defined on E and a bounded closed set $\sigma\in D$, we can find a positive number $A(D,m,K,E,M(z),\sigma)$ such that if f(z) is a holomorphic function in D satisfying the conditions:

1^0 f(z) satisfies the condition K in D,

2^0 $|f(z)|\leqslant M(z)$ for $z\in E$;

then

$$|f(z)|\leqslant A(D,m,K,E,M(z),\sigma) \text{ for } z\in\sigma.$$

Proof. Let $\mathscr{F}$ be the family of all holomorphic functions f(z) in D satisfying the conditions 1^0 and 2^0. This family $\mathscr{F}$ satisfies the conditions of Lemma 9. 5, hence $\mathscr{F}$ is uniformly bounded on the set σ, and accordingly the number A exists.

In the particular case $m=0$, the set E may be taken to be a single point of the domain D. So we have the following corollary:

Corollary 9. 2. Given a bomain D, a condition K of Q_0-normality, a point $z_0\in D$, a positive number M and a dounded closed set $\sigma\subset D$, we can find a positive number $A(D,K,z_0,M,\sigma)$ such that if f(z) is a holomorphic function in D satisfying the conditions:

1^0 f(z) satisfies the condition K in D,

2^0 $|f(z_0)| \leqslant M$;

then

$$|f(z)| \leqslant A(D, K, z_0, M, \sigma) \ for\ z \in \sigma.$$

This corllary implies Schottky's theorem:

Corollary 9. 3. Let $f(z)$ be a holomorphic function in the circle $|z|<1$ satisfying the following conditions:

1^0 $f(z)$ does not take the values 0 and 1 in the circle $|z|<1$.

2^0 $|f(0)| \leqslant M$, M being a positive number.

Then in each circle $|z| \leqslant r$ $(0<r<1)$ we have

$$|f(z)| \leqslant \alpha(M, r),$$

where $\alpha(M, r)$ is a positive number depending only on M and r.

Proof. It is sufficient to define

$$\alpha(M, r) = A(D, K, z_0, M, \sigma),$$

where

D is the circle $|z|<1$,

K is the condition of omitting to take the values $0, 1, \infty$,

$z_0 = 0$,

σ is the circle $|z| \leqslant r$.

Lemma 9. 6. $m \geqslant 1$ and $\nu \geqslant 0$ being two integers, let $\mathscr{F}$ be a Q_m-normal of order ν at most family of holomorphic functions in a domain D. Suppose that there is a set $E \subset D$ satisfying the following conditions:

1^0 E has the property $W_{m-1,\nu+1}$ with respect to D. (see Definition 9. 1)

2^0 At each point of E the family $\mathscr{F}$ is uniformly bounded. Then the family $\mathscr{F}$ is uniformly bounded on each bounded closed set $\sigma \subset D$.

Proof. Consider a bounded closed set $\sigma \subset D$. Assume, on the contrary, that the family $\mathscr{F}$ is not uniformly bounded on σ. Then as in the proof of Lemma 9.5, we get a sequence $S: f_n(z)(n=1,2,\cdots)$ of functions of $\mathscr{F}$, satisfying the condition (9.10). From S we can extract a subsequence $S': f_{n_k}(z)(k=1,2,\cdots)$ which is a C_m-sequence in D and has at most ν nonC_{m-1}-points in D. Denoting by e the set of the nonC_{m-1}-points of S' in D, then S' is a C_{m-1}-sequence in the domain $D_1=D-e$. By the condition 1^0, the set $E^{(m-1)}$ consists of at least $\nu+1$ points, hence there is a point $z_0\in E^{(m-1)}\cap D_1$. Let $\Upsilon: |z-z_0|<r$ be a circle belonging to D_1. Then the set $E_1=E\cap\Upsilon$ has the property W_{m-1} with respect to Υ, hence with respect to D_1. On the other hand, S' considered as a family is Q_{m-1}-normal in D_1. Moreover by the condition 2^0, at each point of E_1, S' is uniformly bounded. Consequently by Lemma 9.5, S' is uniformly bounded on each bounded closed set belonging to D_1. Next by the maximum modulus theorem, we see that S' is uniformly bounded on each bounded closed set $\sigma\subset D$. But this is incompatible with (9.10), so Lemma 9.6 is proved.

Theorem 9.8. Given a domain D, two integers $m\geqslant 1$ and $\nu\geqslant 0$, a condition K of Q_m-normality of order ν at most, a set $E\subset D$ having the property $W_{m-1,\nu+1}$ with respect to D, a positive function $M(z)$ defined on E and a bounded closed set $\sigma\subset D$, we can find a positive number $A(D,m,\nu,K,E,M(z),\sigma)$ such that if $f(z)$ is a holomorphic function in D satisfying the conditions:

1^0 $f(z)$ satisfies the condition K in D,

2^0 $|f(z)|\leqslant M(z)$ for $z\in E$;

then

$$|f(z)|\leqslant A(D,m,\nu,K,E,M(z),\sigma) \text{ for } z\in\sigma.$$

Proof. Let $\mathscr{F}$ be the family of all holomorphic functions $f(z)$ in D satisfying the conditions 1^0 and 2^0. This family $\mathscr{F}$ satisfies the conditions of Lemma 9.6, hence $\mathscr{F}$ is uniformly bounded on the set σ, and accordingly the number A exists.

In Lemmas 9.5 and 9.6 the set E is the same for all functions of the family $\mathscr{F}$. Now we give some extensions of these Lemmas, by making E variable with the functions of $\mathscr{F}$. First we give some definitions.

Definition 9. 3. Let $E_n(n=1,2,\cdots)$ and E be sets of points of the complex plane **C**. We say that E is a limiting set of the sequence $E_n(n=1,2,\cdots)$, if for any point $z_0 \in E$, any positive number ε and any positive integer N, we can find an integer $n \geqslant N$ such that the set $E_n \cap (|z-z_0|<\varepsilon)$ is nonempty.

Definition 9. 4. Let D be a domain and $\{E\}$ a collection of sets of points of D. Let $m \geqslant 0$ be an integer. We say that $\{E\}$ is a (m)-collection with respect to D, if every sequence of sets $E_n(n=1,2,\cdots)$ of $\{E\}$ has a limiting set $\widetilde{E} \subset D$ having the property W_m with respect to D.

In particular, let σ be a bounded closed set of points of D. If each point of σ is considered as a set E, then the collection $\{E\}$ is a (0)-collection with respect to D.

Definition 9. 5. Let D be a domain and $\{E\}$ a collection of sets of points of D. Let $m \geqslant 0$ and $\nu \geqslant 1$ be two integers. We say that $\{E\}$ is a (m,ν)-collection with respect to D, if every sequence of sets $E_n(n=1,2,\cdots)$ of $\{E\}$ has a limiting set $\widetilde{E} \subset D$ having the property $W_{m,\nu}$ with respect to D.

Lemma 9. 7. $m \geqslant 0$ being an integer, let $\mathscr{F}$ be a Q_m-normal family of holomorphic functions in a domain D. Suppose that there are a (m)-collection $\{E\}$ with respect to D and a positive number M satisfying the following condition: For each function $f(z) \in \mathscr{F}$ there is a set $E=E(f)$ of the collection $\{E\}$ such that

$$|f(z)| \leqslant M \ for\ z \in E(f). \tag{9.11}$$

Then the family $\mathscr{F}$ is uniformly bounded on each bounded closed set $\sigma \subset D$.

Proof. As in the proof of Lemma 9. 5, assume that $\mathscr{F}$ is not uniformly bounded on a bounded closed set $\sigma \subset D$. Then there is a sequence of functions S: $f_n(z)(n=1,2,\cdots)$ of $\mathscr{F}$ such that (9. 10) holds for each n. From S we can extract a subsequence S': $f_{n_k}(z)(k=1,2,\cdots)$ which is a C_m-sequence in D. As in the proof of Lemma 9. 5, S' is a C_0-sequence in a domain $D_0=D-e$, where e does not have the property W_m with respect to D. Consider the corresponding sequence of sets $E_k=E(f_{n_k})(k=1,2,\cdots)$ of the collection $\{E\}$, which has a limiting set $\widetilde{E}\subset D$ having the property W_m with respect to D. There is a point $z_0\in D_0\cap\widetilde{E}$. Let Υ : $|z-z_0|<r$ be a circle such that $\overline{\Upsilon}$: $|z-z_0|\leqslant r$ belongs to D_0. By Definition 9. 3, for any positive integer K, there is a positive integer $k\geqslant K$ such that the set $E_k\cap\Upsilon$ is non-empty. This shows that in D_0 the limit function of S′ is a holomorphic function. As in the proof of Lemma 9. 5, this leads to a contradiction with (9. 10).

Theorem 9. 9. Given a domain D, an integer $m\geqslant 0$, a condition K of Q_m-normality, a (m)-collectin $\{E\}$ with respect to D, a positive number M and a bounded closed set $\sigma\subset D$, we can find a positive number $A(D,m,K,\{E\},M,\sigma)$ such that if f(z) is a holomorphic function in D satisfying the conditions:

1^0 f(z) satisfies the condition K in D,

2^0 there is a set E of the collection $\{E\}$ such that $|f(z)|\leqslant M$ for $z\in E$;

then

$$|f(z)|\leqslant A(D,m,K,\{E\},M,\sigma)\ for\ z\in\sigma.$$

To prove Theorem 9. 9, we consider the family $\mathscr{F}$ of all holomorphic functions in the domain D satisfying the conditions 1^0 and 2^0 and then apply Lemma 9. 7.

Corollary 9. 4. Given a domain D, a condition K of Q_0-normality, a bounded closed set $e\subset D$, a positive number M and a bounded closed set $\sigma\subset D$, we can find a positive number $A(D,K,e,M,\sigma)$ such that if f(z) is a holomorphic function in D satisfying the conditions:

1^0 f(z) satisfies the condition K in D,

$2^0 \min_{z\in e}|f(z)|\leqslant M$,

then

$$|f(z)|\leqslant A(D,K,e,M,\sigma)\ for\ z\in\sigma.$$

Lemma 9. 8. $m\geqslant 1$ and $\nu\geqslant 0$ being two integers, let $\mathscr{F}$ be a Q_m-normal of order ν at most family of holomorphic functions in a domain D. Suppose that there is a $(m-1,\nu+1)$-collection $\{E\}$ with respect to D and a positive number M satisfying the following condition : For each function $f(z) \in \mathscr{F}$ there is a set $E = E(f)$ of the collection $\{E\}$ such that (9.11) holds. Then the family $\mathscr{F}$ is uniformly bounded on each bounded closed set $\sigma \subset D$.

Proof. Assume that $\mathscr{F}$ is not uniformly bounded on a bounded closed set $\sigma \subset D$. Then there is a sequence of functions $S: f_n(z)(n=1,2,\cdots)$ of $\mathscr{F}$ such that (9.10) holds for each n. From S we can extract a subsequence $S' : f_{n_k}(z)(k=1,2,\cdots)$ which is a C_m-sequence and has at most ν nonC_{m-1}-points in D. Let e be the set of nonC_{m-1}-points of S' in D. Then in the domain $D_1 = D - e$, S' is a C_{m-1}-sequence. Consider the corresponding sequence of sets $E_k = E(f_{n_k})(k=1,2,\cdots)$ of the collection $\{E\}$, which has a limiting set $\widetilde{E} \subset D$ having the property $W_{m-1,\nu+1}$ with respect to D. Accordingly the set $\widetilde{E}^{(m-1)}$ consists of at least $\nu+1$ points. There is then a point $z_0 \in \widetilde{E}^{(m-1)} \cap D_1$. Let $\Gamma : |z-z_0| < r$ be a circle belonging to D_1. Then the set $\widetilde{E} \cap \Gamma$ has the property W_{m-1} with respect to Γ. On the other hand, let e_1 be the set of nonC_0-points of S' in D_1. Then e_1 does not have the property W_{m-1} with respect to D_1. Consequently there is a point $z_1 \in \widetilde{E} \cap (D_1 - e_1)$. S' is a C_0-sequence in the domain $D_2 = D_1 - e_1$. Let $\Gamma_1 : |z-z_1| < r_1$ be a circle such that $\overline{\Gamma}_1 : |z-z_1| \leqslant r_1$ belongs to D_2. Since $\widetilde{E}$ is a limiting set of the sequence of sets $E_k(k=1,2,\cdots)$, hence by Definition 9.3, for any positive integer K, we can find an integer $k \geqslant K$ such that the set $E_k \cap \Gamma_1$ is non-empty. Consequently by the condition (9.11), we see that the limit function of S' in D_2 is a holomorphic function. By Theorem 9.1, S' converges locally uniformly to a holomorphic function in D. This is incompatible with (9.10).

Theorem 9.10. Given a domain D, two integers $m \geqslant 1$ and $\nu \geqslant 0$, a condition K of Q_m-normality of order ν at most, a $(m-1,\nu+1)$-collection $\{E\}$ with respect to D, a positive number M and a bounded closed set $\sigma \subset D$, we can find a positive number $A(D,m,\nu,K,\{E\},M,\sigma)$ such that if $f(z)$ is a holomorphic function in D satisfying the conditions:

1^0 $f(z)$ satisfies the condition K in D,

2^0 there is a set E of the collection $\{E\}$ such that $|f(z)| \leqslant M$ for $z \in E$;

then

$$|f(z)| \leqslant A(D,m,\nu,K,\{E\},M,\sigma) \quad for\ z \in \sigma.$$

This theorem is deduced from Lemma 9.8 in the usual way.

Corollary 9.5. Given a domain D, an integer $\nu \geqslant 1$, a condition K of Q_1-normality of order ν at most, a bounded closed set $e \subset D$, two positive numbers δ, M and a bounded closed set $\sigma \subset D$, we can find a positive number $A(D, \nu, K, e, \delta, M, \sigma)$ such that if f(z) is a holomorphic function in D satisfying the conditions:

1^0 f(z) satisfies the condition K in D,

2^0 there exist $\nu+1$ points $z_j \in e (j=1,2,\cdots,\nu+1)$ such that

$$|z_j - z_{j'}| \geqslant \delta \ (1 \leqslant j, j' \leqslant \nu + 1, j \neq j') \tag{9.12}$$

and that

$$|f(z_j)| \leqslant M \ (j = 1, 2, \cdots, \nu + 1);$$

then

$$|f(z)| \leqslant A(D, \nu, K, e, \delta, M, \sigma) \ for \ z \in \sigma.$$

In this Corollary it is tacitly assumed that there exist $\nu+1$ points $z_j \in e (j=1, 2, \cdots, \nu+1)$ satisfying the condition (9.12). To deduce this corollary from Theorem 9,10, it is sufficient to define a $(0, \nu+1)$-collection $\{E\}$ with respect to D, by choosing the sets E to be the systems of points $z_j \in e (j=1,2,\cdots,\nu+1)$ satisfying the condition (9.12). (see Lemma 5.10 and its proof)

B) Case of meromorphic functions

In the case of meromorphic functions we can not expect to obtain an inequality of the form

$$|f(z)| \leqslant A \ for \ z \in \sigma,$$

as in the case of holomorphic functions, because the function f(z) may have poles on the set σ. However we shall see that such an inequality can still be found, provided that z is assumed not belonging to certain circles.

Lemma 9.9. $m \geqslant 0$ being an integer, let $\mathscr{F}$ be a Q_m-normal family of meromorphic functions in a domain D. Suppose that there is a set $E \subset D$ satisfying the following conditions:

1^0 E has the property W_m with respect to D.

2^0 At each point of E the family $\mathscr{F}$ is uniformly bounded.

Then for any bounded closed set $\sigma \subset D$ and any positive number δ, we can find a positive number $\alpha(\sigma,\delta)$ such that for each function $f(z) \in \mathscr{F}$, the set

$$\{z \mid z \in \sigma, |f(z)| > \alpha(\sigma,\delta)\}$$

can be contained in a finite number of circles of which the sum of the radii is less than δ.

Proof. Let σ be a bounded closed set of points of D and δ a positive number such that σ can not be contained in a finite number of circles of which the sum of the radii is less than δ. Suppose, on the contrary, that there does not exist a positive number $\alpha(\sigma,\delta)$ having the required property. Then to each positive integer n corresponds a function $f_n(z) \in \mathscr{F}$ such that the set

$$\mu_n = \{z \mid z \in \sigma, |f_n(z)| > n\} \tag{9.13}$$

cannot be contained in a finite number of circles of which the sum of the radii is less than δ. From the sequence $S: f_n(z)\ (n=1,2,\cdots)$ we can extract a subsequence $S': f_{n_k}(z)\ (k=1,2,\cdots)$ which is a C_m-sequence in D. Let e be the set of nonC_0-points of S' in D. In the domain $D_0 = D - e$, S' is a C_0-sequence. As in the proof of Lemma 9.5, we see that the limit function $f(z)$ (with respect to the spherical distance) of S' can not be the constant ∞, and hence $f(z)$ is a meromorphic function in D_0. Now distinguish two cases:

1) The set $\sigma \cap e$ is empty. Then $\sigma \subset D_0$. On σ the function $f(z)$ has at most a finite number of poles $z_j\ (j=1,2,\cdots,h)$. Consider the circles $\gamma_j: |z-z_j| < \delta/(h+1)$ $(j=1,2,\cdots,h)$ and the set $\sigma_1 = \sigma - \bigcup_{j=1}^{h} \gamma_j$. Evidently σ_1 is bounded closed and the function $f(z)$ is finite on σ_1. Hence $f(z)$ is bounded on σ_1. By Theorem 1.2 and the finite covering theorem, S' converges uniformly (with respect to the spherical distance) to $f(z)$ on σ_1. Then it is easy to see that there is a positive integer k_0 such that the sequence $f_{n_k}(z)\ (k \geqslant k_0)$ is uniformly bounded on σ_1. But this is incompatible with the fact that the set $\mu_n \cap \sigma_1$ is non-empty for each n.

2) The set $\sigma\cap e$ is non-empty. By Theorem 8.1, the set e is at most enumerable. So to each point $\zeta\in e$ corresponds a positive integer $n(\zeta)$ such that $n(\zeta_1)\neq n(\zeta_2)$ for $\zeta_1\neq\zeta_2$. Now with each point $\zeta\in\sigma\cap e$ associate the circle $\Gamma_\zeta: |z-\zeta|<\delta/2^{n(\zeta)+1}$. Then since the set $\sigma\cap e$ is obviously bounded closed, there are a finite number of points $\zeta_j(j=1,2,\cdots,p)$ of $\sigma\cap e$, such that

$$\sigma\cap e\subset\bigcup_{j=1}^{p}\Gamma_{\zeta_j}.$$

Note that the sum of the radii of $\Gamma_{\zeta_j}(j=1,2,\cdots,p)$ is

$$\sum_{j=1}^{p}\frac{\delta}{2}\frac{1}{2^{n(\zeta_j)}}<\frac{\delta}{2}\sum_{n=1}^{\infty}\frac{1}{2^n}=\frac{\delta}{2}.$$

The set

$$\sigma_* = \sigma-\bigcup_{j=1}^{p}\Gamma_{\zeta_j}$$

is non-empty and is bounded closed. Moreover $\sigma_*\subset D_0$. Since $\sigma\subset\sigma_*\cup(\bigcup_{j=1}^{p}\Gamma_{\zeta_j})$, the set σ_* can not be contained in a finite number of circles of which the sum of the radii is less than $\delta/2$. Setting

$$\mu_n^* = \{z\,|\,z\in\sigma_*, |f_n(z)|>n\}, \tag{9.14}$$

then $\mu_n=\mu_n^*\cup(\bigcup_{j=1}^{p}\Gamma_{\zeta_j})$ and it follows that μ_n^* can not be contained in a finite number of circles of which the sum of the radii is less than $\delta/2$. Since $\sigma_*\subset D_0$, we can carry on the reasoning as in case 1), in replacing σ by σ_*, δ by $\delta/2$ and (9.13) by (9.14), and get again a contradiction. This completes the proof of Lemma 9.9.

Theorem 9.11. Given a domain D, an integer $m\geqslant 0$, a condition K of Q_m-normality, a set $E\subset D$ having the property W_m with respect to D, a positive function $M(z)$ defined on E, a bounded closed set $\sigma\subset D$ and a positive number δ, we can find a positive number $A(D,m,K,E,M(z),\sigma,\delta)$ such that if $f(z)$ is a meromorphic function in D satisfying the conditions:

1^0 $f(z)$ satisfies the condition K in D,

2^0 $|f(z)|\leqslant M(z)$ for $z\in E$;

then the set

$$\{z \mid z \in \sigma, |f(z)| > A(D, m, K, E, M(z), \sigma, \delta)\}$$

can be contained in a finite number of circles of which the sum of the radii is less than δ.

To prove this theorem, it is sufficient to consider the family $\mathscr{F}$ of all meromorphic functions in D satisfying the conditions 1^0 and 2^0, and to apply Lemma 9. 9.

Lemma 9. 10. $m\geqslant 1$ and $\nu\geqslant 0$ being two integers, let $\mathscr{F}$ be a Q_m-normal of order ν at most family of meromorphic functions in a domain D. Suppose that there is a set $E\subset D$ satisfying the following conditions:

1^0 E has the property $W_{m-1,\nu+1}$ with respect to D.

2^0 At each point of E the family $\mathscr{F}$ is uniformly bounded.

Then for any bounded closed set $\sigma\subset D$ and any positive number δ, we can find a positive number $\alpha(\sigma,\delta)$ such that for each function $f(z)\in\mathscr{F}$, the set

$$\{z \mid z \in \sigma, |f(z)| > \alpha(\sigma,\delta)\}$$

can be contained in a finite number of circles of which the sum of the radii is less than δ.

Proof. Let $\sigma\in D$ be a bounded closed set and δ a positive number such that σ can not be contained in a finite number of circles of which the sum of the radii is less than δ. Suppoes that we can not find a positive number $\alpha(\sigma,\delta)$ having the required property. Then there is a sequence $S: f_n(z)\in\mathscr{F}(n=1,2,\cdots)$ such that for each positive integer n, the set μ_n defined by (9. 13) can not be contained in a finite number of circles of which the sum of the radii is less then δ. From S we can extract a subsequence S' $:f_{n_k}(z)(k=1,2,\cdots)$ which is a C_m-sequence in D and has at most ν nonC_{m-1}-points in D. Let e be the set of nonC_{m-1}-points of S' in D. In the domain $D_1=D$-e, S' is a C_{m-1}-sequence, and hence S' considered as a family, is Q_{m-1}-normal in D_1. On the other hand, there is a point $z_0\in E^{(m-1)}\cap D_1$. Let

$\Upsilon: |z-z_0|<r$ be a circle belonging to D_1. Then the set $E_1=E\cap\Upsilon$ has the property W_{m-1} with respect to D_1. Besides at each point of E_1, S' is uniformly bounded. Now distinguish two cases:

1) $\sigma\subset D_1$. In this case, by Lemma 9. 9, we can find a positive number α_1 such that for each function $f_{n_k}(z)\in S'$, the set

$$\{z\,|\,z\in\sigma, |f_{n_k}(z)|>\alpha_1\}$$

can be contained in a finite number of circles of which the sum of the radii is less than δ. This is incompatible with the fact that the set

$$\mu_{n_k}=\{z\,|\,z\in\sigma, |f_{n_k}(z)|>n_k\}$$

can not be contained in a finite number of circles of which the sum of the radii is less than δ, when $n_k>\alpha_1$.

2) The set $\sigma\cap e$ is non-empty. Let $e=(z_1,z_2,\cdots,z_h)$ and $\sigma_1=\sigma-\bigcup_{j=1}^{h}\Upsilon_j$, where Υ_j is the circle $\Upsilon_j: |z-z_j|<\delta/2(h+1)$. Since $\sigma\subset\sigma_1\cup(\bigcup_{j=1}^{h}\Upsilon_j)$, $\sigma_1\subset D_1$ is a bounded closed set and can not be contained in a finite number of circles of which the sum of the radii is less than $\delta/2$. Again by Lemma 9. 9, we can find a positive number α_2 such that for each function $f_{n_k}(z)\in S'$, the set

$$\{z\,|\,z\in\sigma_1, |f_{n_k}(z)|>\alpha_2\}$$

can be contained in a finite number of circles of which the sum of the radii is less than $\delta/2$. It follows that the set

$$\{z\,|\,z\in\sigma, |f_{n_k}(z)|>\alpha_2\}$$

can be contained in a finite number of circles of which the sum of the radii is less than δ. This is again incompatible with the fact mentioned above. The proof of Lemma 9. 10 is now complete.

From Lemma 9. 10 we deduce in the usual way, the following theorem:

Theorem 9. 12. Given a domain D, two integers $m \geqslant 1$ and $\nu \geqslant 0$, a condition K of Q_m-normality of order ν at most, a set $E \subset D$ having the property $W_{m-1,\nu+1}$ with respect to D, a positive function M(z) defined on E, a bounded closed set $\sigma \subset D$ and a positive number δ, we can find a positive number $A(D,m,\nu,K,E,M(z),\sigma,\delta)$ such that if f(z) is a meromorphic function in D satisfying the conditions:

1^0 f(z) satisfies the condition K in D,

2^0 $|f(z)| \leqslant M(z)$ for $z \in E$;

then the set

$$\{z \mid z \in \sigma, |f(z)| > A(D,m,\nu,K,E,M(z),\sigma,\delta)\}$$

can be contained in a finite number of circles of which the sum of the radii is less than δ.

Lemma 9. 11. $m \geqslant 0$ being an integer, let $\mathscr{F}$ be a Q_m-normal family of meromorphic functions in a domain D. Suppose that there are a (m)-collection {E} with respect to D and a positive number M satisfying the following condition: For each function $f(z) \in \mathscr{F}$ there is a set $E = E(f)$ of the collection {E} such that

$$|f(z)| \leqslant M \text{ for } z \in E(f). \tag{9.15}$$

Then for any bounded closed set $\sigma \subset D$ and any positive number δ, we can find a positive number $\alpha(\sigma,\delta)$ such that for each function $f(z) \in \mathscr{F}$, the set

$$\{z \mid z \in \sigma, |f(z)| > \alpha(\sigma,\delta)\}$$

can be contained in a finite number of circles of which the sum of the radii is less than δ.

Proof. Let $\sigma \subset D$ be a bounded closed set and δ a positive number such that σ can not be contained in a finite number of circles of which the sum of the radii is less than δ. Suppose that we can not find a positive number $\alpha(\sigma,\delta)$ having the required property. Then there is a sequence $S: f_n(z) \in \mathscr{F}$ $(n=1,2,\cdots)$ such that for each positive integer n, the set μ_n defined by (9. 13) can not be contained in a finite number of circles of which the sum of the radii is less than δ. From S we can extract a subsequence S' : $f_{n_k}(z)(k=1,2,\cdots)$ which is a C_m-sequence in D. In the domain $D_0=D-e$, S' is a C_0-sequence, where e is the set of nonC_0-points of S' in D. Consider the corresponding sequence of sets $E_k=E(f_{n_k})(k=1,2,\cdots)$ of the collection $\{E\}$, which has a limiting set $\widetilde{E}\subset D$ having the property W_m with respect to D. There is a point $z_0\in D_0\cap\widetilde{E}$. Let Γ : $|z-z_0|<r$ be a circle such that $\overline{\Gamma}$: $|z-z_0|\leqslant r$ belongs to D_0. By Definition 9. 3, for any positive integer K, there is a positive integer $k\geqslant K$ such that the set $E_k\cap\Gamma$ is non-empty. This and the condition (9. 15) imply that the limit function f(z)(with respect to the spherical distance) of S' in D_0 is not the constant ∞, and hence f(z) is a meromorphic function in D_0. The remaining part of the proof is the same as that of Lemma 9. 9.

From Lemma 9.11 we deduce the following theorem:

Theorem 9. 13. Given a domain D, an integer $m\geqslant 0$, a condition K of Q_m-normality, a (m)-collection $\{E\}$ with respect to D, a positive number M, a bounded closed set $\sigma\subset D$ and a positive number δ, we can find a positive number $A(D,m,K,\{E\},M,\sigma,\delta)$ such that if f(z) is a meromorphic function in D satisfying the conditions:

1^0 f(z) satisfies the condition K in D,

2^0 there is a set E of the collection $\{E\}$ such that $|f(z)|\leqslant M$ for $z\in E$;

then the set

$$\{z\,|\,z\in\sigma,|f(z)|>A(D,m,K,\{E\},M,\sigma,\delta)\}$$

can be contained in a finite number of circles of which the sum of the radii is less than δ.

Corollary 9. 6. Given a domain D, a condition K of Q_0-normality, a bounded closed set $e\subset D$, a positive number M, a bounded closed set $\sigma\subset D$ and a positive number δ, we can find a positive number $A(D,K,e,M,\sigma,\delta)$ such that if f(z) is a meromorphic function in D satisfying the conditions:

1^0 f(z) satisfies the condition K in D,

$2^0 \min_{z\in e}|f(z)|\leqslant M$;

then the set

$$\{z \mid z \in \sigma, |f(z)| > A(D, K, e, M, \sigma, \delta)\}$$

can be contained in a finite number of circles of which the sum of the radii is less than δ.

To deduce this corollary from Theorem 9. 13, it is sufficient to define a (0)-collection $\{E\}$ with respect to D, by considering each point of the set e to be a set E of the collection $\{E\}$.

Corollary 9. 7. Given a domain D, a number $h(0<h<1)$, a bounded closed set $e\subset D$, a positive number M, a bounded closed set $\sigma\subset D$ and a positive number δ, we can find a positive number $A_1(D,h,e,M,\sigma,\delta)$ such that if $f(z)$ is a meromorphic function in D satisfying the conditions:

1^0 there are three values $a_j\in\hat{C}(j=1,2,3)$ such that

$$|a_j, a_{j'}| > h(1 \leqslant j, j' \leqslant 3, j \neq j')$$

and that $f(z)$ does not take the values $a_j(j=1,2,3)$ in D,

$2^0 \min_{z\in e}|f(z)|\leqslant M$;

then the set

$$\{z \mid z \in \sigma, |f(z)| > A_1(D, h, e, M, \sigma, \delta)\}$$

can be contained in a finite number of circles of which the sum of the radii is less than δ.

This corollary can be deduced from Corollary 9. 6. It is sufficient to define

$$A_1(D, h, e, M, \sigma, \delta) = A(D, K, e, M, \sigma, \delta),$$

where K is the condition 1^0 in Corollary 9. 7. K is a condition of Q_0-normality by Theorem 3. 9.

Lemma 9. 12. $m\geqslant1$ and $\nu\geqslant0$ being two integers, Let $\mathscr{F}$ be a Q_m-normal of order ν at most family of meromorphic functions in a domain D. Suppose that there are a $(m-1, \nu+1)$-collection $\{E\}$ with respect to D and a positive number M safisfying the following condition: For each function $f(z) \in \mathscr{F}$ there is a set $E=E(f)$ of the collection $\{E\}$ such that

$$|f(z)| \leqslant M \ for \ z \in E(f). \tag{9. 16}$$

Then for any bounded closed set $\sigma \subset D$ and any positive number δ, we can find a positive number $\alpha(\sigma, \delta)$ such that for each function $f(z) \in \mathscr{F}$, the set

$$\{z \mid z \in \sigma, |f(z)| > \alpha(\sigma, \delta)\}$$

can be contained in a finite number of circles of which the sum of the radii is less than δ.

Proof. Let $\sigma \subset D$ be a bounded closed set and δ a positive number such that σ can not be contained in a finite number of circles of which the sum of the radii is less than δ. Suppose that we can not find a positive number $\alpha(\sigma, \delta)$ having the required property. Then there is a sequence $S: f_n(z) \in \mathscr{F} (n=1,2,\cdots)$ such that for each positive integer n, the set μ_n defined by (9. 13) can not be contained in a finite number of circles of which the sum of the radii is less than δ. From S we can extract a subsequence $S': f_{n_k}(z) (k=1,2,\cdots)$ which is a C_m-sequence in D and has at most ν nonC_{m-1}-points in D. Denoting by e_1 the set of nonC_{m-1}-points of S' in D. Then S' is a C_{m-1}-sequence in the domain $D_1 = D - e_1$. Next let e_2 be the set of nonC_0-points of S' in D_1. Then e_2 does not have the property W_{m-1} with respect to D_1, and S' is a C_0-sequence in the domain $D_2 = D_1 - e_2$. Define $e = e_1 \cup e_2$, then $D_2 = D - e$.

Consider the corresponding sequence of sets $E_k = E(f_{n_k}) (k=1,2,\cdots)$ of the collection $\{E\}$, which has a limiting set $\widetilde{E} \subset D$ having the property $W_{m-1,\nu+1}$ with respect to D. There is then a point $z_0 \in \widetilde{E}^{(m-1)} \cap D_1$. Let $\Gamma: |z-z_0| < r$ be a circle belonging to D_1. Then the set $\widetilde{E} \cap \Gamma$ has the property W_{m-1} with respect to Γ, hence also with respect to D_1. Accordingly there is a point $z_1 \in \widetilde{E} \cap D_2$. Let $\Gamma_1: |z-z_1| < r_1$ be a circle such that $\overline{\Gamma}_1: |z-z_1| \leqslant r_1$ belongs to D_2. Since $\widetilde{E}$ is a limiting set of the sequence of sets $E_k (k=1,2,\cdots)$, hence, by Definition 9. 3, for any positive integer K, we can find an integer $k \geqslant K$ such that the set $E_k \cap \Gamma_1$ is non-empty. This implies, by the condition (9. 16), the limit function f(z) (with respect to the spherical distance) of S' in D_2 is not the constant ∞ and hence $f(z)$ is a meromorphic function in D_2. The remaining part of the proof is the same as that of Lemma 9.9.

From Lemma 9.12 we deduce the following theorem:

Theorem 9.14. Given a domain D, two integers $m \geqslant 1$ and $\nu \geqslant 0$, a condition K of Q_m-normality of order ν at most, a $(m-1, \nu+1)$-collection $\{E\}$ with respect to D, a positive number M, a bounded closed set $\sigma \subset D$ and a positive number δ, we can find a positive number $A(D, m, \nu, K, \{E\}, M, \sigma, \delta)$ such that if $f(z)$ is a meromorphic function in D satisfying the conditions:

1^0 $f(z)$ satisfies the condition K in D,

2^0 There is a set E of the collection $\{E\}$ such that $|f(z)| \leqslant M$ for $z \in E$;

then the set

$$\{z \mid z \in \sigma, |f(z)| > A(D, m, \nu, K, \{E\}, M, \sigma, \delta)\}$$

can be contained in a finite number of circles of which the sum of the radii is less than δ.

Corollary 9.8. Given a domain D, an integer $\nu \geqslant 1$, a condition K of Q_1-normality of order ν at most, a bounded closed set $e \subset D$ and a positive number η, a positive number M, a bounded closed set $\sigma \subset D$ and a positive number δ, we can find a positive number $A(D, \nu, K, e, \eta, M, \sigma, \delta)$ such that if $f(z)$ is a meromorphic function in D satisfying the conditions:

1^0 $f(z)$ satisfies the condition K in D,

2^0 there exist $\nu+1$ points $z_j \in e$ $(j=1,2,\cdots,\nu+1)$ such that

$$|z_j - z_{j'}| \geqslant \eta \quad (1 \leqslant j, j' \leqslant \nu+1, j \neq j') \tag{9.17}$$

and that

$$|f(z_j)| \leqslant M \quad (j = 1, 2, \cdots, \nu+1);$$

then the set

$$\{z \mid z \in \sigma, |f(z)| > A(D, \nu, K, e, \eta, M, \sigma, \delta)\}$$

can be contained in a finite number of circles of which the sum of the radii is less than δ.

In this corollary and the corollary below, it is tacitly assumed that there exist $\nu+1$ points $z_j \in e (j=1,2,\cdots,\nu+1)$ satisfying the condition (9.17).

To deduce Corollary 9.8 from Theorem 9.14, it is sufficient to define a $(o, \nu+1)$-collection $\{E\}$ with respect to D, by choosing the sets E to be the systems of points $z_j \in e (j=1,2,\cdots,\nu+1)$ satisfying the condition (9.17). (see the proof of Corollary 9.5)

Corollary 9.9. Given a domain D, an integer $\nu \geqslant 1$, a number $h (0<h<1)$, a bounded closed set $e \subset D$ and a positive number η, a positive number M, a bounded closed set $\sigma \subset D$ and a positive number δ, we can find a positive number $A_1(D, \nu, h, e, \eta, M, \sigma, \delta)$ such that if f(z) is a meromorphic function in D satisfying the conditions:

1^0 there are three values $a_j \in \hat{\mathbf{C}} (j=1,2,3)$ such that

$$|a_j, a_{j'}| > h \ (1 \leqslant j, j' \leqslant 3, j \neq j')$$

and that for each $j (1 \leqslant j \leqslant 3)$, the equation $f(z)=a_j$ has at most ν distinct roots in D,

2^0 there exist $\nu+1$ points $z_j \in e (j=1,2,\cdots,\nu+1)$ such that

$$|z_j - z_{j'}| \geqslant \eta \ (1 \leqslant j, j' \leqslant \nu+1, j \neq j')$$

and that

$$|f(z_j)| \leqslant M \ (j=1,2,\cdots,\nu+1);$$

then the set

$$\{z \mid z \in \sigma, |f(z)| > A_1(D, \nu, h, e, \eta, M, \sigma, \delta)\}$$

can be contained in a finite number of circles of which the sum of the radii is less than δ.

This corollary can be deduced from Corollary 9. 8, by defining

$A_1(D,\nu,h,e,\eta,M,\sigma,\delta)=A(D,\nu,K,e,\eta,M,\sigma,\delta)$,

where K is the condition stated in 1^0 of Corollary 9. 9. K is a condition of Q_1-normality of order ν at most, by Theorem 5. 4.

Finally we prove some theorems on the limitation of the spherical derivative of a meromorphic function.

Lemma 9. 13. $m\geqslant 0$ being an integer, let $\mathscr{F}$ be a Q_m-normal family of meromorphic functions in a domain D, and let $\{E\}$ be a (m)-collection with respect to D. Then we can find two positive numbers M and r such that for any function $f(z)\in\mathscr{F}$ and any set E of the collection $\{E\}$, there is a point $z_0\in E$ satisfying the condition:

$$\partial(z,f)\leqslant M \text{ for } z\in D\cap(|z-z_0|<r).$$

Proof. Assuming that such positive numbers M and r do not exist, then to each positive integer n correspond a function $f_n(z)\in\mathscr{F}$ and a set E_n of the collection $\{E\}$ such that there does not exist a point $z_n\in E_n$ satisfying the condition:

$$\partial(z,f_n)\leqslant n \text{ for } z\in D\cap(|z-z_n|<\frac{1}{n}). \tag{9. 18}$$

From the sequence $S: f_n(z)\,(n=1,2,\cdots)$ we can extract a subsequence $S': f_{n_k}(z)$ $(k=1,2,\cdots)$ which is a C_m-sequence in D, and hence is a C_0-sequence in a domain $D_0=D\text{-}e$, where e does not have the property W_m with respect to D. The sequence of sets $E_{n_k}\,(k=1,2,\cdots)$ has a limiting set $\widetilde{E}\subset D$ having the property W_m with respect to D. Accordingly there is a point $z_0\in\widetilde{E}\cap D_0$. Since S' constitutes a Q_0-normal family in D_0, hence by Theorem 1. 6, we can find a circle $\Gamma: |z-z_0|<\rho$ belonging to D_0 and a number $\alpha>0$ such that for $k\geqslant 1$, we have

$$\partial(z,f_{n_k})\leqslant\alpha \text{ for } z\in\Gamma. \tag{9. 19}$$

Choosing a positive integer K such that

$$n_K > a, \ \frac{1}{n_K} < \frac{\rho}{2}, \tag{9.20}$$

by Definition 9.3, there is a positive integer $k \geqslant K$ such that the set $E_{n_k} \cap (|z-z_0| < \frac{\rho}{2})$ is non-empty. Let ζ_k be a point of this set. Then $\zeta_k \in E_{n_k}$ and the circle $|z-\zeta_k| < 1/n_k$ belongs to T Consequently by (9.19) and (9.20), we have

$$\partial(z, f_{n_k}) < n_k \ for \ |z-\zeta_k| < \frac{1}{n_k}.$$

This is incompatible with the fact that for each n, there does not exist a point $z_n \in E_n$ satisfying (9.18).

Theorem 9.15. Given a domain D, and integer $m \geqslant 0$, a condition K of Q_m-normality and a (m)-collection $\{E\}$ with respect to D, we can find two positive numbers $M = M(D, m, K, \{E\})$ and $r = r(D, m, K, \{E\})$ having the following property: If f(z) is a meromorphic function in D satisfying the condition K in D, then for any set E of the collection $\{E\}$, there is a point $z_0 \in E$ such that

$$\partial(z, f) \leqslant M \ for \ z \in D \cap (|z-z_0| < r).$$

To prove this theorem, we consider the family $\mathscr{F}$ of all meromorphic functions in D satisfying the condition K in D, and then apply Lemma 9.13 to $\mathscr{F}$ and $\{E\}$.

Lemma 9.14. $m \geqslant 1$ and $\nu \geqslant 0$ being two integers, let $\mathscr{F}$ be a Q_m-normal of order ν at most family of meromorphic functions in a domain D, and let $\{E\}$ be a $(m-1, \nu+1)$-collection with respect to D. Then we can find two positive numbers M and r such that for any function $f(z) \in \mathscr{F}$ and any set E of the collection $\{E\}$, there is a point $z_0 \in E$ satisfying the condition:

$$\partial(z, f) \leqslant M \ for \ z \in D \cap (|z-z_0| < r).$$

Proof. Assuming that such positive numbers M and r do not exist, then to each positive integer n correspond a function $f_n(z)\in\mathscr{F}$ and a set E_n of the collection $\{E\}$ such that there does not exist a point $z_n\in E_n$ satisfying the condition (9.18). From the sequence $S: f_n(z)(n=1,2,\cdots)$ we can extract a subsequence $S': f_{n_k}(z)(k=1,2,\cdots)$ which is a C_m-sequence in D and has at most ν nonC_{m-1}-points in D. S' is a C_{m-1}-sequence in the domain $D_1=D-e_1$, where e_1 is the set of nonC_{m-1}-points of S in D. Let e_2 be the set of nonC_0-points of S' in D_1, then e_2 does not have the property W_{m-1} with respect to D_1, and S' is a C_0-sequence in the domain $D_2=D_1-e_2$. The sequence of sets $E_{n_k}(k=1,2,\cdots)$ has a limiting set $\widetilde{E}\subset D$ having the property $W_{m-1,\nu+1}$ with respect to D. As in the proof of Lemma 9.12, we see that there is a point $z_1\in\widetilde{E}\cap D_2$. The remaining part of the proof is carried out in the same way as that of Lemma 9.13.

From Lemma 9.14 we deduce the following theorem:

Theorem 9.16. Given a domain D, two integers $m\geqslant 1$ and $\nu\geqslant 0$, a condition K of Q_m-normality of order ν at most and a $(m-1,\nu+1)$-collection $\{E\}$ with respect to D, we can find two positive numbers $M=M(D,m,\nu,K,\{E\})$ and $r=r(D,m,\nu,K,\{E\})$ having the following property: If $f(z)$ is a meromorphic function in D satisfying the condition K in D, then for any set E of the collection $\{E\}$, there is a point $z_0\in E$ such that

$$\partial(z,f)\leqslant M \text{ for } z\in D\cap(|z-z_0|<r).$$

Theorems 9.15 and 9.16 are generalizations of Theorem 3.4.

9.4. LIMITATION OF THE NUMBER OF THE ZEROS OF MEROMORPHIC FUNCTIONS

We first state the following lemma which is an immediate consequence of Theorem 5.19:

Lemma 9.15. Let $\mathscr{F}$ be a Q_0-normal family of meromorphic functions in a domain D. Then for any bounded domain Δ such that $\bar{\Delta}=\Delta+\partial\Delta\subset D$ and any number $0<\varepsilon<1$, we can find a positive integer $N(\Delta,\varepsilon)$ having the following property: If $f(z)$ is a function of $\mathscr{F}$, then for every value $w\in\hat{\mathbf{C}}$, we have

$$n(f,w,\Delta) \leqslant N(\Delta,\varepsilon),$$

except at most the values w which belong to a spherical circle $|w, w_0(f)| < \varepsilon$.

Here and in what follows, we shall use the notation $n(f,w,\Delta)$ to denote the number of the roots of the equation $f(z)=w$ in an open set $\Delta \subset D$, with due count of order of multiplicity.

From this lemma we deduce the following theorem:

Theorem 9.17. Given a domain D, a condition K of Q_0-normality, a domain Δ such that $\bar{\Delta}=\Delta+\partial\Delta\subset D$ and a number $0<\varepsilon<1$, we can find a positive integer $N(D,K,\Delta,\varepsilon)$ such that if $f(z)$ is a meromorphic function in D satisfying the condition K in D, then for every value $w\in\hat{C}$, we have

$$n(f,w,\Delta) \leqslant N(D,K,\Delta,\varepsilon),$$

except at most the values w which belong to a spherical circle $|w, w_0(f)| < \varepsilon$.

To prove this theorem, it is sufficient to apply Lemma 9.15 to the family $\mathscr{F}$ of meromorphic functions in D and satisfying the condition K in D.

Lemma 9.16. $m\geqslant 0$ being an integer, let $\mathscr{F}$ be a Q_m-normal family of meromorphic functions in a domain D, let $\{E\}$ be a (m)-collection with respect to D and let $0<\varepsilon<1$ be a number. Then we can find a positive integer N and a positive number r such that for any function $f(z)\in\mathscr{F}$, and any set E of the collection $\{E\}$, there is a point $z_0\in E$ such that for every value $w\in\hat{C}$, we have

$$n(f,w,\Omega) \leqslant N,\ \Omega = D \cap (|z-z_0| < r),$$

except at most the values w which belong to a spherical circle $|w, w_0(f,E)| < \varepsilon$.

Proof. Assuming that such N and r do not exist, then to each positive integer n correspond a function $f_n(z)\in\mathscr{F}$ and a set E_n of the collection $\{E\}$ such that there does not exist a point $z_n\in E_n$ having the following property: For every value $w\in\hat{C}$, we have

$$n(f_n,w,\Omega_n) \leqslant n,\ \Omega_n = D \cap (|z-z_n| < \frac{1}{n}), \qquad (9.21)$$

except at most the values w which belong to a spherical circle $|w, w_n| < \varepsilon$.

From the sequence $S: f_n(z)(n=1,2,\cdots)$ we can extract a subsequence $S': f_{n_k}(z)(k=1,2,\cdots)$ which is a C_m-sequence in D, and hence is a C_0-sequence in a domain $D_0 = D - e$, where e does not have the property W_m with respect to D. The sequence of sets $E_{n_k}(k=1,2,\cdots)$ has a limiting set $\tilde{E} \subset D$ having the property W_m with respect to D. Accordingly there is a point $z_0 \in \tilde{E} \cap D_0$. Let $\Gamma: |z-z_0| < \rho$ be a circle such that $\bar{\Gamma}: |z-z_0| \leqslant \rho$ belongs to D_0. Since S' constitutes a Q_0-normal family in D_0, by Lemma 9.15, we can find a positive integer P such that for $k \geqslant 1$ we have

$$n(f_{n_k}, w, \Gamma) \leqslant P$$

which holds for every value $w \in \hat{C}$, except at most the values w which belong to a spherical circle $|w, w'_k| < \varepsilon$.

Choosing a positive integer K such that

$$n_K > P, \quad \frac{1}{n_K} < \frac{\rho}{2},$$

by Definition 9.3, there is a positive integer $k \geqslant K$ such that the set $E_{n_k} \cap (|z-z_0| < \frac{\rho}{2})$ is non-empty. Let ζ_k be a point of this set. Then $\zeta_k \in E_{n_k}$ and the circle $c: |z-\zeta_k| < 1/n_k$ belongs to Γ. Hence for every value $w \in \hat{C}$, we have

$$n(f_{n_k}, w, c) \leqslant P < n_k,$$

except at most the values w which belong to a spherical circle $|w, w'_k| < \varepsilon$. This is incompatible with the fact that for each n there does not exist a point $z_n \in E_n$ having the property (9.21).

Theorem 9.18. Given a domain D, an integer $m \geqslant 0$, a condition K of Q_m-normality, a (m)-collection $\{E\}$ with respect to D and a number $0 < \varepsilon < 1$, we can find a positive integer $N = N(D, m, K, \{E\}, \varepsilon)$ and a positive number $r = r(D, m, K, \{E\}, \varepsilon)$ having the following property: If f(z) is a meromorphic function in D satisfying the condition K in D, then for any set E of the collection $\{E\}$, there is a point $z_0 \in E$ such that for every value $w \in \hat{C}$, we have

$$n(f, w, \Omega) \leqslant N, \; \Omega = D \cap (|z-z_0| < r),$$

except at most the values w which belong to a spherical circle $|w, w_0(f, E)| < \varepsilon$.

To prove this theorem, it is sufficient to apply Lemma 9.16 to the family $\mathscr{F}$ of the meromorphic functions satisfying the condition K in D and to the collection $\{E\}$ and the number ε.

Lemma 9.17. $m \geqslant 1$ and $\nu \geqslant 0$ being two integers, let $\mathscr{F}$ be a Q_m-normal of order ν at most family of meromorphic functions in a domain D, let $\{E\}$ be a $(m-1, \nu+1)$-collection with respect to D and let $0 < \varepsilon < 1$ be a number. Then we can find a positive integer N and a positive number r such that for any function $f(z) \in \mathscr{F}$ and any set E of the collection $\{E\}$, there is a point $z_0 \in E$ such that for every value $w \in \hat{C}$, we have

$$n(f, w, \Omega) \leqslant N, \quad \Omega = D \cap (|z-z_0| < r),$$

except at most the values w which belong to a spherical circle $|w, w_0(f, E)| < \varepsilon$.

Proof. Assuming that such N and r do not exist, then to each positive integer n correspond a function $f_n(z) \in \mathscr{F}$ and a set E_n of the collection $\{E\}$ such that there does not exist a point $z_n \in E_n$ having the property that (9.21) is true for every value $w \in \hat{C}$, except at most those in a spherical circle $|w, w_n| < \varepsilon$. From the sequence $S: f_n(z) (n = 1, 2, \cdots)$ we can extract a subsequence $S': f_{n_k}(z)$ $(k = 1, 2, \cdots)$ which is a C_m-sequence in D and has at most ν nonC_{m-1}-points in D. S' is a C_{m-1}-sequence in the domain $D_1 = D - e_1$, where e_1 is the set of nonC_{m-1}-points of S' in D. Let e_2 be the set of nonC_0-points of S' in D_1, then e_2 does not have the property W_{m-1} with respect to D_1 and S' is a C_0-sequence in the domain $D_2 = D_1 - e_2$.

The sequence of sets $E_{n_k} (k = 1, 2, \cdots)$ has a limiting set $\tilde{E} \subset D$ having the property $W_{m-1, \nu+1}$ with respect to D. As in the proof of Lemma 9.12, we see that there is a point $z_1 \in \tilde{E} \cap D_2$. Since S' constitutes a Q_0-normal family in D_2, the remaining part of the proof is carried out in the same way as in the proof of Lemma 9.16.

Theorem 9. 19. Given a domain D, two integers $m \geqslant 1$, $\nu \geqslant 0$, a condition K of Q_m-normality of order ν at most, a $(m-1, \nu+1)$-collection $\{E\}$ with respect to D and a number $0 < \varepsilon < 1$, we can find a positive integer $N = N(D, m, \nu, K, \{E\}, \varepsilon)$ and a positive number $r = r(D, m, \nu, K, \{E\}, \varepsilon)$ having the following property: If $f(z)$ is a meromorphic function in D satisfying the condition K in D, then for any set E of the collection $\{E\}$, there is a point $z_0 \in E$ such that for every value $w \in \hat{C}$, we have

$$n(f, w, \Omega) \leqslant N, \ \Omega = D \cap (|z - z_0| < r),$$

except at most the values w which belong to a spherical circle $|w, w_0(f, E)| < \varepsilon$.

As for Theorem 9. 18, we deduce Theorem 9. 19 from Lemma 9. 17.

9. 5. DOMAINS COVERED BY THE VALUES OF MEROMORPHIC FUNCTIONS

As simple consequences of Theorems 8. 11, 8. 12, 8. 15, 8. 16 and Theorems 8. 19, 8. 20, we are going to prove the following theorems which in a certain sense, are extensions of theorems on the domains covered by the values of holomorphic or meromorphic functions proved in Chapters 2, 3, 5, 6, and 7.

Definition 9. 6. Let $E_j (j = 1, 2, 3)$ be three sets of points of the extended complex plane $\hat{C}$. We say that $E_j (j = 1, 2, 3)$ are uniformly mutually disjoint, if there is a number $\delta (0 < \delta < 1)$ such that for $j \neq j'$ $(1 \leqslant j, j' \leqslant 3)$ and $w \in E_j$, $w' \in E_{j'}$, we have aways $|w, w'| > \delta$.

Theorem 9. 20. Let D be a domain, $\nu \geqslant 0$ an integer and $\sigma^{(1)}: z_j (1 \leqslant j \leqslant \nu + 1)$ a system of points of D, satisfying the condition (Γ_1) with respect to D. (see Definition 8. 8). Let $E_j (j = 1, 2, 3)$ be three uniformly mutually disjoint sets of points of $\hat{C}$. Then we can find a positive number r such that the circles $\overline{\Gamma}_j: |z - z_j| \leqslant r (j = 1, 2, \cdots, \nu + 1)$ belong to D and a positive number A, both depending only on D, ν, $\sigma^{(1)}$ and $E_j (j = 1, 2, 3)$, having the following property: If $f(z)$ is a meromorphic function in D such that

$$\min_{1\leqslant j\leqslant \nu+1}(\max_{z\in \overline{\Gamma}_j}\partial(z,f))\geqslant A,$$

then there is an $j(1\leqslant j\leqslant 3)$ such that for each point $w\in E_j$, $f(z)$ has at least $\nu+1$ distinct $(w)_1$-circles in D.

Proof. Assume that such r and A do not exist. Taking a positive number ρ such that the circles $|z-z_j|\leqslant\rho(j=1,2,\cdots,\nu+1)$ belong to D, then to each positive integer n corresponds a meromorphic function $f_n(z)$ in D such that

$$\min_{1\leqslant j\leqslant \nu+1}(\max_{|z-z_j|\leqslant\frac{\rho}{n}}\partial(z,f_n))\geqslant n \tag{9.22}$$

and that there are three points $w_j^{(n)}\in E_j(j=1,2,3)$ such that for each $j(1\leqslant j\leqslant 3)$, $f_n(z)$ has at most ν distinct $(w_j^{(n)})_1$-circles in D. By Theorem 8.20, the family $S:f_n(z)(n=1,2,\cdots)$ is Q_1-normal of order ν at most in D. Next by Theorem 8.11, the family S satisfies the condition (M)(see Definition 8.9) with respect to the system of points $\sigma^{(1)}:z_j(1\leqslant j\leqslant\nu+1)$. This is incompatible with (9.22).

Theorem 9.21. Let D be a domain and $\sigma^{(1)}:z_j(j\geqslant 1)$ be a sequence of points of D, satisfying the condition (Γ_1') with respect to D. (see Definition 8.8) Let $E_j(j=1,2,3)$ be three uniformly mutually disjoint sets of points of $\hat{\mathbf{C}}$. Then we can find a positive integer p, a positive number r such that the circles $\overline{\Gamma}_j:|z-z_j|\leqslant r(j=1,2,\cdots,p)$ belong to D and a positive number A, all depending only on D, $\sigma^{(1)}$ and $E_j(j=1,2,3)$, having the following property: If f(z) is a meromorphic function in D such that

$$\min_{1\leqslant j\leqslant p}(\max_{z\in \overline{\Gamma}_j}\partial(z,f))\geqslant A,$$

then there is an $j(1\leqslant j\leqslant 3)$ such that for each point $w\in E_j$, $f(z)$ has a $(w)_2$-circle in D.

Proof. Assume that such p, r and A do not exist. Taking a sequence of positive numbers $\rho_p(p=1,2,\cdots)$ such that

$$\lim_{p\to+\infty}\rho_p = 0$$

and that for each $p(p\geqslant 1)$ the circles $\overline{\Gamma}_j^{(p)}: |z-z_j|\leqslant\rho_p(j=1,2,\cdots,p)$ belong to D, then to each positive integer p corresponds a meromorphic function $f_p(z)$ in D such that

$$\min_{1\leqslant j\leqslant p}(\max_{z\in\overline{\Gamma}_j^{(p)}}\partial(z,f_p))\geqslant p \tag{9.23}$$

and that there are three points $w_j^{(p)}\in E_j(j=1,2,3)$ such that for each $j(1\leqslant j\leqslant 3)$ $f_p(z)$ has no $(w_j^{(p)})_2$-circle in D. By Theorem 8.19, the family $S: f_p(z)(p=1,2,\cdots)$ is Q_1-normal in D. Next by Theorem 8.12, we can find an integer $p_0\geqslant 1$ such that the family S satisfies the condition (M) with respect to the system $z_j(1\leqslant j\leqslant p_0)$. Accordingly by Definition 8.9, we can find circles $\overline{C}_j: |z-z_j|\leqslant r_j(j=1,2,\cdots,p_0)$ belonging to D and a number $\alpha>0$ such that for each $p\geqslant 1$, we have

$$\min_{1\leqslant j\leqslant p_0}(\max_{z\in\overline{C}_j}\partial(z,f_p))\leqslant\alpha.$$

Now choosing a positive integer P such that

$$P>p_0,\quad P>\alpha,\quad \rho_P<r_j\quad(j=1,2,\cdots,p_0), \tag{9.24}$$

then

$$\begin{aligned}\min_{1\leqslant j\leqslant P}(\max_{z\in\overline{\Gamma}_j^{(P)}}\partial(z,f_P)) &\leqslant \min_{1\leqslant j\leqslant p_0}(\max_{z\in\overline{\Gamma}_j^{(P)}}\partial(z,f_P))\\ &\leqslant \min_{1\leqslant j\leqslant p_0}(\max_{z\in\overline{C}_j}\partial(z,f_P))\leqslant\alpha.\end{aligned}$$

this is incompatible with (9.23) and (9.24).

Similarly basing upon Theorems 8.15, 8.16 and Theorems 8.19, 8.20, we can prove the following two theorems:

Theorem 9.22. Let D be a domain, $m\geqslant 2$, $\nu\geqslant 0$ two integers and $\sigma^{(m)}$: $z_{j_1j_2\cdots j_m}(1\leqslant j_1\leqslant \nu+1, j_2\geqslant 1,\cdots,j_m\geqslant 1)$ be a m-ple sequence of points of D, satisfying the condition (Γ_m) with respect to D. (see Definition 8.8) Let $E_j(j=1,2,3)$

be three uniformly mutually disjoint sets of points of $\hat{C}$. Then we can find a positive integer p, a positive number r such that the circles $\overline{\Gamma}_{j_1j_2\cdots j_m}: |z-z_{j_1j_2\cdots j_m}|\leqslant r\,(1\leqslant j_1\leqslant \nu+1, 1\leqslant j_2\leqslant p, \cdots, 1\leqslant j_m\leqslant p)$ belong to D and a positive number A, all depending only on D, m, ν, $\sigma^{(m)}$ and $E_j(j=1,2,3)$, having the following property: If f(z) is a meromorphic function in D such that

$$\min_{1\leqslant j_1\leqslant \nu+1, 1\leqslant j_2\leqslant p, \cdots, 1\leqslant j_m\leqslant p}\left(\max_{z\in \overline{\Gamma}_{j_1j_2\cdots j_m}} \partial(z,f)\right)\geqslant A,$$

then there is an $j(1\leqslant j\leqslant 3)$ such that for each point $w\in E_j$, f(z) has at least $\nu+1$ distinct $(w)_m$-circles in D.

Theorem 9. 23. Let D be a domain, $m\geqslant 2$ an integer and $\sigma^{(m)}: z_{j_1j_2\cdots j_m}(j_1\geqslant 1, j_2\geqslant 1, \cdots, j_m\geqslant 1)$ be a m-ple sequence of points of D, satisfying the condition (Γ'_m) with respect to D. (see Definition 8. 8). Let $E_j(j=1,2,3)$ be three uniformly mutually disjoint sets of points of $\hat{C}$. Then we can find a positive integer p, a positive number r such that the circles $\overline{\Gamma}_{j_1j_2\cdots j_m}: |z-z_{j_1j_2\cdots j_m}|\leqslant r\,(1\leqslant j_1\leqslant p, 1\leqslant j_2\leqslant p, \cdots, 1\leqslant j_m\leqslant p)$ belong to D and a postive number A, all depending only on D, m, $\sigma^{(m)}$ and $E_j(j=1,2,3)$, having the following property: If f(z) is a meromorphic function in D such that

$$\min_{1\leqslant j_1\leqslant p, 1\leqslant j_2\leqslant p, \cdots, 1\leqslant j_m\leqslant p}\left(\max_{z\in \overline{\Gamma}_{j_1j_2\cdots j_m}} \partial(z,f)\right)\geqslant A,$$

then there is an $j(1\leqslant j\leqslant 3)$ such that for each point $w\in E_j$, f(z) has a $(w)_{m+1}$-circle in D.

Lemma 9. 18. Given any integers $m\geqslant 1$ and $n\geqslant 1$, we can find in the interval $I: 0<x<1$, a m-ple sequence of points $\sigma^{(m)}: x_{j_1j_2\cdots j_m}(1\leqslant j_1\leqslant n, j_2\geqslant 1, \cdots, j_m\geqslant 1)$ satisfying the condition (Γ_m) with respect to the interval I, and a m-ple sequence of points $\sigma_1^{(m)}: x_{j_1j_2\cdots j_m}(j_1\geqslant 1, j_2\geqslant 1, \cdots, j_m\geqslant 1)$ satisfying the condition (Γ'_m) with respect to the interval I. (see Definition 8. 8, for I)

Proof. For the case $m=1$, $\sigma^{(1)}: x_{j_1}(1\leqslant j_1\leqslant n)$ is a system of points. In this case, the existence of $\sigma^{(1)}$ is evident. Now assume that for an integer $m\geqslant 1$, we can find in the interval I a m-ple sequence of points $\sigma^{(m)}: x_{j_1j_2\cdots j_m}(1\leqslant j_1\leqslant n, j_2\geqslant 1,$

$\cdots, j_m \geqslant 1)$ satisfying the condition (Γ_m) with respect to I. Let us define a $(m+1)$-ple sequence of points $\sigma^{(m+1)}: x_{j_1 j_2 \cdots j_{m+1}} (1 \leqslant j_1 \leqslant n, j_2 \geqslant 1, \cdots, j_{m+1} \geqslant 1)$ of I as follows: To each point $x_{j_1 j_2 \cdots j_m}$ associate an increasing sequence of points $X_j^{(j_1 j_2 \cdots j_m)} (j \geqslant 1)$ of the interval $0 < x < x_{j_1 j_2 \cdots j_m}$ such that

$$\lim_{j \to +\infty} X_j^{(j_1 j_2 \cdots j_m)} = x_{j_1 j_2 \cdots j_m}$$

and then define

$$x_{j_1 j_2 \cdots j_{m+1}} = X_{j_{m+1}}^{(j_1 j_2 \cdots j_m)},$$

then evidently $\sigma^{(m+1)}$ satisfies the condition (Γ_{m+1}) with respect to I.

On the other hand, for the case $m=1$, $\sigma_1^{(1)}: x_{j_1} (j_1 \geqslant 1)$ is a sequence of points of I. In this case, it is suficient to take an increasing sequence $\sigma_1^{(1)}: x_{j_1} (j_1 \geqslant 1)$ of points of I, such that the limit of $\sigma_1^{(1)}$ belongs to I. Then evidently $\sigma_1^{(1)}$ satisfies the condition (Γ_1') with respect to I. Next by mathematical induction, we prove as in the above, the existence of $\sigma_1^{(m)}$ for each $m \geqslant 1$.

Definition 9. 7. Let $\varphi(x) (0 \leqslant x \leqslant 1)$ be a complex valued continuous function in the interval $0 \leqslant x \leqslant 1$, such that for $x_1 \neq x_2 (0 \leqslant x_1, x_2 \leqslant 1)$ we have $\varphi(x_1) \neq \varphi(x_2)$. Then we call $\varphi(x) (0 \leqslant x \leqslant 1)$ a simple curve and denote it by C.

Lemma 9. 19. Let D be a domain and $C: \varphi(x) (0 \leqslant x \leqslant 1)$ be a simple curve belonging to D. Let $m \geqslant 1$ and $n \geqslant 1$ be two integers. Then the following two assertions are true:

1^0 If $\sigma^{(m)}: x_{j_1 j_2 \cdots j_m} (1 \leqslant j_1 \leqslant n, j_2 \geqslant 1, \cdots, j_m \geqslant 1)$ is a m-ple sequence of points of the interval $I: 0 < x < 1$, satisfying the condition (Γ_m) with respect to I, then the m-ple sequence of poitns $\tilde{\sigma}^{(m)}: z_{j_1 j_2 \cdots j_m} = \varphi(x_{j_1 j_2 \cdots j_m}) (1 \leqslant j_1 \leqslant n, j_2 \geqslant 1, \cdots, j_m \geqslant 1)$ of C satisfies the condition (Γ_m) with respect to D.

2^0 If $\sigma_1^{(m)}: x_{j_1 j_2 \cdots j_m} (j_1 \geqslant 1, j_2 \geqslant 1, \cdots, j_m \geqslant 1)$ is a m-ple sequence of points of the interval $I: 0 < x < 1$, satisfying the condition (Γ_m') with respect to I, then the m-ple sequence of points $\tilde{\sigma}_1^{(m)}: z_{j_1 j_2 \cdots j_m} = \varphi(x_{j_1 j_2 \cdots j_m}) (j_1 \geqslant 1, j_2 \geqslant 1, \cdots, j_m \geqslant 1)$ of C satisfies the condition (Γ_m') with respect to D.

To prove 1^0, consider first the case $m=1$. In this case, $\sigma^{(1)}: x_{j_1} (1 \leqslant j_1 \leqslant n)$ is a system fo points, and the assertion 1^0 is evidently true. Now assume that 1^0 is true for an integer $m\geqslant 1$. Consider a $(m+1)$-ple sequence $\sigma^{(m+1)}: x_{j_1j_2\cdots j_{m+1}}(1\leqslant j_1\leqslant n, j_2\geqslant 1,\cdots,j_{m+1}\geqslant 1)$ of points of I, satisfying the condition (Γ_{m+1}) with respect to I, and the corresponding $(m+1)$-ple sequence $\tilde{\sigma}^{(m+1)}: z_{j_1j_2\cdots j_{m+1}}=\varphi(x_{j_1j_2\cdots j_{m+1}})(1\leqslant j_1\leqslant n, j_2\geqslant 1,\cdots,j_{m+1}\geqslant 1)$ of points of C. Then by Definition 8.8 (for I), for each system $j_1, j_2,\cdots,j_m(1\leqslant j_1\leqslant n, j_2\geqslant 1,\cdots,j_m\geqslant 1)$, the sequence of points $x_{j_1j_2\cdots j_mj_{m+1}}$ $(j_{m+1}\geqslant 1)$ are distinct and the limit

$$\lim_{j_{m+1}\to+\infty} x_{j_1j_2\cdots j_mj_{m+1}} = \xi_{j_1j_2\cdots j_m}$$

exists and belongs to I. Hence the sequence of points $z_{j_1j_2\cdots j_mj_{m+1}}=\varphi(x_{j_1j_2\cdots j_mj_{m+1}})(j_{m+1}\geqslant 1)$ are distinct and the limit

$$\lim_{j_{m+1}\to+\infty} z_{j_1j_2\cdots j_mj_{m+1}} = \varphi(\xi_{j_1j_2\cdots j_m})$$

exists ad belongs to C. Moreover by Definition 8.8(for I), the m-ple sequence $\xi_{j_1j_2\cdots j_m}(1\leqslant j_1\leqslant n, j_2\geqslant 1,\cdots,j_m\geqslant 1)$ satisfies the condition (Γ_m) with respect to I. Hence by assumption, the m-ple sequence $\varphi(\xi_{j_1j_2\cdots j_m})(1\leqslant j_1\leqslant n, j_2\geqslant 1,\cdots,j_m\geqslant 1)$ satisfies the condition (Γ_m) with respect to D. Consequently by Definition 8.8, the $(m+1)$-ple sequence $\tilde{\sigma}^{(m+1)}$ satisfies the condition (Γ_{m+1}) with respect to D.

To prove 2^0, consider first the case $m=1$. In this case, $\sigma_1^{(1)}: x_{j_1}(j_1\geqslant 1)$ is a sequence of distinct points of I, such that

$$\lim_{j_1\to+\infty} x_{j_1} = \xi$$

exists and belongs to I. Hence the sequence $\tilde{\sigma}_1^{(1)}: z_{j_1}=\varphi(x_{j_1})(j_1\geqslant 1)$ are distinct points of C, such that

$$\lim_{j_1\to+\infty} z_{j_1} = \varphi(\xi)$$

exists and belongs to C. So in this case, 2^0 is true. As for 1^0, we complete the proof of 2^0 by mathematical induction.

Theorem 9.24. Let D be a domain, $m\geqslant 1$, $\nu\geqslant 0$ two integers and $C:\varphi(x)$

$(0\leqslant x\leqslant 1)$ a simple curve belonging to D. Let $E_j(j=1,2,3)$ be three uniformly mutually disjoint sets of points of $\hat{C}$. Then we can find a positive number A depending only on D, m, ν, C and $E_j(j=1,2,3)$ having the following property: If f (z) is a meromorphic function in D such that

$$\min_{z\in C}\partial(z,f)\geqslant A, \tag{9.25}$$

then there is an $j(1\leqslant j\leqslant 3)$ such that for each point $w\in E_j$, f(z) has at least $\nu+1$ distinct $(w)_m$-circles in D.

Proof. First of all, by Lemmas 9.18 and 9.19, we can find a m-ple sequence of points $\tilde{\sigma}^{(m)}: z_{j_1j_2\cdots j_m}(1\leqslant j_1\leqslant \nu+1, j_2\geqslant 1,\cdots,j_m\geqslant 1)$ of C, satisfying the condition (Γ_m) with respect to D. Next by Theorems 9.20 and 9.22, we can find a finite number of terms $\zeta_i(i=1,2,\cdots,q)$ of $\tilde{\sigma}^{(m)}$, a positive number r such that the circles $\bar{\Gamma}_i: |z-\zeta_i|\leqslant r(i=1,2,\cdots,q)$ belong to D and a positive number A, having the following property: If f(z) is a meromorphic function in D such that

$$\min_{1\leqslant j\leqslant q}(\max_{z\in\bar{\Gamma}_i}\partial(z,f))\geqslant A, \tag{9.26}$$

then there is an $j(1\leqslant j\leqslant 3)$ such that for each point $w\in E_j$, f(z) has at least $\nu+1$ distinct $(w)_m$-circles in D. Since (9.25) implies (9.26), hence this positive number A has the required property.

Similarly we can prove the following theorem:

Theorem 9.25. Let D be a domain, $m\geqslant 1,\nu\geqslant 0,k\geqslant 1$ three integers and $C:\varphi(x)(0\leqslant x\leqslant 1)$ a simple curve belonging to D. Let $M,\delta(0<\delta<M)$ be two numbers. Then we can find a positive number A depending only on D, m, ν, k, C and M, δ having the following property: If f(z) is a meromorphic function in D such that

$$\min_{z\in C}\partial(z,f)\geqslant A, \tag{9.27}$$

then one at least of the following two assertions holds:

1° For each point w of the circle $|w|<M$, f(z) has at least $\nu+1$ distinct

$(w)_m$-circles in D.

2^0 For each point w of the domain $\delta<|w|<M$, $f^{(k)}(z)$ has at least $\nu+1$ distinct $(w)_m$-circles in D.

Proof. Assume on the contrary, that there does not exist a positive number A having the required property. Then to each positive integer n corresponds a meromorphic function $f_n(z)$ in D, such that

$$\min_{z\in C}\partial(z,f_n)\geqslant n \tag{9.28}$$

and that there are two points w_n' and w_n'' respectively of the circle $|w|<M$ and the domain $\delta<|w|<M$, for which $f_n(z)$ has at most ν distinct $(w_n')_m$-circles in D and $f_n^{(k)}(z)$ has at most ν distinct $(w_n'')_m$-circles in D. By Theorem 8.22, the family of functions $f_n(z)(n=1,2,\cdots)$ is Q_m-normal of order 2ν at most in D. By Lemmas 9.18 and 9.19, we can find a m-ple sequence of points $\tilde{\sigma}^{(m)}: z_{j_1j_2\cdots j_m}(1\leqslant j_1\leqslant 2\nu+1$, $j_2\geqslant 1,\cdots,j_m\geqslant 1)$ of the curve C, satisfying the condition (Γ_m) with respect to D. Next by Theorems 8.11 and 8.15, we can find a finite number of points $\zeta_i(i=1,2,\cdots,q)$ of $\tilde{\sigma}^{(m)}$ and circles $\overline{\gamma}_i: |z-\zeta_i|\leqslant r_i(i=1,2,\cdots,q)$ belonging to D and a number $\alpha>0$, such that for each n we have

$$\min_{1\leqslant i\leqslant q}(\max_{z\in\overline{\gamma}_i}\partial(z,f_n))\leqslant\alpha. \tag{9.29}$$

Evidently (9.28) and (9.29) imply $n\leqslant\alpha$ and hence we get a contradiction.

As an example, consider the entire function

$$f_\lambda(z)=e^{\lambda z},$$

where $\lambda>0$ is a constant. We have

$$\partial(z,f_\lambda)=\frac{|f_\lambda'(z)|}{1+|f_\lambda(z)|^2}=\frac{\lambda|e^{\lambda z}|}{1+|e^{\lambda z}|^2}.$$

So for real y we have

$$\partial(iy, f_\lambda) = \frac{\lambda}{2}$$

which can be as large as we please, provided that λ is sufficiently large. Accordingly by Theorem 9.24, for any integer $m \geqslant 1$, $f_\lambda(z)$ should have $(w)_m$-circles ($w \neq 0, \infty$), when λ is sufficiently large. We are going to show that this is really true.

In fact, consider a circle $\Gamma: |z-iy_0| < \delta$, where y_0 is real and $\delta < \eta_1$, η_1 being given in Definition 8.13 and a domain $E: 1/R < |w| < R\,(R > 1)$. Let us show that when $\lambda > \lambda_1$, $\lambda_1 = (\log R + 2\pi)/\delta$, the function $f_\lambda(z)$ takes in Γ every value $w \in E$. Accordingly, by Definition 8.13, for every $w \in E$, the circle Γ is a $(w)_1$-circle of $f_\lambda(z)$. To prove this consider a point $w = \rho e^{i\theta} \in E\,(1/R < \rho < R,\ \theta$ real) and a number $\lambda > \lambda_1$. Let k be the integer such that

$$2k\pi \leqslant \lambda y_0 - \theta < 2(k+1)\pi,$$

and set

$$\zeta = \frac{1}{\lambda}(\log \rho + i(\theta + 2k\pi)).$$

Then

$$|\zeta - iy_0| \leqslant \frac{1}{\lambda}(\log R + |2k\pi - (\lambda y_0 - \theta)|)$$
$$< \frac{1}{\lambda}(\log R + 2\pi) < \delta.$$

On the other hand, we have

$$f_\lambda(\zeta) = e^{\lambda\zeta} = \rho e^{i\theta} = w.$$

This proves our assertion.

Consider again the circle Γ, but now assume that $\delta < \eta_2$, η_2 being given in Definition 8.13. Let $\Gamma'_j: |z-iy_j| < \delta'$ ($\delta' < \eta_1, y_j$ real)$(j = 1, 2, \cdots, I_2)$ be I_2 circles such that the circles $\overline{\Gamma}'_j: |z-iy_j| \leqslant \delta'$ $(j = 1, 2, \cdots, I_2)$ belong to Γ and are mutually disjoint. By the assertion just proved, we see that we can find a number $\lambda_2>0$ such that for $\lambda>\lambda_2$ and $w\in E$, each of the circles Γ'_j $(j=1,2,\cdots,I_2)$ is a $(w)_1$-circle of the function $f_\lambda(z)$ in Γ. Consequently Γ is a $(w)_2$-circle of $f_\lambda(z)$.

In general, basing upon Definition 8.13, it is easy to prove by mathematical induction, the following assertion:

Given an integer $m\geqslant 1$, a circle Γ: $|z-iy_0|<\delta$ $(0<\delta<\eta_m$, η_m being given in Definition 8.13) and a domain E: $1/R<|w|<R$ $(R>1)$, we can find a number $\Lambda>0$ such that for $\lambda>\Lambda$ and $w\in E$, Γ is a $(w)_m$-circle of the function $f_\lambda(z)$.

APPENDIX A

A GENERAL THEOREM ON HOLOMORPHIC FUNCTIONS IN THE UNIT CIRCLE

The purpose of this appendix is to prove by the method of Wiman-Valiron in a simplified form due to Macintyre[21, 22], a general theorem on holomorphic functions in the unit circle and to deduce from it some consequences. For this we need to prove first a theorem on convex functions.

1. A THEOREM ON CONVEX FUNCTIONS

Definition 1. $f(x)$ being an increasing convex function in an interval, we denote by $f'_+(x)$ and $f'_-(x)$ the right-hand derivative and the left hand derivative of $f(x)$ respectively.

Definition 2. Let $\alpha>0$ and $0<\lambda<1$ be two numbers. Let $U(x)$ be a positive non-decreasing function for $x\geqslant 0$ such that $\int_0^\infty dx/U(x)$ converges. We define

$$H_{\alpha,\lambda,U}=\frac{2\alpha}{1-\lambda}+\int_0^\infty\frac{dx}{U(x)}$$

and

$$\Phi_{\alpha,\lambda,U}(X)=\frac{H_{\alpha,\lambda,U}}{X}\frac{2\alpha}{1-\lambda}+\int_X^\infty\frac{dx}{U(x)}.$$

The function $\Phi_{\alpha,\lambda,U}(X)$ is positive, decreasing and continuous for $X>0$ and satisfies the conditions

$$\lim_{X\to 0}\Phi_{\alpha,\lambda,U}(X)=+\infty,\quad \lim_{X\to+\infty}\Phi_{\alpha,\lambda,U}(X)=0.$$

We denote by $X=\Psi_{\alpha,\lambda,U}(Y)$ the inverse function of $Y=\Phi_{\alpha,\lambda,U}(X)$. The function $\Psi_{\alpha,\lambda,U}(Y)$ is positive, decreasing and continuous for $Y>0$ and satisfies the conditions:

$$\lim_{Y\to+\infty}\Psi_{\alpha,\lambda,U}(Y)=0,\quad \lim_{Y\to 0}\Psi_{\alpha,\lambda,U}(Y)=+\infty.$$

Theorem 1. Let $m(t)$ be an increasing convex function for $t<0$ such that $m'_+(t)\geqslant 1$ for $t<0$. Let α,λ and $U(x)$ have the same meaning as in Definition 2. Assume that there exists a number $t_0<0$ such that

$$m(t_0)\geqslant\Psi_{\alpha,\lambda,U}(-t_0). \tag{1}$$

Under these conditions, there exist three numbers $t_j(j=1,2,3)$ such that

1) $-H_{\alpha,\lambda,U}\leqslant t_1<t_2<t_3<0$;

2) $m(t_2)-m(t_1)=m(t_3)-m(t_2)=\alpha$;

3) $\lambda(t_2-t_1)<t_3-t_2\leqslant t_2-t_1$;

4) $m(t_2)\geqslant\alpha$;

5) $m'_+(t_2)\leqslant U\{m(t_2)\}$.

For the proof of this theorem we need some lemmas.

Lemma 1. Let $d_n(n\geqslant 0)$ be a non-increasing sequence of positive numbers such that $\sum_{n=0}^{\infty}d_n$ converges. Let $0<\lambda<1$ be a number. Let $V(y)$ be a positive non-decreasing function for $y\geqslant 0$ such that $\int_0^\infty dy/V(y)$ converges. Assume that

$$\sum_{n=0}^{\infty}d_n\geqslant\frac{d_0}{1-\lambda}+\int_0^\infty\frac{dy}{V(y)}.$$

Under these conditions, there exists an integer $n_0 \geqslant 1$ such that

$$d_{n_0} > \lambda d_{n_0-1},\ d_{n_0} \geqslant \frac{1}{V(n_0)}. \tag{2}$$

Proof. We have

$$\begin{aligned}(1-\lambda)\sum_{n=0}^{\infty} d_n &\geqslant d_0 + (1-\lambda)\int_0^{\infty} \frac{dy}{V(y)} \\ &\geqslant d_0 + (1-\lambda)\sum_{n=1}^{\infty} \frac{1}{V(n)},\end{aligned}$$

$$\sum_{n=0}^{\infty} d_n \geqslant d_0 + \lambda \sum_{n=0}^{\infty} d_n + (1-\lambda)\sum_{n=1}^{\infty} \frac{1}{V(n)},$$

$$\begin{aligned}\sum_{n=1}^{\infty} d_n &\geqslant \lambda \sum_{n=1}^{\infty} d_{n-1} + (1-\lambda)\sum_{n=1}^{\infty} \frac{1}{V(n)} \\ &= \sum_{n=1}^{\infty} \left\{ \lambda d_{n-1} + (1-\lambda)\frac{1}{V(n)} \right\}.\end{aligned}$$

Hence there is an integer $n_0 \geqslant 1$ such that

$$d_{n_0} \geqslant \lambda d_{n_0-1} + (1-\lambda)\frac{1}{V(n_0)}$$

which togather with the inequality $d_{n_0-1} \geqslant d_{n_0}$, gives (2).

Lemma 2. Let $d_n (0 \leqslant n \leqslant N, N \geqslant 1)$ be a non-increasing finite sequence of positive numbers. Let λ and $V(y)$ have the same meaning as in Lemma 1. Assume that

$$\sum_{n=0}^{N} d_n \geqslant \frac{d_0}{1-\lambda} + \int_0^{\infty} \frac{dy}{V(y)}.$$

Under these conditions, there exists an integer $1 \leqslant n_0 \leqslant N$ such that

$$d_{n_0} > \lambda d_{n_0-1},\ d_{n_0} > \frac{1}{V(n_0)}. \tag{3}$$

Proof. We have

$$(1-\lambda)\sum_{n=0}^{N} d_n \geqslant d_0 + (1-\lambda)\int_0^{\infty}\frac{dy}{V(y)}$$

$$> d_0 + (1-\lambda)\sum_{n=1}^{N}\frac{1}{V(n)},$$

$$\sum_{n=0}^{N} d_n > d_0 + \lambda\sum_{n=0}^{N} d_n + (1-\lambda)\sum_{n=1}^{N}\frac{1}{V(n)},$$

$$\sum_{n=1}^{N} d_n > \lambda\sum_{n=1}^{N} d_{n-1} + (1-\lambda)\sum_{n=1}^{N}\frac{1}{V(n)}$$

$$= \sum_{n=1}^{N}\left\{\lambda d_{n-1} + (1-\lambda)\frac{1}{V(n)}\right\}.$$

Hence there is an integer $1\leqslant n_0\leqslant N$ such that

$$d_{n_0} > \lambda d_{n_0-1} + (1-\lambda)\frac{1}{V(n_0)}$$

which togather with the inequality $d_{n_0-1}\geqslant d_{n_0}$, gives (3).

Lemma 3. Let $h(s)$ be an increasing convex function for $a\leqslant s<b$ such that $h(a)\geqslant 0$ and $h'_+(a)>0$. Let α,λ and $U(x)$ have the same meaning as in Definition 2. Assume that

$$b-a \geqslant \frac{1}{h'_+(a)}\frac{2\alpha}{1-\lambda} + \int_{h(a)}^{\infty}\frac{dx}{U(x)} \tag{4}$$

Under these conditions, there exist three numbers $s_j (j=1,2,3)$ such that

1) $a\leqslant s_1<s_2<s_3<b$;

2) $h(s_2)-h(s_1)=h(s_3)-h(s_2)=\alpha$;

3) $\lambda(s_2-s_1)<s_3-s_2\leqslant s_2-s_1$;

4) $h(s_2)\geqslant\alpha$;

5) $h'_+(s_2)\leqslant U\{h(s_2)\}$.

Proof. Distinguish two cases as follows:

Case 1. The function h(s) is unbounded for $a \leqslant s < b$. Denote by $\sigma_n (n \geqslant 0)$ the sequence of values of s defined by the equality

$$h(\sigma_n) = h(a) + n\alpha$$

and put

$$d_n = \sigma_{n+1} - \sigma_n \qquad (n \geqslant 0).$$

We have

$$d_n > 0, \ \sum_{n=0}^{\infty} d_n = b - a, \quad \frac{\alpha}{h'_+(a)} \geqslant d_0. \tag{5}$$

Moreover by the convexity of the function h(s), the sequence d_n is non-increasing. Define

$$V(y) = \frac{1}{\alpha} U\{h(a) + \alpha y\} \quad (y \geqslant 0). \tag{6}$$

V(y) is positive non-decreasing for $y \geqslant 0$. We have

$$\int_{h(a)}^{\infty} \frac{dx}{U(x)} = \int_0^{\infty} \frac{dy}{V(y)}. \tag{7}$$

The relations (4), (5) and (7) yield

$$\sum_{n=0}^{\infty} d_n \geqslant \frac{2d_0}{1-\lambda} + \int_0^{\infty} \frac{dy}{V(y)}.$$

Hence by Lemma 1, there exists an integer $n_0 \geqslant 1$ such that

$$d_{n_0} > \lambda d_{n_0 - 1}, \quad d_{n_0} \geqslant \frac{1}{V(n_0)}. \tag{8}$$

By means of the convexity of the function h(s) and (8), we see that the three values $s_1=\sigma_{n_0-1}, s_2=\sigma_{n_0}, s_3=\sigma_{n_0+1}$ satisfy the conditions 1)—5) in Lemma 3.

Case 2. The function $h(s)$ is bounded for $a\leqslant s<b$. Define

$$h(b)=\lim_{s\to b} h(s).$$

Then by the convexity of the function $h(s)$ and (4), we get

$$h(b)-h(a)\geqslant(b-a)h'_+(a)>\frac{2\alpha}{1-\lambda}>2\alpha.$$

Let $P\geqslant 2$ be the integer such that

$$P<\frac{h(b)-h(a)}{\alpha}\leqslant P+1.$$

Let $\sigma_n(0\leqslant n\leqslant P)$ be the finite sequence of values of s defined by the equality

$$h(\sigma_n)=h(a)+n\alpha.$$

Put

$$N=P-1,\quad d_n=\sigma_{n+1}-\sigma_n\qquad(0\leqslant n\leqslant N).$$

We have

$$d_n>0,\quad \frac{\alpha}{h'_+(a)}\geqslant d_0,\tag{9}$$

and from the inequality

$$h'_+(a)\leqslant\frac{h(\sigma_P)-h(a)}{\sigma_P-a}\leqslant\frac{h(b)-h(\sigma_P)}{b-\sigma_P}\leqslant\frac{\alpha}{b-\sigma_P}$$

we have

$$b - a = (b - \sigma_p) + (\sigma_p - a) \leqslant \frac{a}{h'_+(a)} + \sum_{n=0}^{N} d_n. \tag{10}$$

Moreover by the convexity of the function $h(s)$, the finite sequence d_n is non-increasing. The relations (4),(7),(9) and (10) yield

$$\sum_{n=0}^{N} d_n > \frac{d_0}{1-\lambda} + \int_0^\infty \frac{dy}{V(y)}.$$

Hence by Lemma 2, there exists an integer $1 \leqslant n_0 \leqslant N$ such that

$$d_{n_0} > \lambda d_{n_0-1},\ d_{n_0} > \frac{1}{V(n_0)}. \tag{11}$$

Then, as in case 1, by means of the convexity of the function $h(s)$ and (11), we see that the three values $s_1 = \sigma_{n_0-1}, s_2 = \sigma_{n_0}, s_3 = \sigma_{n_0+1}$ satisfy the conditions 1)—5) in Lemma 3.

Now let us come back to the proof of Theorem 1. Put $H = H_{a,\lambda,U}$ and distinguish two cases according to $m(-H) \geqslant 0$ or < 0.

First assume that $m(-H) \geqslant 0$. In this case, consider the increasing convex function $m(t)$ for $-H \leqslant t < 0$, such that $m(-H) \geqslant 0$ and $m'_+(-H) \geqslant 1$; the numbers α, λ and the function $U(x)$. We have

$$H \geqslant \frac{1}{m'_+(-H)} \frac{2a}{1-\lambda} + \int_{m(-H)}^\infty \frac{dx}{U(x)}.$$

Hence by Lemma 3, there exist three numbers $t_j (j = 1,2,3)$ satisfying the conditions 1)—5) in Theorem 1.

Now assume that $m(-H) < 0$. The inequality (1) implies

$$-t_0 \geqslant \Phi_{a,\lambda,U}\{m(t_0)\} = \frac{H}{m(t_0)} \frac{2a}{1-\lambda} + \int_{m(t_0)}^\infty \frac{dx}{U(x)}. \tag{12}$$

The function $m(t)$ being increasing, the inequalities

$$m(-H) < 0, \quad m(t_0) > 0$$

imply

$$-H < t_0. \tag{13}$$

Then by the convexity of the function $m(t)$, we get

$$m'_+(t_0) \geqslant \frac{m(t_0) - m(-H)}{t_0 + H} > \frac{m(t_0)}{H}. \tag{14}$$

The inequalities (12) and (14) yield

$$-t_0 > \frac{1}{m'_+(t_0)} \frac{2\alpha}{1-\lambda} + \int_{m(t_0)}^{\infty} \frac{dx}{U(x)}. \tag{15}$$

Consider the increasing convex function $m(t)$ for $t_0 \leqslant t < 0$ such that $m(t_0) > 0$ and $m'_+(t_0) \geqslant 1$; the numbers α, λ and the function $U(x)$. By (15) and (13) we can apply Lemma 3 and conclude that there exist three numbers $t_j (j=1,2,3)$ satisfying the conditions 1)—5) in Theorem 1.

Corollary 1. Let $m(t)$ be an increasing convex function for $t<0$ such that $m'_+(t) \geqslant 1$ for $t<0$. Let $\alpha>0$ and $0<\lambda<1$ be two numbers. Define

$$h_{\alpha,\lambda} = \frac{2\alpha}{1-\lambda} + 2, \qquad \psi_{\alpha,\lambda}(Y) = 1 + \frac{(h_{\alpha,\lambda} - 1)^2}{Y} \quad (Y > 0). \tag{16}$$

Assume that there exists a number $t_0 < 0$ such that

$$m(t_0) \geqslant \psi_{\alpha,\lambda}(-t_0). \tag{17}$$

Under these conditions, there exist three numbers $t_j (j=1,2,3)$ such that

1) $-h_{\alpha,\lambda} \leqslant t_1 < t_2 < t_3 < 0$;

2) $m(t_2) - m(t_1) = m(t_3) - m(t_2) = \alpha$;

3) $\lambda(t_2 - t_1) < t_3 - t_2 \leqslant t_2 - t_1$;

4) $m(t_2) \geqslant \alpha$;

5) $m'_+(t_2) \leqslant 1 + \{m(t_2)\}^2$.

Proof. Let U(x) be the function defined for $x \geqslant 0$ as follows:

$$U(x) = 1 \text{ for } 0 \leqslant x < 1, \; U(x) = x^2 \text{ for } x \geqslant 1.$$

Then it is easy to see that

$$H_{a,\lambda,U} = h_{a,\lambda}, \quad \Phi_{a,\lambda,U}(X) = \frac{(h_{a,\lambda} - 1)^2}{X} \quad (X \geqslant 1)$$

and

$$\Psi_{a,\lambda,U}(Y) = \frac{(h_{a,\lambda} - 1)^2}{Y} \quad (0 < Y \leqslant (h_{a,\lambda} - 1)^2)$$

$$\Psi_{a,\lambda,U}(Y) < 1 \quad (Y > (h_{a,\lambda} - 1)^2).$$

Hence

$$\Psi_{a,\lambda,U}(Y) < \psi_{a,\lambda}(Y) \quad (Y > 0).$$

Besides we have

$$U(x) \leqslant 1 + x^2 \quad (x \geqslant 0).$$

Consequently Corollary 1 follows from Theorem 1.

2. A GENERAL THEOREM ON HOLOMORPHIC FUNCTIONS IN THE UNIT CIRCLE

Definition 3. f(z) being a holomorphic function in the circle $|z| < 1$, we define

$$M(r, f) = \max_{|z|=r} |f(z)| \quad (0 < r < 1).$$

Theorem 2. Let f(z) be a holomorphic function in the circle $|z|<1$ and S a positive number. Assume that there exists a number $0<r_0<1$ such that

$$M(r_0,f) \geqslant 4e(S+|f(0)|)e^{\frac{a}{-\log r_0}}.$$

Under these conditions, we can find two numbers ρ, N such that

$$e^{-b}<\rho<1, \; M=M(\rho,f)>S, \; 1\leqslant N<1+(\log \frac{M}{S})^2$$

and two holomorphic functions $u(\zeta)$, $v(\zeta)$ in the rectangle

$$(R): -2<Re(\zeta)<-\frac{1}{2}, \qquad 0<Im(\zeta)<4\pi$$

having the following properties:

1^0 In (R), $u(\zeta)$ is univalent and $Re\{u(\zeta)\}<0$. For each number $0\leqslant\omega<2\pi$, $e^{u(\zeta)}$ is univalent in the rectangle

$$(R_\omega): -2<Re(\zeta)<-\frac{1}{2}, \qquad \omega<Im(\zeta)<\omega+2\pi.$$

In (R), we have

$$f\{\rho e^{u(\zeta)}\}=Me^{\zeta}.$$

2^0 In (R), we have

$$|v(\zeta)|<2\pi+\frac{1}{4}$$

and

$$f'\{\rho e^{u(\zeta)}\}=\frac{N}{\rho}Me^{(1-\frac{1}{N})\zeta+v(\zeta)}.$$

3^0 For $n \geqslant 1$, we have

$$M(\rho, f^{(n)}) < n!e^c M(\frac{N}{\rho})^n.$$

In this theorem, a, b, c are absolute positive constants whose values will be written out explicitly.

Proof. Define

$$f_1(z) = \frac{1}{A} f(z), \quad A = 2(S + |f(0)|). \tag{18}$$

We have

$$|f_1(0)| < 1, \quad M(r, f_1) = \frac{1}{A} M(r, f), \tag{19}$$
$$M(r_0, f_1) \geqslant (2e) e^{\frac{a}{-\log r_0}}.$$

Next define

$$f_2(z) = f_1(z) - f_1(0). \tag{20}$$

We have

$$f_2(0) = 0, \quad M(r, f_2) \geqslant M(r, f_1) - 1,$$
$$M(r_0, f_2) > e^{1 + \frac{a}{-\log r_0}}.$$

In what follows, we first study the function $f_2(z)$, then the function $f_1(z)$ and finally the function $f(z)$.

Study of the function $f_2(z)$.

By a theorem of Hadamard, the function

$$m(t) = \log M(e^t, f_2) \quad (t < 0) \tag{21}$$

is increasing and convex for $t<0$. Since $f_2(z)=zg(z)$, where $g(z)$ is a holomorphic function in the circle $|z|<1$, we have

$$m(t) = t + \log M(e^t, g) \quad (t < 0)$$

which implies that

$$m'_+(t) \geqslant m'_-(t) \geqslant 1 \quad (t < 0). \tag{22}$$

Moreover we have

$$m(t_0) > 1 + \frac{a}{-t_0} \quad (t_0 = \log r_0). \tag{23}$$

Now let us choose

$$\alpha = 3e4^4, \ \lambda = \frac{\alpha}{\alpha + 1} \tag{24}$$

and

$$a = (h_{\alpha,\lambda} - 1)^2, \tag{25}$$

where $h_{\alpha,\lambda}$ is defined by (16). Then (23) becomes

$$m(t_0) > \psi_{\alpha,\lambda}(-t_0),$$

where $\psi_{\alpha,\lambda}(Y)$ is defined by (16). Applying Corollary 1 to $m(t)$, α, λ, we conclude that there exist three numbers $t_j(j=1,2,3)$ satisfying the conditions 1)—5) in Corollary 1. Put

$$\rho = e^{t_2}. \tag{26}$$

By the condition 1) in Corollary 1, we have

$$e^{-b} < \rho < 1 \tag{27}$$

with

$$b = h_{a,\lambda}. \tag{28}$$

Taking a point z_0 of the circle $|z| = \rho$, such that

$$|f_2(z_0)| = M(\rho, f_2) \geqslant e^a \tag{29}$$

and introducing the number

$$N = \frac{z_0 f_2'(z_0)}{f_2(z_0)}, \tag{30}$$

we are going to show that N is real and

$$m_-'(t_2) \leqslant N \leqslant m_+'(t_2). \tag{31}$$

In fact consider a circle $\Upsilon : |z - z_0| < \delta$ in which $f_2(z) \neq 0$ and let $l(z)$ be a branch of the logarithm of $f_2(z)$, holomorphic in Υ. Then putting

$$z_0 = \rho e^{i\theta_0},$$

we have

$$iN = \left\{ \frac{d}{d\theta} l(\rho e^{i\theta}) \right\}_{\theta=\theta_0},$$

$$Re(iN) = \left\{ \frac{d}{d\theta} \log |f_2(\rho e^{i\theta})| \right\}_{\theta=\theta_0}.$$

Since the function $\log |f_2(\rho e^{i\theta})|$ has a maximum at $\theta = \theta_0$, hence $Re(iN) = 0$ and N is real. On the other hand, we have

$$N = \rho \left\{ \frac{d}{dr} l(re^{i\theta_0}) \right\}_{r=\rho},$$

$$N = \rho\left\{\frac{d}{dr}\log|f_2(re^{i\theta_0})|\right\}_{r=\rho},$$

which may be written in the form

$$N = \left\{\frac{d\log|f_2(re^{i\theta_0})|}{d\log r}\right\}_{r=\rho}.$$

Consequently if we define

$$Q(r) = \frac{\log|f_2(re^{i\theta_0})| - \log|f_2(\rho e^{i\theta_0})|}{\log r - \log\rho}.$$

then

$$N = \lim_{r\to\rho} Q(r).$$

Finally from the inequalities

$$Q(r) \leqslant \frac{\log M(r,f_2) - \log M(\rho,f_2)}{\log r - \log\rho} \quad (\rho < r < 1),$$

$$Q(r) \geqslant \frac{\log M(r,f_2) - \log M(\rho,f_2)}{\log r - \log\rho} \quad (0 < r < \rho),$$

we deduce (31). The inequalities (22), (31) and the condition 5) in Corollary 1 yield

$$1 \leqslant N \leqslant 1 + \{\log M(\rho,f_2)\}^2. \tag{32}$$

Consider the function

$$\varphi(\tau) = \frac{f_2(z_0e^{\tau})}{f_2(z_0)}e^{-N\tau} \tag{33}$$

holomorphic in the half plane $\mathrm{Re}(\tau) < -t_2$. Putting

$$\sigma' = t_2 - t_1, \qquad \sigma = t_3 - t_2, \tag{34}$$

we are going to show that

$$|\varphi(\tau)| < e \text{ for } -\sigma' \leqslant Re(\tau) \leqslant \sigma. \tag{35}$$

In fact, we have

$$|\varphi(\tau)| \leqslant e^{m(t_2+u)-m(t_2)-Nu} \quad (u = Re(\tau)).$$

Since (31) implies

$$\frac{m(t_2 - \sigma') - m(t_2)}{-\sigma'} \leqslant N \leqslant \frac{m(t_2 + \sigma) - m(t_2)}{\sigma},$$

if follows that if $0 < u < \sigma$, we have, by the condition 2) in Corollary 1,

$$\begin{aligned} &m(t_2 + u) - m(t_2) - Nu = u\left\{\frac{m(t_2 + u) - m(t_2)}{u} - N\right\} \\ &\leqslant u\left\{\frac{m(t_2 + \sigma) - m(t_2)}{\sigma} - N\right\} \leqslant \sigma\left\{\frac{m(t_2 + \sigma) - m(t_2)}{\sigma} - N\right\} \\ &= m(t_2 + \sigma) - m(t_2) - \sigma N = a - \sigma N, \end{aligned}$$

and if $-\sigma' < u < 0$, we have

$$\begin{aligned} &m(t_2 + u) - m(t_2) - Nu = u\left\{\frac{m(t_2 + u) - m(t_2)}{u} - N\right\} \\ &\leqslant u\left\{\frac{m(t_2) - m(t_2 - \sigma')}{\sigma'} - N\right\} \leqslant -\sigma'\left\{\frac{m(t_2) - m(t_2 - \sigma')}{\sigma'} - N\right\} \\ &= m(t_2 - \sigma') - m(t_2) + \sigma' N = \sigma' N - a. \end{aligned}$$

Consequently

$$\begin{aligned} |\varphi(\tau)| &\leqslant e^{a-\sigma N} \quad \text{for } 0 \leqslant Re\tau \leqslant \sigma, \\ |\varphi(\tau)| &\leqslant e^{\sigma' N - a} \quad \text{for } -\sigma' \leqslant Re\tau \leqslant 0. \end{aligned}$$

Finally by the conditions 2) and 3) in Corollary 1, we have

$$\frac{\alpha}{\sigma'} \leqslant N \leqslant \frac{\alpha}{\sigma}, \quad \lambda\sigma' < \sigma \leqslant \sigma', \tag{36}$$

and we get, by (24),

$$\alpha - \sigma N \leqslant \alpha(1 - \frac{\sigma}{\sigma'}) < \alpha(1 - \lambda) < 1$$

$$\sigma' N - \alpha \leqslant \alpha(\frac{\sigma'}{\sigma} - 1) < \alpha(\frac{1}{\lambda} - 1) = 1.$$

Hence we have (35).

We have

$$\varphi(0) = 1, \ \varphi'(0) = 0. \tag{37}$$

From (35) and (37), we can apply Schwarz's lemma to the function $\varphi(\tau) - 1$ and get

$$|\varphi(\tau) - 1| \leqslant \frac{e+1}{\sigma^2}|\tau|^2 \qquad for\ |\tau| \leqslant \sigma.$$

In particular, in the circle

$$(C): |\tau| < \frac{\sigma}{16\sqrt{3}},$$

we have

$$|\varphi(\tau) - 1| < \frac{4}{3(16)^2} = \frac{1}{12}\frac{1}{16},$$

and from (33),

$$\frac{f_2(z_0 e^\tau)}{f_2(z_0)} = e^{N\tau}\{1 + \psi(\tau)\} \tag{38}$$

with

$$|\psi(\tau)| < \frac{1}{12}\frac{1}{16} \tag{39}$$

in (C).

Now consider the function

$$\varphi_1(\tau) = \frac{z_0 e^{\tau} f_2'(z_0 e^{\tau})}{N f_2(z_0)} e^{-N\tau}. \tag{40}$$

We see easily that

$$\varphi_1(\tau) = \varphi(\tau) + \frac{\varphi'(\tau)}{N}. \tag{41}$$

By Cauchy's inequality and (35), we have

$$\frac{\sigma}{2}|\varphi'(\tau)| < e \text{ for } -\frac{\sigma}{2} \leqslant Re(\tau) \leqslant \frac{\sigma}{2}.$$

Hence

$$\frac{|\varphi'(\tau)|}{N} < \frac{2e}{N\sigma}.$$

By (36),

$$N\sigma > \lambda\sigma' N \geqslant \lambda a = a\frac{a}{a+1} > \frac{a}{2}. \tag{42}$$

Consequently

$$\frac{|\varphi'(\tau)|}{N} < \frac{4e}{a} = \frac{1}{12}\frac{1}{16}.$$

It follows that we have

$$\frac{z_0 e^{\tau} f_2'(z_0 e^{\tau})}{N f_2(z_0)} = e^{N\tau}\{1 + \psi_1(\tau)\} \tag{43}$$

with

$$|\psi_1(\tau)| < \frac{1}{6}\frac{1}{16} \tag{44}$$

in (C).

By the condition 2) in Corollary 1, we have

$$M(\rho', f_2) = e^{\alpha} M(\rho, f_2) \quad (\rho = e^{t_2}, \rho' = e^{t_3}).$$

n$\geqslant$1 being an integer, we have, by Cauchy's inequalities,

$$M(\rho, f_2^{(n)}) \leqslant n!\frac{M(\rho', f_2)}{(\rho' - \rho)^n} = n!\frac{e^{\alpha} M(\rho, f_2)}{(\rho' - \rho)^n}.$$

From (34) and (42), we have

$$\rho' - \rho = \rho(e^{\sigma} - 1) > \rho\sigma, \ N\sigma > \frac{\alpha}{2},$$

hence

$$M(\rho, f_2^{(n)}) < n! e^{\alpha} \frac{1}{(\frac{\alpha}{2})^n} M(\rho, f_2)(\frac{N}{\rho})^n. \tag{45}$$

Study of the function $f_1(z)$.

By (19) and (20), we have

$$f_1(z) = f_2(z) + f_1(0), \ |f_1(0)| < 1, \tag{46}$$

and by (29) and the condition 4) in Corollary 1, we have

$$|f_2(z_0)| \geqslant e^a, \ f_1(z_0) = f_2(z_0)(1+\beta), \ |\beta| < \frac{1}{e^a}. \tag{47}$$

On the other hand from (38),(39),(29) and (36), we have in the circle (C),

$$|f_2(z_0 e^\tau)| = |f_2(z_0)| e^{NRe\tau} |1+\psi(\tau)| > \frac{1}{2} e^a e^{-\frac{Na}{2}} \geqslant \frac{1}{2} e^{\frac{a}{2}}. \tag{48}$$

Hence in the circle (C), we have

$$f_1(z_0 e^\tau) = f_2(z_0 e^\tau)\{1 + b(\tau)\}, |b(\tau)| < \frac{2}{e^{\frac{a}{2}}}. \tag{49}$$

It follows from (38),(47) and (49), that

$$\frac{f_1(z_0 e^\tau)}{f_1(z_0)} = e^{N\tau}\{1+\psi(\tau)\}\frac{1+b(\tau)}{1+\beta}.$$

Consequently if we write

$$\frac{f_1(z_0 e^\tau)}{f_1(z_0)} = e^{N\tau}\{1+\delta(\tau)\}, \tag{50}$$

then we have

$$\delta(\tau) = \frac{b(\tau)-\beta}{1+\beta} + \psi(\tau) + \psi(\tau)\frac{b(\tau)-\beta}{1+\beta}. \tag{51}$$

We are going to show that in the circle (C), we have

$$|\delta(\tau)| < \frac{1}{2}\,\frac{1}{16}. \tag{52}$$

In fact we have

$$\left|\frac{b(\tau)-\beta}{1+\beta}\right| \leqslant 2|b(\tau)-\beta| < 2\left(\frac{2}{e^{\frac{a}{2}}}+\frac{1}{e^{a}}\right) < \frac{6}{e^{\frac{a}{2}}},$$

$$e^{\frac{a}{2}} > \frac{a}{2} = 2e \times 12 \times 16,$$

$$\left|\frac{b(\tau)-\beta}{1+\beta}\right| < \frac{1}{4e}\,\frac{1}{16}. \tag{53}$$

From (51),(53) and (39), we see that (52) holds.

By (46) we have

$$f_1'(z) = f_2'(z). \tag{54}$$

Hence from (43) and (47), we have

$$\frac{z_0 e^{\tau} f_1'(z_0 e^{\tau})}{N f_1(z_0)} = e^{N\tau}\,\frac{1+\psi_1(\tau)}{1+\beta}.$$

Then from (44) and (47), we see easily that if we write

$$\frac{z_0 e^{\tau} f_1'(z_0 e^{\tau})}{N f_1(z_0)} = e^{N\tau}\{1+\delta_1(\tau)\}, \tag{55}$$

we have

$$|\delta_1(\tau)| < \frac{1}{2}\,\frac{1}{16} \tag{56}$$

in the circle (C).

$n \geqslant 1$ being an integer, we deduce from (45) and (46),

$$M(\rho, f_1^{(n)}) < n!\,e^{c} M(\rho, f_1)\left(\frac{N}{\rho}\right)^{n}, \tag{57}$$

with

$$c = a. \tag{58}$$

By (46) and (29), we have

$$|f_2(z_0)| - 1 < |f_1(z_0)| < |f_2(z_0)| + 1, \tag{59}$$

$$M(\rho, f_2) - 1 < M(\rho, f_1) < M(\rho, f_2) + 1 \tag{60}$$

and

$$1 \geqslant \frac{|f_1(z_0)|}{M(\rho, f_1)} > \frac{|f_2(z_0)| - 1}{|f_2(z_0)| + 1} = 1 - \frac{2}{|f_2(z_0)| + 1} \geqslant 1 - \frac{2}{e^a + 1}.$$

Then making use of the inequality

$$\log(1 - x) > -\frac{x}{1 - x} \qquad (0 < x < 1)$$

we find

$$\left| \log \frac{|f_1(z_0)|}{M(\rho, f_1)} \right| < \frac{2}{e^a - 1} < \frac{2}{\alpha} < \frac{1}{16}. \tag{61}$$

Study of the function f(z).

From (18), (29) and (60), we have

$$\begin{aligned} M = M(\rho, f) = A M(\rho, f_1) > A\{M(\rho, f_2) - 1\} \geqslant A(e^a - 1) \\ > \alpha A > S. \end{aligned} \tag{62}$$

From (60) and (62), we have

$$M(\rho, f_2) < 2M(\rho, f_1) = \frac{2M}{A} \leqslant \frac{M}{S},$$

hence from (32), we have

$$1 \leqslant N < 1 + (\log \frac{M}{S})^2. \tag{63}$$

On the other hand, from (18), (50), (52), (55), (56), (57) and (61), we deduce

$$\frac{f(z_0e^\tau)}{f(z_0)} = e^{N\tau}\{1 + \delta(\tau)\}, \tag{64}$$

$$\frac{z_0e^\tau f'(z_0e^\tau)}{Nf(z_0)} = e^{N\tau}\{1 + \delta_1(\tau)\}, \tag{65}$$

with

$$|\delta(\tau)| < \frac{1}{2}\frac{1}{16}, \qquad |\delta_1(\tau)| < \frac{1}{2}\frac{1}{16} \tag{66}$$

in the circle (C), and

$$M(\rho, f^{(n)}) < n!e^c M(\frac{N}{\rho})^n, \tag{67}$$

$$\left| log \frac{|f(z_0)|}{M(\rho, f)} \right| < \frac{1}{16}. \tag{68}$$

Let log(1+Z) be the branch of the logarithm of 1+Z, which is holomorphic for $|Z|<1$ and equal to zero at the point $Z=0$. Then we have

$$|log(1+Z)| \leqslant 2|Z| \; for \; |Z| < \frac{1}{2}.$$

Hence, by (66) and (68), the function

$$\lambda(\tau) = log \frac{|f(z_0)|}{M} + log\{1 + \delta(\tau)\}$$

is holomorphic in the circle (C) and satisfies the inequality

$$|\lambda(\tau)| < \frac{1}{8}. \tag{69}$$

Next, putting

$$z_0 = \rho e^{i\theta_0}(0 \leqslant \theta_0 < 2\pi),\ f(z_0) = |f(z_0)|e^{i\Gamma} \quad (0 \leqslant \Gamma < 2\pi),$$

we can write the formula (64) in the form

$$f(\rho e^{\tau + i\theta_0}) = Me^{\Lambda(\tau)}, \Lambda(\tau) = N\tau + i\Gamma + \lambda(\tau) \tag{70}$$

in the circle (C).

We are going to show that the circle (C): $|\tau| < \dfrac{\sigma}{16\sqrt{3}}$ contains the rectangle

$$(A):\ -\frac{4}{N} \leqslant Re(\tau) \leqslant 0,\ \frac{-\Gamma - 2\pi}{N} \leqslant Im(\tau) \leqslant \frac{-\Gamma + 6\pi}{N}.$$

In fact, for $\tau \in (A)$, we have

$$|\tau| \leqslant |Re(\tau)| + |Im(\tau)| \leqslant \frac{4}{N} + \frac{6\pi}{N},$$

and hence it is sufficient to verify that

$$16\sqrt{3}(4 + 6\pi) < N\sigma.$$

It is easy to see that this is true, by making use of (42) and (24). The function

$$\zeta = N\tau + i\Gamma$$

maps the rectangle (A) onto the rectangle

$$(R'):\ -4 \leqslant Re(\zeta) \leqslant 0,\ -2\pi \leqslant Im(\zeta) \leqslant 6\pi;$$

the interior and the boundary of (A) correspond respectively to the interior and boundary of (R'). Moreover if ζ' is a point on the boundary of (R') and ζ is a point of the rectangle

$$(R): -2 < Re(\zeta) < -\frac{1}{2}, \qquad 0 < Im(\zeta) < 4\pi,$$

we have

$$|\zeta' - \zeta| > \frac{1}{2}.$$

Consequently if ζ is a point of (R), then on the boundary of (A), we have

$$|N\tau + i\Gamma - \zeta| > \frac{1}{2},$$

hence by (69), we can apply Rouché's theorem and conclude that the function

$$\Lambda(\tau) - \zeta = (N\tau + i\Gamma - \zeta) + \lambda(\tau)$$

has a zero $\tau = \tau(\zeta)$ in the interior of (A). The function $\tau = \tau(\zeta)$ is holomorphic and univalent in (R) and

$$\Lambda\{\tau(\zeta)\} = \zeta$$

in (R). It follows that in (R) the function

$$u(\zeta) = \tau(\zeta) + i\theta_0 \tag{71}$$

is holomorphic and univalent with

$$Re\{u(\zeta)\} = Re\{\tau(\zeta)\} < 0.$$

Moreover by (70) we have

$$f\{\rho e^{u(\zeta)}\} = Me^{\zeta} \tag{72}$$

in (R).

Consider a number $0 \leqslant \omega < 2\pi$ and the rectangle

$$(R_\omega): -2 < Re(\zeta) < -\frac{1}{2}, \qquad \omega < Im(\zeta) < \omega + 2\pi.$$

We are going to show that the function $e^{u(\zeta)}$ is univalent in (R_ω). In fact, assume on the contrary that there exist two points $\zeta_1, \zeta_2 (\zeta_1 \neq \zeta_2)$ of (R_ω) such that

$$e^{u(\zeta_1)} = e^{u(\zeta_2)}.$$

Then by (71) and (72), we have

$$e^{\tau(\zeta_1)} = e^{\tau(\zeta_2)}, \quad \tau(\zeta_1) - \tau(\zeta_2) = 2p\pi i \quad (p \text{ integer}, p \neq 0),$$
$$e^{\zeta_1} = e^{\zeta_2}, \ \zeta_1 - \zeta_2 = 2q\pi i \quad (q \text{ integer}).$$

Next from (69) and (70), we have

$$\zeta_1 = N\tau(\zeta_1) + i\Gamma + \lambda\{\tau(\zeta_1)\}, \zeta_2 = N\tau(\zeta_2) + i\Gamma + \lambda\{\tau(\zeta_2)\},$$
$$N|\tau(\zeta_1) - \tau(\zeta_2)| < |\zeta_1 - \zeta_2| + \frac{1}{4}.$$

On the other hand, since ζ_1 and ζ_2 belong to (R_ω), we have

$$2|q|\pi = |Im(\zeta_1) - Im(\zeta_2)| < 2\pi, \qquad q = 0,$$

then by (32), we have

$$N \geqslant 1, \ N2|p|\pi < 2|q|\pi + \frac{1}{4} = \frac{1}{4}$$

and hence $|p| < 1$. So we get a contradiction.

Now consider (65) and define

$$\lambda_1(\tau) = log \frac{|f(z_0)|}{M} + log\{1 + \delta_1(\tau)\}$$

which is holomorphic in the circle (C) and satisfies the inequality

$$|\lambda_1(\tau)| < \frac{1}{8}. \tag{73}$$

The formula (65) can be written in the form

$$f'(\rho e^{\tau + i\theta_0}) = \frac{N}{\rho} M e^{(N-1)\tau + i(\Gamma - \theta_0) + \lambda_1(\tau)}.$$

Then making use of the formula

$$\Lambda(\tau) = N\tau + i\Gamma + \lambda(\tau),$$

we have

$$f'(\rho e^{\tau + i\theta_0}) = \frac{N}{\rho} M e^{(1-\frac{1}{N})\Lambda(\tau) + \mu(\tau)}, \tag{74}$$

where

$$\begin{aligned}\mu(\tau) &= -(1 - \frac{1}{N})\lambda(\tau) + \lambda_1(\tau) - (1 - \frac{1}{N})i\Gamma + i(\Gamma - \theta_0)\\ &= -(1 - \frac{1}{N})\lambda(\tau) + \lambda_1(\tau) + i(\frac{\Gamma}{N} - \theta_0)\end{aligned}$$

is holomorphic in the circle (C) and satisfies the inequality

$$|\mu(\tau)| < \frac{1}{4} + 2\pi.$$

In (74) setting $\tau = \tau(\zeta)$, we get, in (R),

$$f'(\rho e^{u(\zeta)}) = \frac{N}{\rho} M e^{(1-\frac{1}{N})\zeta + v(\zeta)}, \tag{75}$$

where

$$v(\zeta) = \mu\{\tau(\zeta)\} \tag{76}$$

is holomorphic in (R) and satisfies the inequality

$$|v(\zeta)| < \frac{1}{4} + 2\pi$$

in (R).

Theorem 2 is now completely proved. In what follows, we deduce some consequences of Theorem 2, in which the symbol a always denotes the absolute positive constant in Theorem 2.

Corollary 2. Let f(z) be a holomorphic function in the circle $|z|<1$ and S a positive number. Assume that there exists a number $0<r_0<1$ such that

$$M(r_0, f) \geqslant 4e(S + |f(0)|)e^{\frac{a}{-\log r_0}},$$

Then we can find a number M such that

$$M > S$$

and that for each number $0 \leqslant \omega < 2\pi$, the domain

$$(\Gamma_\omega): \quad Me^{-2} < |w| < Me^{-\frac{1}{2}}, \quad \omega < \arg w < \omega + 2\pi$$

is a simply covered image domain of f(z) for the circle $|z|<1$. (see Definition 2. 1)

Proof. By Theorem 2, we can find ρ, N, $u(\zeta)$, $v(\zeta)$ having the properties in Theorem 2. The function $\rho e^{u(\zeta)}$ maps the domain (R_ω) onto a domain d_ω belonging to the circle $|z|<\rho$. Then by the formula

$$f\{\rho e^{u(\zeta)}\} = Me^{\zeta},$$

evidently f(z) maps d_ω onto the domain (Γ_ω).

Definition 4. Let F(z) be a function which is holomorphic and does not take the value zero in the circle $|z|<1$. We denote by Log F(z) the branch of the logarithm of F(z), which is holomorphic in the circle $|z|<1$ and is such that $0\leqslant \operatorname{Im}\{\operatorname{Log}F(0)\}<2\pi$.

Corollary 3. Let F(z) be a function which is holomorphic and does not take the value zero in the circle $|z|<1$. Let R be a positive number. Assume that there exists a number $0<r_0<1$ such that

$$M(r_0, Log F) \geqslant (4e)^2(R + |Log F(0)|)e^{\frac{\sigma}{-\log r_0}}. \tag{77}$$

Then we can find a number $B>R$ and a holomorphic and univalent function $\varphi(w)$ in the circle $|w-w_0|<R$ $(w_0=Bi)$, such that

$$|\varphi(w)| < 1,\ F\{\varphi(w)\} = e^w \tag{78}$$

in the circle $|w-w_0|<R$.

Proof. Putting $f(z)=\operatorname{Log}F(z)$ and $S=4eR$, then (77) implies

$$M(r_0, f) \geqslant (4e)(S + |f(0)|)e^{\frac{\sigma}{-\log r_0}}.$$

Hence, by Theorem 2, we can find two numbers ρ, M such that

$$0 < \rho < 1,\ M > S$$

and a holomorphic function $u(\zeta)$ in the rectangle

$$-2 < Re(\zeta) < -\frac{1}{2},\ 0 < Im(\zeta) < 4\pi$$

having the property 1^0 in Theorem 2. The function $w = Me^{\zeta}$ is holomorphic and univalent in the rectangle

$$(R_0): \; -2 < Re(\zeta) < -\frac{1}{2}, \; 0 < Im(\zeta) < 2\pi$$

and maps (R_0) onto the domain

$$(\Gamma_0): \; Me^{-2} < |w| < Me^{-\frac{1}{2}}, \; 0 < \arg w < 2\pi.$$

The function $w = Me^{\zeta}$ has then an inverse function $\zeta = L(w)$ which is holomorphic and univalent in (Γ_0). The function

$$\varphi(w) = \rho e^{u(\zeta)}, \; \zeta = L(w)$$

is then holomorphic and univalent in (Γ_0) and we have

$$|\varphi(w)| < 1, \; f\{\varphi(w)\} = w, \; F\{\varphi(w)\} = e^{w}$$

in (Γ_0). Evidently (Γ_0) contains the circle

$$|w - Bi| < P, \; B = \frac{M}{2}(e^{-\frac{1}{2}} + e^{-2}), \; P = \frac{M}{2}(e^{-\frac{1}{2}} - e^{-2})$$

and we have

$$B > P > \frac{M}{2}e^{-\frac{1}{2}}(1 - e^{-1}) > \frac{1}{4}Me^{-\frac{1}{2}} > \frac{1}{4}e^{-1}S = R.$$

Corollary 3 is therefore proved.

Corollary 4. Let $F(z)$ be a function which is holomorphic and does not take the value zero in the circle $|z| < 1$. Let R be a positive number. Assume that there exists a number $0 < r_0 < 1$ such that

$$\log M(r_0, F) \geqslant 2\pi(4e)^2(1 + R + \overset{+}{\log} |F(0)|)e^{\frac{\sigma}{-\log r_0}}. \tag{79}$$

Then the conclusion of Corollary 3 holds.

Proof. If $|F(0)| \geqslant 1$, then (79) implies (77), hence the conclusion of Corollary 3 holds. If $|F(0)| < 1$, then (79) becomes

$$\log M(r_0, F) \geqslant 2\pi(4e)^2(1 + R)e^{\frac{\sigma}{-\log r_0}}. \tag{80}$$

Let $z_0 = r_0 e^{i\theta_0}$ be a point of the circle $|z| = r_0$, such that

$$|F(z_0)| = M(r_0, F) > 1.$$

Then there is a point $z_1 = r_1 e^{i\theta_0}$ $(0 < r_1 < r_0)$ such that

$$|F(z_1)| = 1.$$

Consider the function

$$F_1(\zeta) = F\{z_1 + \zeta(1 - |z_1|)\} \quad (|\zeta| < 1).$$

We have

$$|F_1(0)| = 1, \; F_1(\zeta_0) = F(z_0),$$

where

$$\zeta_0 = \frac{z_0 - z_1}{1 - |z_1|} = \frac{r_0 - r_1}{1 - r_1} e^{i\theta_0}, \; |\zeta_0| < r_0.$$

Then by (80), we have

$$M(r_0, \operatorname{Log} F_1) \geqslant \log M(r_0, F_1) \geqslant \log |F_1(\zeta_0)|$$
$$> (4e)^2(R + |\operatorname{Log} F_1(0)|)e^{\frac{\sigma}{-\log r_0}}.$$

Consequently by Corollary 3, we can find a number $B>R$ and a holomorphic and univalent function $\varphi_1(w)$ in the circle $|w-w_0|<R\,(w_0=Bi)$, such that

$$|\varphi_1(w)|<1,\ F_1\{\varphi_1(w)\}=e^w$$

in the circle $|w-w_0|<R$. The function

$$\varphi(w)=z_1+\varphi_1(w)(1-|z_1|)$$

is holomorphic and univalent in the circle $|w-w_0|<R$ and is such that

$$|\varphi(w)|<1,\ F\{\varphi(w)\}=e^w$$

in the circle $|w-w_0|<R$.

Corollary 5. Let $F(z)$ be a function which is holomorphic and does not take the value zero in the circle $|z|<1$. Let $K>1$ be a number. Assume that there exists a number $0<r_0<1$ such that

$$\log M(r_0,F)\geqslant 8\pi^2(4e)^2(2+\log K+\overset{+}{\log}\,|F(0)|)e^{\frac{a}{-\log r_0}}. \tag{81}$$

Then for each number $0\leqslant\omega<2\pi$, the domain

$$(C_\omega):\quad \frac{1}{K}<|Z|<K,\ \omega<\arg Z<\omega+2\pi$$

is a simply covered image domain of $F(z)$ for the circle $|z|<1$.

Proof. Putting $R=\log K+4\pi$, then (81) implies (79), hence the conclusion of Corollary 3 holds. We are going to show that the circle $|w-w_0|<R\,(w_0=Bi)$ contains the rectangle

$$(T):\ -\log K<Re(w)<\log K,\ 2l\pi<Im(w)<2l\pi+4\pi,$$

with

$$l = \left[\frac{B}{2\pi}\right].$$

In fact, if w is a point of (T), then

$$|w - Bi| \leqslant |Re(w)| + |Im(w) - B| < log K + 4\pi = R.$$

Consider a number $0 \leqslant \omega < 2\pi$ and the rectangle

$$(T_\omega): - log K < Re(w) < log K, \ \omega + 2l\pi < Im(w) < (\omega + 2l\pi) + 2\pi$$

which belongs to (T). The function $\varphi(w)$ maps (T_ω) onto a domain D_ω belonging to the circle $|z|<1$. By the formula (78), it is easy to see that the function F (z) is univalent in D_ω and maps D_ω onto (C_ω).

Similarly we can prove the following corollary:

Corollary 6. Let F(z) be a function which is holomorphic and does not take the value zero in the circle $|z|<1$. Let R be a positive number. Assume that there exists a number $0<r_0<1$ such that

$$log M(r_0, \frac{1}{F}) \geqslant 2\pi(4e)^2(1 + R + \overset{+}{log} \frac{1}{|F(0)|})e^{\frac{a}{-log r_0}},$$

Then the conclusion of Corollary 3 holds.

Proof. If $|F(0)| \leqslant 1$, the inequality in Corollary 6 implies (77), hence the conclusion of Corollary 3 holds. If $|F(0)|>1$, then the inequality in Corollary 6 becomes

$$log M(r_0, \frac{1}{F}) \geqslant 2\pi(4e)^2(1 + R)e^{\frac{a}{-log r_0}}.$$

Let $z_0 = r_0 e^{i\theta_0}$ be a point of the circle $|z| = r_0$, such that

$$\frac{1}{|F(z_0)|} = M(r_0, \frac{1}{F}) > 1.$$

Then there is a point $z_1 = r_1 e^{i\theta_0}$ $(0 < r_1 < r_0)$ such that

$$|F(z_1)| = 1.$$

As in the proof of Corollary 4, consider the function

$$F_1(\zeta) = F\{z_1 + \zeta(1 - |z_1|)\} \quad (|\zeta| < 1).$$

We have

$$|F_1(0)| = 1,\ F_1(\zeta_0) = F(z_0),\ |z_0| < r_0.$$

Then we have

$$\begin{aligned} M(r_0, Log\, F_1) &\geqslant log\, M(r_0, \frac{1}{F_1}) \geqslant log\, \frac{1}{|F_1(\zeta_0)|} \\ &= log\, M(r_0, \frac{1}{F}) \\ &\geqslant 2\pi(4e)^2(1 + R)e^{\frac{a}{-log\, r_0}} \\ &> (4e)^2(R + |Log\, F_1(0)|)e^{\frac{a}{-log\, r_0}}. \end{aligned}$$

Then as in the proof of Corollary 4, we see that the conclusion of Corollary 3 holds.

From Corollary 6 we deduce, as in the proof of Corollary 5, the following corollary:

Corollary 7. Let $F(z)$ be a function which is holomorphic and does not take the value zero in the circle $|z| < 1$. Let $K > 1$ be a number. Assume that there exists a number $0 < r_0 < 1$ such that

$$log\, M(r_0, \frac{1}{F}) \geqslant 8\pi^2(4e)^2(2 + log\, K + \overset{+}{log}\, \frac{1}{|F(0)|})e^{\frac{a}{-log\, r_0}}.$$

Then the conclusion of Corollary 5 holds.

Consider a function $F(z)$ which is holomorphic and does not take the value zero in the circle $|z|<1$. Putting $F=e^f$, where $f=\operatorname{Log}F$ is holomorphic in the circle $|z|<1$, the successive derivatives of F can be expressed as

$$F' = e^f f',$$
$$F'' = e^f(f'^2+f''),$$
$$F''' = e^f(f'^3+3f'f''+f'''),$$

and in general

$$F^{(\nu)} = e^f\{f'^\nu + P_\nu(f', f'', \cdots, f^{(\nu)})\} \quad (\nu \geqslant 2), \tag{82}$$

where P_ν is a polynomial of $f', f'', \cdots, f^{(\nu)}$ of degree $\nu-1$ whose coefficients are positive integers. Grouping together the similar terms of P_ν, we may assume that no two terms of P_ν are similar. Denote by I_ν the largest coefficient of P_ν and by J_ν the number of the terms of P_ν, and define

$$q(z) = P_\nu(f', f'', \cdots, f^{(\nu)}). \tag{83}$$

Corollary 8. Let $F(z)$ be a function which is holomorphic and does not take the value zero in the circle $|z|<1$. Let ν be a positive integer and R a positive number. Assume that there exists a number $0<r_0<1$ such that

$$M(r_0, \operatorname{Log}F) \geqslant c_\nu(1+R+|\operatorname{Log}F(0)|)e^{\frac{a}{-\operatorname{Log}r_0}}, \tag{84}$$

where c_ν is a positive constant depending only on ν. Then we can find a number $B>R$ and a holomorphic and univalent function $\varphi(w)$ in the circle $|w-w_0|<R$ $(w_0=Bi)$, such that

$$|\varphi(w)|<1, \quad F^{(\nu)}\{\varphi(w)\} = e^w \tag{85}$$

in the circle $|w-w_0|<R$.

Proof. For the sake of definiteness we assume $\nu \geqslant 2$. The case of $\nu = 1$ is much simpler. Consider the function $f(z) = \mathrm{Log}F(z)$ and a positive number S. Assume that there exists a number $0 < r_0 < 1$ such that

$$M(r_0, f) \geqslant 4e(S + |f(0)|)e^{\frac{a}{-\log r_0}}. \tag{86}$$

Then by Theorem 2, we can find $\rho, N, u(\zeta), v(\zeta)$ having the properties in Theorem 2. If

$$S > e, \tag{87}$$

we have

$$1 \leqslant N < 2(\log M)^2$$

and by the property 3^0 in Theorem 2, we have for $1 \leqslant n \leqslant \nu$,

$$M(\rho, f^{(n)}) < A_\nu M(\log M)^{2\nu}, A_\nu = \nu! 2^\nu e^{c+\nu b}. \tag{88}$$

Then from (83) we see that in the circle $|z| \leqslant \rho$, we have

$$|q(z)| < B_\nu M^{\nu-1}(\log M)^{2\nu(\nu-1)}, B_\nu = I_\nu J_\nu A_\nu^{\nu-1}. \tag{89}$$

On the other hand, by the property 2^0 in Theorem 2, we see that in the rectangle (R) we have

$$|f'\{\rho e^{u(\zeta)}\}| > \alpha M, \ \alpha = e^{-(2+2\pi+\frac{1}{4})}. \tag{90}$$

Consequently in (R) we have

$$\left| \frac{q\{\rho e^{u(\zeta)}\}}{(f'\{\rho e^{u(\zeta)}\})^\nu} \right| < \frac{B_\nu}{\alpha^\nu} \frac{(\log M)^{2\nu(\nu-1)}}{M}. \tag{91}$$

Evidently we can find a positive number K_ν depending only on ν, such that when $M > K_\nu$, the right member of (91) is less than 1/2. Hence if

$$S > K_\nu, \tag{92}$$

then in (R) we have, by (82),

$$F^{(\nu)}\{\rho e^{u(\zeta)}\} = e^{Me^\zeta}(f'\{\rho e^{u(\zeta)}\})^\nu(1 + \mu(\zeta)), \tag{93}$$

where $\mu(\zeta)$ is a holomorphic function in (R) with

$$|\mu(\zeta)| < \frac{1}{2} \tag{94}$$

in (R). From (93) and the property 2^0 in Theorem 2, we have

$$F^{(\nu)}\{\rho e^{u(\zeta)}\} = e^{Me^\zeta + g(\zeta)} \tag{95}$$

in (R), where

$$g(\zeta) = \nu\{log(\frac{N}{\rho}M) + (1 - \frac{1}{N})\zeta + v(\zeta)\} + Log\{1 + \mu(\zeta)\}$$

is a holomorphic function in (R) and we have in (R),

$$|g(\zeta)| < 3\nu log\, M + \beta\nu, \tag{96}$$

where $\beta > 0$ is an absolute constant.

Now consider the function $Z = Me^\zeta$ which is holomorphic and univalent in the rectangle

$$(R_0):\ -2 < Re(\zeta) < -\frac{1}{2},\quad 0 < Im(\zeta) < 2\pi$$

and maps (R_0) onto the domain

$$(\Gamma_0):\ Me^{-2} < |Z| < Me^{-\frac{1}{2}},\ 0 < arg\, Z < 2\pi.$$

Let $\zeta = L(Z)$ b the inverse function of the function $Z = Me^{\zeta}$; $L(Z)$ is holomorphic and univalent in (Γ_0). In (95) put $\zeta = L(Z)$, we get

$$F^{(\nu)}\{\psi(Z)\} = e^{Z+h(Z)} \tag{97}$$

in (Γ_0), where

$$\psi(Z) = \rho e^{u(L(Z))} \tag{98}$$

is a holomorphic and univalent function in (Γ_0) with

$$|\psi(Z)| < 1 \tag{99}$$

in (Γ_0). On the other hand,

$$h(Z) = g\{L(Z)\} \tag{100}$$

is a holomorphic function in (Γ_0) with

$$|h(Z)| < 3\nu \log M + \beta\nu \tag{101}$$

in (Γ_0).

Evidently (Γ_0) contains the circle

$$|Z - Bi| \leqslant Q, \quad B = \frac{M}{2}(e^{-\frac{1}{2}} + e^{-2}), \quad Q = \frac{M}{4}(e^{-\frac{1}{2}} - e^{-2}).$$

Consider a point w of the circle Υ : $|w - Bi| < \frac{Q}{2}$ and a point Z on the circle $|Z - Bi| = Q$. Then

$$|Z - w| > \frac{Q}{2}.$$

On the other hand, when $M > K'_\nu$, where K'_ν is a positive number depending only on ν, we have

$$3v\log M + \beta v < \frac{Q}{2}.$$

Hence if

$$S > K_v',$$

then for each point $w \in \Upsilon$, we have

$$|h(Z)| < |Z - w|$$

on the circle $|Z - Bi| = Q$. Consequently by Rouché's theorem, the function $Z + h(Z) - w$ has just one zero $Z(w)$ in the circle $|Z - Bi| < Q$. The function $Z = Z(w)$ is holomorphic and univalent in the circle Υ. In (97) putting $Z = Z(w)$, we get

$$F^{(v)}\{\varphi(w)\} = e^{w},$$

in Υ, where the function

$$\varphi(w) = \psi\{Z(w)\}$$

is holomorphic and univalent in Υ with

$$|\varphi(w)| < 1$$

in Υ.

Finally let $m_v = \max(e, K_v, K_v')$ and taking

$$S = m_v + 16eR, \tag{102}$$

then

$$B > \frac{Q}{2} > \frac{M}{16e} > \frac{S}{16e} > R.$$

Consequently if we take in (84),

$$c_\nu = 4e(m_\nu + 16e),$$

then (84) implies (86) with S defined by (102), and hence the conclusion of Corollary 8 holds.

In what follows, c_ν always denotes the positive constant in (84).

Corollary 9. Let $F(z)$ be a function which is holomorphic and does not take the value zero in the circle $|z|<1$. Let ν be a positive integer and R a positive number. Assume that there exists a number $0<r_0<1$ such that

$$\log M(r_0,F) \geqslant (1+2\pi)\nu c_\nu(1+R+\overset{+}{\log}\ |F(0)| + \log\frac{1}{1-r_0})e^{\frac{a}{-\log r_0}}. \tag{103}$$

Then the conclusion of Corollary 8 holds.

Proof. If $|F(0)|\geqslant 1$, then (103) implies (84), hence the conclusion of Corollary 8 holds. If $|F(0)|<1$, then (103) becomes

$$\log M(r_0,F) \geqslant (1+2\pi)\nu c_\nu(1+R+\log\frac{1}{1-r_0})e^{\frac{a}{-\log r_0}}. \tag{104}$$

$z_0=r_0e^{i\theta_0}$ being a point of the circle $|z|=r_0$, such that

$$|F(z_0)| = M(r_0,F) > 1,$$

there exists a point $z_1=r_1e^{i\theta_0}(0<r_1<r_0)$, such that

$$|F(z_1)| = 1.$$

Consider the function

$$F_1(\zeta) = \frac{1}{(1-|z_1|)^\nu}F\{z_1+\zeta(1-|z_1|)\} \quad (|\zeta|<1).$$

We have

$$|F_1(0)| = \frac{1}{(1-|z_1|)^\nu}, \quad F_1(\zeta_0) = \frac{1}{(1-|z_1|)^\nu}F(z_0),$$

where

$$\zeta_0 = \frac{z_0 - z_1}{1-|z_1|} = \frac{r_0 - r_1}{1-r_1}e^{i\theta_0}, |\zeta_0| < r_0.$$

Then by (104), we have

$$M(r_0, Log F_1) \geqslant log\, M(r_0, F_1) \geqslant log\, |F_1(\zeta_0)| > log\, M(r_0, F)$$
$$> c_\nu(1 + R + |Log F_1(0)|)e^{\frac{a}{-log r_0}}.$$

Hence by Corollary 8, we can find a number $B>R$ and a holomorphic and univalent function $\varphi_1(w)$ in the circle $|w-w_0|<R$ $(w_0=Bi)$, such that

$$|\varphi_1(w)| < 1, \; F_1^{(\nu)}\{\varphi_1(w)\} = e^w$$

in the circle $|w-w_0|<R$. The fucntion $\varphi(w)=z_1+\varphi_1(w)(1-|z_1|)$ is holomorphic and univalent in the circle $|w-w_0|<R$ and we have

$$|\varphi(w)| < 1, \; F^{(\nu)}\{\varphi(w)\} = e^w$$

in the circle $|w-w_0|<R$.

Corollary 10. Let $F(z)$ be a function which is holomorphic and does not take the value zero in the circle $|z|<1$. Let ν be a positive integer and $K>1$ a number. Assume that there exists a number $0<r_0<1$ such that

$$log\, M(r_0, F) \geqslant c'_\nu(2 + log\, K + \overset{+}{log}\, |F(0)| + log\, \frac{1}{1-r_0})e^{\frac{a}{-log r_0}}, \quad (104)$$

where $c'_v = 4\pi(1+2\pi)\nu c_v$. Then for each number $0 \leqslant \omega < 2\pi$, the domain

$$(C_\omega):\quad \frac{1}{K} < |Z| < K,\ \omega < \arg Z < \omega + 2\pi$$

is a simply covered image domain of $F^{(\nu)}(z)$ for the circle $|z|<1$.

This corollary is deduced from Corollary 9 by the method used in the proof of Corollary 5.

Similarly we can prove the following corollary:

Corollary 11. Let $F(z)$ be a function which is holomorphic and does not take the value zero in the circle $|z|<1$. Let ν be a positive integer and R a positive number. Assume that there exists a number $0<r_0<1$ such that

$$\log M\left(r_0, \frac{1}{F}\right) \geqslant (1+2\pi)\nu c_\nu\left(1 + R + \overset{+}{\log}\frac{1}{|F(0)|} + \log\frac{1}{1-r_0}\right)e^{\frac{a}{-\log r_0}}.$$

Then the conclusion of Corollary 8 holds.

Proof. If $|F(0)| \leqslant 1$, then the inequality in Corollary 11 implies (84), hence the conclusion of Corollary 8 holds. If $|F(0)|>1$, then the inequality in Corollary 11 becomes

$$\log M\left(r_0, \frac{1}{F}\right) \geqslant (1+2\pi)\nu c_\nu\left(1 + R + \log\frac{1}{1-r_0}\right)e^{\frac{a}{-\log r_0}}.$$

Let $z_0 = r_0 e^{i\theta_0}$ be a point of the circle $|z|=r_0$, such that

$$\frac{1}{|F(z_0)|} = M\left(r_0, \frac{1}{F}\right) > 1.$$

Then there is a point $z_1 = r_1 e^{i\theta_0}$ $(0<r_1<r_0)$ such that

$$|F(z_1)| = 1.$$

As in the proof of Corollary 9, consider the function

$$F_1(\zeta) = \frac{1}{(1 - |z_1|)^\nu} F\{z_1 + \zeta(1 - |z_1|)\} \quad (|\zeta| < 1).$$

We have

$$|F_1(0)| = \frac{1}{(1 - |z_1|)^\nu}, \ F_1(\zeta_0) = \frac{1}{(1 - |z_1|)^\nu} F(z_0), \ |\zeta_0| < r_0.$$

Then we have

$$\begin{aligned}
M(r_0, Log\, F_1) &\geqslant log\, M(r_0, \frac{1}{F_1}) \geqslant log \frac{1}{|F_1(\zeta_0)|} = \nu log\,(1 - |z_1|) \\
&= \nu log\,(1 - |z_1|) + log \frac{1}{|F(z_0)|} > \nu log\,(1 - r_0) + log\, M(r_0, \frac{1}{F}) \\
&> \nu log\,(1 - r_0) + c_\nu(1 + 2\pi + R + \nu log \frac{1}{1 - r_0}) e^{\frac{a}{-log\, r_0}} \\
&\qquad + c_\nu(2\pi\nu)(log \frac{1}{1 - r_0}) e^{\frac{a}{-log\, r_0}} \\
&> c_\nu(1 + 2\pi + R + \nu log \frac{1}{1 - r_0}) e^{\frac{a}{-log\, r_0}} \\
&> c_\nu(1 + R + |Log\, F_1(0)|) e^{\frac{a}{-log\, r_0}}.
\end{aligned}$$

Then we complete the proof of Corollary 11 as in the proof of Corollary 9.

As for Corollary 10, from Corollary 11 we can deduce the following corollary:

Corollary 12. Let $F(z)$ be a function which is holomorphic and does not take the value zero in the circle $|z| < 1$. Let ν be a positive integer and $K > 1$ a number. Assume that there exists a number $0 < r_0 < 1$ such that

$$log\, M(r_0, \frac{1}{F}) \geqslant c'_\nu(2 + log\, K + \overset{+}{log} \frac{1}{|F(0)|} + log \frac{1}{1 - r_0}) e^{\frac{a}{-log\, r_0}},$$

where $c'_\nu = 4\pi(1 + 2\pi)\nu c_\nu$. Then the conclusion of Corollary 10 holds.

3. GENERALIZATION OF THEOREM 2

Theorem 2 can be further generalized as follows:

Theorem 3. Let $f(z)$ be a holomorphic function in the circle $|z|<1$, k a positive integer and S a positive number. Assume that there exists a number $0<r_0<1$ such that

$$M(r_0,f) \geqslant 4e\left(S + \sum_{n=0}^{k-1} \frac{|f^{(n)}(0)|}{n!}\right) e^{\frac{a_1}{-\log r_0}}.$$

Under these conditions, we can find two numbers ρ, N such that

$$e^{-b_1} < \rho < 1,\ M = M(\rho,f) > S,\ k \leqslant N < 1 + \left(\log \frac{M}{S}\right)^2$$

and $k+1$ holomorphic functions $u(\zeta)$, $v_p(\zeta)$ $(p=1,2,\cdots,k)$ in the rectangle

$$(R):\ -2 < Re(\zeta) < -\frac{1}{2},\quad 0 < Im(\zeta) < 2(k+1)\pi$$

having the following properties:

1^0 In (R), $u(\zeta)$ is univalent and $Re\{u(\zeta)\}<0$. For each number $0\leqslant\omega<2\pi$, $e^{u(\zeta)}$ is univalent in the rectangle

$$(R_\omega):\ -2 < Re(\zeta) < -\frac{1}{2},\ \omega < Im(\zeta) < \omega + 2k\pi.$$

In (R) we have

$$f\{\rho e^{u(\zeta)}\} = Me^{\zeta}.$$

2^0 For $1\leqslant p\leqslant k$, in (R) we have

$$|v_p(\zeta)| < 2\pi + \frac{1}{2}$$

and

$$f^{(p)}\{\rho e^{u(\zeta)}\} = \frac{N(N-1)\cdots(N-p+1)}{\rho^p} M e^{(1-\frac{p}{N})\zeta + v_p(\zeta)}.$$

3^0 For $n \geqslant 1$, we have

$$M(\rho, f^{(n)}) < n! e^{c_k} M (\frac{N}{\rho})^n.$$

In this theorem, a_k, b_k, c_k are positive constants depending only on k.

For the proof of Theorem 3, which is quite similar to that of Theorem 2, see [10], where Theorem 3 in a more complete form is proved. In what follows, we deduce some consequences of Theorem 3, in which the symbol a_k always denotes the positive constant in Theorem 3.

Corollary 13. Let f(z) be a holomorphic function in the circle $|z| < 1$, k a positive integer and S a positive number. Assume that there exists a number $0 < r_0 < 1$ such that

$$M(r_0, f) \geqslant 4e(S + \sum_{n=0}^{k-1} \frac{|f^{(n)}(0)|}{n!}) e^{\frac{a_k}{-\log r_0}}.$$

Then we can find a number M such that

$$M > S$$

and that for each number $0 \leqslant \omega < 2\pi$, there are k disjoint simply connected domains $D_j(\omega)$ $(j = 1, 2, \cdots, k)$ belonging to the circle $|z| < 1$, such that for each j $(1 \leqslant j \leqslant k)$, f(z) is univalent in $D_j(\omega)$ and maps $D_j(\omega)$ onto the domain

$$(\Gamma_\omega): \quad Me^{-2} < |w| < Me^{-\frac{1}{2}}, \quad \omega < \arg w < \omega + 2\pi.$$

Proof. By Theorem 3, we can find ρ and $u(\zeta)$ having the properties in that theorem. We are going to show that the number

$$M = M(\rho, f) > S$$

satisfies the required condition. In fact, consider a number $0 \leqslant \omega < 2\pi$ and an integer $1 \leqslant j \leqslant k$. Since the function $\rho e^{u(\zeta)}$ is univalent in the domain

$$-2 < Re(\zeta) < -\frac{1}{2},\ \omega + 2(j-1)\pi < Im(\zeta) < \omega + 2j\pi,$$

$\rho e^{u(\zeta)}$ maps this domain onto a simply connected domain $D_j(\omega)$ belonging to the circle $|z| < 1$. By the formula

$$f\{\rho e^{u(\zeta)}\} = Me^{\zeta},$$

we see easily that $f(z)$ is univalent in $D_j(\omega)$ and maps $D_j(\omega)$ onto (Γ_ω). Since $\rho e^{u(\zeta)}$ is univalent in (R_ω), evidently $D_{j_1}(\omega)$ and $D_{j_2}(\omega)$ have no common point, when $j_1 \neq j_2$.

Corollary 14. Let $f(z)$ be a holomorphic function in the circle $|z| < 1$. Let k, p, d be positive integers and R a positive number. Let $F(z)$ be a function of the following form:

$$F(z) = e^{f(z)}\{Q_0(z) + \sum_{i=1}^{v} Q_i(z)\}, \qquad (105)$$

where

$$Q_0(z) = a_0 \prod_{j=0}^{k} (f^{(j)})^{m_j},\ Q_i(z) = a_i \prod_{j=0}^{p} (f^{(j)})^{m_{ij}} \quad (i = 1, 2, \cdots, n) \qquad (106)$$

in which $a_0 \neq 0$, $a_i (i = 1, 2, \cdots, v)$ are constants and $m_j (j = 0, 1, \cdots, k)$, $m_{ij} (i = 1, 2, \cdots, v, j = 0, 1, \cdots, p)$ are non-negative integers with

$$\sum_{j=0}^{k} m_j = d,\ \sum_{j=0}^{p} m_{ij} < d \quad (i = 1, 2, \cdots, v). \qquad (107)$$

Assume that there exists a number $0<r_0<1$ such that

$$M(r_0,f) \geqslant \alpha(A + R + \sum_{n=0}^{k-1} \frac{|f^{(n)}(0)|}{n!})e^{\frac{a_1}{-\log r_0}}, \tag{108}$$

where α is a positive constant depending only on k, p, d and

$$A = 1 + |\log|a_0|| + (\sum_{i=1}^{v}|\frac{a_i}{a_0}|)^2. \tag{109}$$

Then we can find a number $B>R$ and a holomorphic and univalent function $\varphi(w)$ in the circle $|w-w_0|<R$ $(w_0=Bi)$, such that

$$|\varphi(w)| < 1, \; F\{\varphi(w)\} = e^w \tag{110}$$

in the circle $|w-w_0|<R$.

Proof. The proof of this corollary is quite similar to that of Corollary 8. The main difference is that now we base upon Theorem 3 instead of Theorem 2. Following the procedure of the proof of Corollary 8, consider the function $f(z)$ and a positive number S. Assume that there exists a number $0<r_0<1$ such that

$$M(r_0,f) \geqslant 4e(S + \sum_{n=0}^{k-1} \frac{|f^{(n)}(0)|}{n!})e^{\frac{a_1}{-\log r_0}}.$$

Then by Theorem 3, we can find $\rho, \mathbf{N}, u(\zeta), v_p(\zeta)$ $(p=1,2,\cdots,k)$ having the properties in Theorem 3. If

$$S > e, \tag{111}$$

we have

$$k \leqslant N < 2(\log M)^2 \tag{112}$$

and by the property 3^0 in Theorem 3, we have for $0\leqslant n\leqslant p$,

$$M(\rho, f^{(n)}) < A_{k,p} M(\log M)^{2p}, \ A_{k,p} = p\,!\,2^p e^{(c_k + p b_k)}.$$

Then from (106) and (107), we see that in the circle $|z| \leqslant \rho$, we have

$$|Q_i(z)| \leqslant |a_i| B_{k,p,d} M^{d-1} (\log M)^{2p(d-1)}, \ B_{k,p,d} = A_{k,p}^{d-1} \tag{113}$$
$$(i = 1, 2, \cdots, v).$$

Hence if we put

$$Q(z) = \sum_{i=1}^{v} Q_i(z),$$

then in the circle $|z| \leqslant \rho$, we have

$$|Q(z)| \leqslant B_{k,p,d} M^{d-1} (\log M)^{2p(d-1)} \sum_{i=1}^{v} |a_i|. \tag{114}$$

On the other hand, from the properties 1^0 and 2^0 in Theorem 3, in (R) we have for $0 \leqslant j \leqslant k$,

$$|f^{(j)}\{\rho e^{u(\zeta)}\}| > M e^{-(2+2\pi+\frac{1}{2})}.$$

Hence in (R) we have

$$|Q_0(\rho e^{u(\zeta)})| > |a_0| M^d e^{-(2+2\pi+\frac{1}{2})d}. \tag{115}$$

From (114) and (115), in (R) we have

$$\left| \frac{Q\{\rho e^{u(\zeta)}\}}{Q_0\{\rho e^{u(\zeta)}\}} \right| < B'_{k,p,d} \frac{(\log M)^{2p(d-1)}}{M} \sum_{i=1}^{v} \left| \frac{a_i}{a_0} \right|, \quad B'_{k,p,d} = B_{k,p,d} e^{(2+2\pi+\frac{1}{2})d}.$$

Let $\lambda_{k,p,d}$ be a positive number such that when $M > \lambda_{k,p,d}$, we have

$$B'_{k,p,d} \frac{(\log M)^{2p(d-1)}}{M^{\frac{1}{2}}} < \frac{1}{2}.$$

Then we see that if

$$S > \lambda_{k,p,d},\ S > (\sum_{i=1}^{\nu}\left|\frac{a_i}{a_0}\right|)^2, \tag{116}$$

we have in (R),

$$\left|\frac{Q\{\rho e^{u(\zeta)}\}}{Q_0\{\rho e^{u(\zeta)}\}}\right| < \frac{1}{2}. \tag{117}$$

By the properties 1^0 and 2^0 in Theorem 3, we see that for $0 \leqslant j \leqslant k$, we have in (R),

$$f^{(j)}\{\rho e^{u(\zeta)}\} = e^{\log M + V_j(\zeta)}, \tag{118}$$

where $V_j(\zeta)$ is a holomorphic function in (R) such that

$$|V_j(\zeta)| < \Gamma_k + k\log N, \tag{119}$$

in (R), Γ_k being a positive constant depending only on k. Consequently from (118), (119) and (106), we have in (R),

$$Q_0\{\rho e^{u(\zeta)}\} = e^{Log\, a_0 + d\log M + V(\zeta)}, \tag{120}$$

where $V(\zeta)$ is a holomorphic function in (R) such that

$$|V(\zeta)| < d(\Gamma_k + k\log N) \tag{121}$$

in (R).

From (105), (117), (120) and the property 1^0 in Theorem 3, in (R) we have

$$F\{\rho e^{u(\zeta)}\} = e^{Me^{\zeta} + Log\, a_0 + d\log M + V(\zeta) + l(\zeta)}, \tag{122}$$

where $l(\zeta)$ is a holomorphic function in (R) such that

$$|l(\zeta)| < 1 \tag{123}$$

in (R). (122) can be written in the form

$$F\{\rho e^{u(\zeta)}\} = e^{Me^{\zeta}+g(\zeta)}, \tag{124}$$

where

$$g(\zeta) = Log\, a_0 + d\, log\, M + V(\zeta) + l(\zeta)$$

is a holomorphic function in (R) and by (121),(123) and (112), we have

$$|g(\zeta)| < d(1+2k)log\, M + |log\, |a_0|| + \Gamma'_{k,d},$$
$$\Gamma'_{k,d} = 2\pi + 1 + d(\Gamma_k + k\, log\, 2) \tag{125}$$

in (R).

Comparing (124) with (95), evidently we can continue the proof in the same way as in the proof of Corollary 8. We see that if

$$S > \lambda'_{k,d},\ S > \frac{16}{e^{-\frac{1}{2}} - e^{-2}}|log\, |a_0||,\ S > 16eR, \tag{126}$$

then we get the conclusion of Corollary 14. Hence in (108), it is sufficient to take

$$a = 4e\Lambda_{k,p,d},$$

where

$$\Lambda_{k,p,d} = max(\lambda_{k,p,d}, \lambda'_{k,d}, 32e).$$

APPENDIX B

SOME THEOREMS ON MEROMORPHIC FUNCTIONS

1. A THEOREM ON MEROMORPHIC FUNCTIONS IN THE PLANE

For the sake of convenience, we first give the following definition: (see Definition 3. 3)

Definition. Let $f(z)$ be a meromorphic function in $\mathbf{C}$ and $a \in \hat{\mathbf{C}}$ a value. Let $m \geqslant 2$ be an integer or $m = +\infty$. We say that a is an exceptional value of weight $1-\frac{1}{m}$ of $f(z)$ in $\mathbf{C}$, if in $\mathbf{C}$ the equation $f(z)=a$ has no root whose order of multiplicity is less than m. In particular, if $m=+\infty$, an exceptional value a of weight 1 of $f(z)$ in $\mathbf{C}$ is such that the equation $f(z)=a$ has no root in $\mathbf{D}$.

Theorem 1. Let $a_j \in \hat{\mathbf{C}} (j=1,2,\cdots,q; 3 \leqslant q \leqslant 5)$ be q distinct values and $m_j \geqslant 2 (j=1,2,\cdots,q)$ be such that for each $1 \leqslant j \leqslant q$, m_j is an integer or $+\infty$, and that

$$\sum_{j=1}^{q} (1-\frac{1}{m_j}) > 2. \tag{1}$$

Then there does not exist a non-constant meromorphic function $f(z)$ in $\mathbf{C}$ such that for each $1 \leqslant j \leqslant q$, the value a_j is an exceptional value of weight $1-\frac{1}{m_j}$ of $f(z)$ in $\mathbf{C}$.

Proof. Assume on the contrary that there exists a non-constant meromorphic function $f(z)$ in $\mathbf{C}$ such that for each $1 \leqslant j \leqslant q$, the value a_j is an exceptional value of weight $1-\frac{1}{m_j}$ of f(z) in C. By the second fundamental theorem in Nevanlinna's theory of meromorphic functions[13], we have the inequality

$$(q-2)T(r,f) \leqslant \sum_{j=1}^{q} \overline{N}(r,a_j) + S(r,f). \tag{2}$$

In general $\overline{N}(r,a)$ is defined by the formula

$$\overline{N}(r,a) = \int_0^r \frac{\overline{n}(t,a)-\overline{n}(o,a)}{t}dt + \overline{n}(o,a)\log r, \tag{3}$$

where $\overline{n}(t,a)$ denotes the number of the roots in the circle $|z| \leqslant t$ of the equation $f(z)=a$, each root being counted once. On the other hand, $S(r,f)$ denotes a function of r, such that there is a set σ of values of r of finite linear measure, such that

$$\lim_{\substack{r \to +\infty \\ r \notin \sigma}} \frac{S(r,f)}{T(r,f)} = 0. \tag{4}$$

By the assumption, evidently for each $1 \leqslant j \leqslant q$, we have

$$\overline{N}(r,a_j) \leqslant \frac{1}{m_j} N(r,a_j). \tag{5}$$

In general $N(r,a)$ is defined by the formula

$$N(r,a) = \int_0^r \frac{n(t,a)-n(o,a)}{t}dt + n(o,a)\log r. \tag{6}$$

where $n(t,a)$ denotes the number of the roots in the circle $|z| \leqslant t$ of the equation $f(z)=a$, with due count of order of multiplicity. By the first fundamental theorem in Nevanlinna's theory of meromorphic functions, we have

$$N(r,a_j) \leqslant T(r,f) + O(1). \tag{7}$$

From (2), (5), (7), we deduce

$$\left\{\sum_{j=1}^{q}(1-\frac{1}{m_j})-2\right\}T(r,f)\leqslant 0(1)+S(r,f).$$

This inequality is impossible, in view of (4) and (1). So we get a contradiction.

2. A THEOREM ON MEROMORPHIC FUNCTIONS IN THE UNIT CIRCLE

Theorem 2. Let $k\geqslant 1$ be an integer. Then we can find a positive constant A_k depending only on k having the following property: If $f(z)$ is a meromorphic function in the circle $|z|<1$, such that each of the equations

$$f(z)=0,\ f^{(k)}(z)=1 \tag{8}$$

has no root in the circle $|z|<1$, then in the circle $|z|<\frac{1}{32}$ one of the inequalities

$$|f(z)|<A_k,\ \left|\frac{1}{f(z)}\right|<A_k$$

holds.

This theorem was first proved by Gu[17](for the circle $|z|<\frac{1}{64}$). Its proof was later simplified by Yang[38]. In what follows we give a complete proof of Theorem 2. For this we need some lemmas.

Lemma 1. Given an integer $k\geqslant 1$, we can find positive constants $A_j(j=1, 2,\cdots,7)$ having the following property: If $f(z)$ is a meromorphic function in a domain $|z|<R\ (0<R\leqslant+\infty)$ which has, in the neighborhood of the point $z=0$, a development of the form:

$$f(z)=c_s z^s+c_{s+1}z^{s+1}+\cdots,\ c_s\neq 0,\ s\geqslant 0, \tag{9}$$

than for $0<r<\rho<R$ we have

$$m(r,\frac{f^{(k)}}{f})<A_1\overset{+}{\log}\,T(\rho,f)+A_2\overset{+}{\log}\,\frac{1}{\rho-r}+A_3\overset{+}{\log}\,\rho+A_4\overset{+}{\log}\,\frac{1}{r}$$

$$+ A_5 \overset{+}{\log} \overset{+}{\log} \frac{1}{|c_s|} + A_6 \overset{+}{\log} s + A_7. \tag{10}$$

Proof. For $k=1$, Lemma 1 is known[13]. We are going to prove it by mathematical induction. Assume that it is true for $k=1,2,\cdots,p$, and let us show that it holds for $k=p+1$. In fact, let $f(z)$ be a meromorphic function in a domain $|z|<R$ having a development of the form (9) in the neighborhood of the point $z=0$. Consider the function

$$g(z) = z^p f(z) = c_s z^{s+p} + c_{s+1} z^{s+p+1} + \cdots. \tag{11}$$

We have

$$g^{(p+1)} = z^p f^{(p+1)} + a_1 z^{p-1} f^{(p)} + a_2 z^{(p-2)} f^{(p-1)} + \cdots + a_p f',$$

where $a_j (j=1,2,\cdots,p)$ are positive integers. Hence

$$\frac{g^{(p+1)}}{g} = \frac{f^{(p+1)}}{f} + a_1 \frac{1}{z} \frac{f^{(p)}}{f} + a_2 \frac{1}{z^2} \frac{f^{(p-1)}}{f} + \cdots + a_p \frac{1}{z^p} \frac{f'}{f}$$

and we have

$$m\left(r, \frac{f^{(p+1)}}{f}\right) \leqslant m\left(r, \frac{g^{(p+1)}}{g}\right) + \sum_{j=1}^{p} m\left(r, \frac{f^{(j)}}{f}\right) + \mu \overset{+}{\log} \frac{1}{r} + \mu', \tag{12}$$

where $\mu = \sum_{j=1}^{p} j$ and $\mu' = \sum_{j=1}^{p} \log |a_j| + \log(p+1)$. Since by assumption Lemma 1 is true for $k=1,2,\cdots,p$, it follows from (12) that it is sufficient to show that for $0<r<\rho<R$, we have an inequality of the following form:

$$m\left(r, \frac{g^{(p+1)}}{g}\right) < B_1 \overset{+}{\log} T(\rho, g) + B_2 \overset{+}{\log} \frac{1}{\rho - r} + B_3 \overset{+}{\log} \rho + B_4 \overset{+}{\log} \frac{1}{r}$$
$$+ B_5 \overset{+}{\log} \overset{+}{\log} \frac{1}{|c_s|} + B_6 \overset{+}{\log} (s+p) + B_7, \tag{13}$$

where $B_j (j=1,2,\cdots,7)$ are positive constants, in view of the inequalities

$$T(\rho,g) \leqslant p \overset{+}{\log} \rho + T(\rho,f), \quad \overset{+}{\log}(s+p) \leqslant \overset{+}{\log} s + \log(2p).$$

In fact by the identity

$$\frac{g^{(p+1)}}{g} = \frac{g^{(p+1)}}{g^{(p)}} \frac{g^{(p)}}{g},$$

we have for $0<r<\rho<R$, $\rho' = (r+\rho)/2$,

$$m(r,\frac{g^{(p+1)}}{g}) \leqslant m(r,\frac{g^{(p+1)}}{g^{(p)}}) + m(r,\frac{g^{(p)}}{g}), \tag{14}$$

$$m(r,\frac{g^{(p+1)}}{g^{(p)}}) < a_1 \overset{+}{\log} T(\rho',g^{(p)}) + a_2 \overset{+}{\log} \frac{1}{\rho'-r} + a_3 \overset{+}{\log} \rho' + a_4 \overset{+}{\log} \frac{1}{r}$$
$$+ a_5 \overset{+}{\log} \overset{+}{\log} \frac{1}{|\lambda c_s|} + a_6 \overset{+}{\log} s + a_7, \tag{15}$$

where $\lambda=(s+p)(s+p-1)\cdots(s+1)$ and $a_j(j=1,2,\cdots,7)$ are positive numerical constants, and by assumption,

$$m(r,\frac{g^{(p)}}{g}) < B'_1 \overset{+}{\log} T(\rho,g) + B'_2 \overset{+}{\log} \frac{1}{\rho-r} + B'_3 \overset{+}{\log} \rho + B'_4 \overset{+}{\log} \frac{1}{r}$$
$$+ B'_5 \overset{+}{\log} \overset{+}{\log} \frac{1}{|c_s|} + B'_6 \overset{+}{\log}(s+p) + B'_7 \tag{16}$$

where $B'_j(j=1,2,\cdots,7)$ are positive constants. Consider (15). We have

$$T(\rho',g^{(p)}) = m(\rho',g^{(p)}) + N(\rho',g^{(p)}) \leqslant m(\rho',g) + m(\rho',\frac{g^{(p)}}{g})$$
$$+ (p+1)N(\rho',g)$$
$$\leqslant (p+1)T(\rho,g) + m(\rho',\frac{g^{(p)}}{g}),$$
$$\overset{+}{\log} T(\rho',g^{(p)}) \leqslant \overset{+}{\log} T(\rho,g) + \overset{+}{\log} m(\rho',\frac{g^{(p)}}{g}) + \log 2(p+1),$$
$$\overset{+}{\log} m(\rho',\frac{g^{(p)}}{g}) \leqslant m(\rho',\frac{g^{(p)}}{g}) < H + B'_2 \overset{+}{\log} 2,$$

where H denotes the right member of (16). Hence

$$\overset{+}{\log} T(\rho', g^{(p)}) < \overset{+}{\log} T(\rho, g) + H + \beta, \tag{17}$$

where β is a positive constant. (14), (15), (16), (17) yield (13).

Evidently in Lemma 1 the positive constants $A_j (j=1,2,\cdots,7)$ depend only on k.

Lemma 2. Let $k \geqslant 1$ be an integer and $f(z)$ a meromorphic function in a domain $|z| < R (0 < R \leqslant +\infty)$ such that $f(0) \neq 0, \infty$. Then for $0 < r < \rho < R$ we have

$$m(r, \frac{f^{(k)}}{f}) < A_1 \overset{+}{\log} T(\rho, f) + A_2 \overset{+}{\log} \frac{1}{\rho - r} + A_3 \overset{+}{\log} \rho + A_4 \overset{+}{\log} \frac{1}{r} + A_5 \overset{+}{\log} \overset{+}{\log} \frac{1}{|f(0)|} + A_6, \tag{18}$$

where $A_j (j=1,2,\cdots,6)$ are positive constants depending only on k.

This is a particular case of Lemma 1.

Lemma 3. Let $k \geqslant 1$ be an integer and $f(z)$ a meromorphic function in a domain $|z| < R (0 < R \leqslant +\infty)$ such that

$$f(0) \neq 0, \infty, f^{(k)}(0) \neq 0.$$

Then for $0 < r < \rho < R$ we have

$$m(r, \frac{f^{(k+1)}}{f^{(k)}}) < B_1 \overset{+}{\log} T(\rho, f) + B_2 \overset{+}{\log} \frac{1}{\rho - r} + B_3 \overset{+}{\log} \rho + B_4 \overset{+}{\log} \frac{1}{r} + B_5 \overset{+}{\log} \overset{+}{\log} \frac{1}{|f(0)|} + B_6 \overset{+}{\log} \overset{+}{\log} \frac{1}{|f^{(k)}(0)|} + B_7, \tag{19}$$

where $B_j (j=1,2,\cdots,7)$ are positive constants depending only on k.

Proof. First by Lemma 2, for $0 < r < \rho < R$, $\rho' = (r+\rho)/2$, we have

$$m\left(r,\frac{f^{(k+1)}}{f^{(k)}}\right) < a_1 \overset{+}{\log} T(\rho',f^{(k)}) + a_2 \overset{+}{\log}\frac{1}{\rho'-r} + a_3 \overset{+}{\log}\rho' + a_4 \overset{+}{\log}\frac{1}{r}$$
$$+ a_5 \overset{+}{\log}\overset{+}{\log}\frac{1}{|f^{(k)}(0)|} + a_6, \tag{20}$$

where $a_j (j=1,2,\cdots,6)$ are positive numerical constants. Next as (17), we find

$$\overset{+}{\log} T(\rho',f^{(k)}) < \overset{+}{\log} T(\rho,f) + L + \alpha, \tag{21}$$

where L denotes the right member of (18) and α is a positive constant depending only on k. From (20) and (21) we deduce (19).

Similarly we prove the following lemma:

Lemma 4. Let $k \geqslant 1$ be an integer and $f(z)$ a meromorphic function in a domain $|z|<R\,(0<R\leqslant+\infty)$ such that

$$f(0) \neq 0,\infty,\ f^{(k)}(0) \neq 1.$$

Then for $0<r<\rho<R$ we have

$$m\left(r,\frac{f^{(k+1)}}{f^{(k)}-1}\right) < B_1 \overset{+}{\log} T(\rho,f) + B_2 \overset{+}{\log}\frac{1}{\rho-r} + B_3 \overset{+}{\log}\rho + B_4 \overset{+}{\log}\frac{1}{r}$$
$$+ B_5 \overset{+}{\log}\overset{+}{\log}\frac{1}{|f(0)|} + B_6 \overset{+}{\log}\overset{+}{\log}\frac{1}{|f^{(k)}(0)-1|} + B_7',$$

where $B_j (j=1,2,\cdots,6)$ and B_7' are positive constants depending only on k.

Lemma 5. Let $f(z)$ be a holomorphic function in a domain $|z|<R\,(0<R\leqslant+\infty)$. Then for $0<r<\rho<R$ we have

$$\overset{+}{\log} M(r,f) \leqslant \frac{\rho+r}{\rho-r} T(\rho,f).$$

This lemma is known[13].

Lemma 6. Let $U(r)$ be a non-negative and non-decreasing function in an interval $r_0 < r < R$ $(r_0 \geqslant 0, r_0 < R \leqslant +\infty)$. Let a and b be two positive numbers such that $b \geqslant 2a$ and $b \geqslant 8a^2$. Assume that for $r_0 < r < \rho < R$ we have

$$U(r) < a \overset{+}{\log} U(\rho) + a\log \frac{\rho}{\rho - r} + b. \tag{22}$$

Then for $r_0 < r < \rho < R$ we have

$$U(r) < 2a\log \frac{\rho}{\rho - r} + 2b. \tag{23}$$

Proof. In the case $r_0 = 0$, this lemma is known[13]. Consider the case $r_0 > 0$. Let $V(r)$ be the non-negetive and non-decreaseing function defined for $0 < r < R$ as follows:

$$V(r) = 0 \; for \; 0 < r \leqslant r_0, \; V(r) = U(r) \; for \; r_0 < r < R.$$

Then evidently (22) holds for $0 < r < \rho < R$, if we replace $U(r)$ and $U(\rho)$ respectively by $V(r)$ and $V(\rho)$. Consequently (23) holds for $0 < r < \rho < R$, if we replace $U(r)$ by $V(r)$. This implies (23) for $r_0 < r < \rho < R$.

Lemma 7. Let $U(r)$ be a non-nagative and non-decreasing function in an interval $0 < r < R_0$ $(0 < R_0 \leqslant 1)$, such that for $0 < r < \rho < R_0$ we have

$$U(r) < A_1 \overset{+}{\log} U(\rho) + A_2\log \frac{1}{\rho - r} + A_3\log \frac{1}{r} + A_4, \tag{24}$$

where A_j $(j = 1, 2, 3, 4)$ are positive constants. Let r_0 $(0 < r_0 < R_0)$ be a number. Then for $r_0 < r < \rho < R_0$ we have

$$U(r) < 2A\log \frac{\rho}{\rho - r} + 2B, \tag{25}$$

where

$$A = max(A_1, A_2),\ B = (A_2 + A_3)log\frac{1}{r_0} + A_4 + max(2A, 8A^2).$$

Proof. For $r_0 < r < \rho < R_0$, (24) implies

$$U(r) < A\overset{+}{log}\,U(\rho) + A\,log\frac{\rho}{\rho - r} + B.$$

Hence by Lemma 6, for $r_0 < r < \rho < R_0$ we have (25).

Lemma 8. Let $a > e$ and $x > 0$ be two numbers. Then we have

$$log\,x + a\,\overset{+}{log}\,\overset{+}{log}\frac{1}{x} \leqslant a(log\,a - 1) + \overset{+}{log}\,x.$$

This lemma is known[13].

Lemma 9. Let $f(z)$ be a meromorphic function in a circle $|z| < R\,(0 < R < +\infty)$ and $k \geqslant 1$ an integer, such that each of the equations

$$f(z) = 0,\ f^{(k)}(z) = 1 \tag{26}$$

has no root in the circle $|z| < R$. Put

$$\varphi(z) = f^{(k)}(z),\ \psi(z) = (k+2)\frac{\varphi'(z)}{\varphi(z) - 1} - (k+1)\frac{\varphi''(z)}{\varphi'(z)} \tag{27}$$

and assume

$$f(0) \neq \infty,\ \varphi'(0) \neq 0,\ \psi(0) \neq 0. \tag{28}$$

Then for $0 < r < R$ we have

$$T(r,f) \leqslant (2 + \frac{1}{k})\{m(r, \frac{\varphi}{f}) + m(r, \frac{\varphi'}{f})\} + (2 + \frac{2}{k})m(r, \frac{\varphi'}{\varphi - 1}) + \frac{1}{k}m(r, \frac{\varphi''}{\varphi'})$$

$$+ (2+\frac{1}{k})\log\left|\frac{f(0)\{(\varphi(0)-1\}}{\varphi'(0)}\right| + \frac{1}{k}\log\frac{1}{|\psi(0)|} + a_k, \tag{29}$$

where a_k is a positive constant depending only on k.

Concerning this lemma see [15].

Proof. We start from the following identity of Bureau:

$$\frac{1}{f} = \frac{\varphi}{f} - \frac{\varphi-1}{\varphi'}\frac{\varphi'}{f}.$$

We have

$$m(r,\frac{1}{f}) \leqslant m(r,\frac{\varphi}{f}) + m(r,\frac{\varphi-1}{\varphi'}) + m(r,\frac{\varphi'}{f}) + \log 2. \tag{30}$$

Since

$$T(r,f) = m(r,\frac{1}{f}) + N(r,\frac{1}{f}) + \log|f(0)|$$
$$= m(r,\frac{1}{f}) + \log|f(0)|,$$
$$m(r,\frac{\varphi-1}{\varphi'}) = m(r,\frac{\varphi'}{\varphi-1}) + N(r,\frac{\varphi'}{\varphi-1}) - N(r,\frac{\varphi-1}{\varphi'}) + \log\left|\frac{\varphi(0)-1}{\varphi'(0)}\right|,$$

and by a formula of Nevanlinna[28],

$$N(r,\frac{\varphi'}{\varphi-1}) - N(r,\frac{\varphi-1}{\varphi'}) = N(r,\varphi') - N(r,\frac{1}{\varphi'}) + N(r,\frac{1}{\varphi-1}) - N(r,\varphi-1)$$
$$= \overline{N}(r,f) - N(r,\frac{1}{\varphi'}),$$

hence

$$T(r,f) \leqslant \overline{N}(r,f) - N(r,\frac{1}{\varphi'}) + S(r), \tag{31}$$

where

$$S(r) = m(r,\frac{\varphi}{f}) + m(r,\frac{\varphi'}{f}) + m(r,\frac{\varphi'}{\varphi - 1}) + \log\left|\frac{f(0)\{\varphi(0)-1\}}{\varphi'(0)}\right| + \log 2. \tag{32}$$

It remains to estimate $\overline{N}(r,f)$. First of all, if we denote by $N_{1)}(r,f)$ the counting function for the simple poles of $f(z)$ and by $\overline{N}_{(2}(r,f)$ the counting function for the multiple poles of $f(z)$, each multiple pole being counted only once, then evidently

$$\overline{N}(r,f) = N_{1)}(r,f) + \overline{N}_{(2}(r,f). \tag{33}$$

Put

$$g(z) = \varphi(z) - 1 = f^{(k)}(z) - 1, \ \psi(z) = (k+2)\frac{g'(z)}{g(z)} - (k+1)\frac{g''(z)}{g'(z)} \tag{34}$$

and consider a simple pole z_0 of $f(z)$. Then in a circle $\Gamma: |z - z_0| < \rho$, we have

$$f = \frac{c}{z - z_0} + P, \qquad c \neq 0,$$

where $P = P(z)$ is a holomorphic function in Γ. Then in Γ we have successively

$$\varphi = f^{(k)} = (-1)^k \frac{ck!}{(z - z_0)^{k+1}} + P^{(k)},$$

$$g = (-1)^k \frac{ck!}{(z - z_0)^{k+1}}(1 + Q), \ Q = c'(z - z_0)^{k+1}(P^{(k)} - 1),$$

$$g' = (-1)^{k+1} \frac{c(k+1)!}{(z - z_0)^{k+2}} + P^{(k+1)} = (-1)^{k+1} \frac{c(k+1)!}{(z - z_0)^{k+2}}(1 + Q_1),$$

$$g'' = (-1)^{k+2} \frac{c(k+2)!}{(z - z_0)^{k+3}} + P^{(k+2)} = (-1)^{k+2} \frac{c(k+2)!}{(z - z_0)^{k+3}}(1 + Q_2),$$

$$Q_1 = c_1(z - z_0)^{k+2}P^{(k+1)}, \quad Q_2 = c_2(z - z_0)^{k+3}P^{(k+2)},$$

hence

$$\frac{g'}{g} = (-1)\frac{k+1}{z-z_0}\frac{1+Q_1}{1+Q} = (-1)\frac{k+1}{z-z_0}(1+\frac{Q_1-Q}{1+Q})$$

$$= (-1)\frac{k+1}{z-z_0}\{1+(z-z_0)^{k+1}R_1\} = (-1)\frac{k+1}{z-z_0} + (z-z_0)^k c_1' R_1,$$

$$\frac{g''}{g'} = (-1)\frac{k+2}{z-z_0} + (z-z_0)^{k+1}c_2' R_2,$$

$$\psi = (z-z_0)^k\{(k+2)c_1'R_1 - (k+1)c_2'(z-z_0)R_2\}.$$

Consequently z_0 is a zero of $\psi(z)$, the order of multiplicity of z_0 is at least equal to k. It follows that

$$N_{1)}(r,f) \leqslant \frac{1}{k}N(r,\frac{1}{\psi}) \leqslant \frac{1}{k}\{N(r,\psi) + m(r,\psi) + \log\frac{1}{|\psi(0)|}\}. \quad (35)$$

Let us estimate $N(r,\psi)$. By (39), the poles of $\psi(z)$ are all simple poles which are the zeros and poles of $g(z)$ and $g'(z)$. But by hypothesis of Lemma 9, $g(z)$ has no zero and the poles of $g(z)$ and $g'(z)$ are the poles of $f(z)$. On the other hand, we have shown that the simple poles of $f(z)$ are zeros of $\psi(z)$. It follows that

$$N(r,\psi) \leqslant \overline{N}_{(2}(r,f) + \overline{N}(r,\frac{1}{g'}) = \overline{N}_{(2}(r,f) + \overline{N}(r,\frac{1}{\varphi'}). \quad (36)$$

From (35) and (36) we get

$$N_{1)}(r,f) \leqslant \frac{1}{k}\{\overline{N}_{(2}(r,f) + \overline{N}(r,\frac{1}{\varphi'})\} + \frac{1}{k}\{m(r,\psi) + \log\frac{1}{|\psi(0)|}\}. \quad (37)$$

On the other hand, by (31) we have

$$N_{1)}(r,f) + 2\overline{N}_{(2}(r,f) \leqslant N(r,f) \leqslant T(r,f) \leqslant \overline{N}(r,f) - N(r,\frac{1}{\varphi'}) + S(r)$$

$$= N_{1)}(r,f) + \overline{N}_{(2}(r,f) - N(r,\frac{1}{\varphi'}) + S(r),$$

hence

$$\overline{N}_{(2}(r,f) \leqslant - N(r,\frac{1}{\varphi'}) + S(r). \tag{38}$$

From (37) and (38) we have

$$N_{1)}(r,f) \leqslant \frac{1}{k}S(r) + \frac{1}{k}\{m(r,\psi) + log\ \frac{1}{|\psi(0)|}\}, \tag{39}$$

and then

$$\overline{N}(r,f) = N_{1)}(r,f) + \overline{N}_{(2}(r,f) \leqslant - N(r,\frac{1}{\varphi'}) + (1 + \frac{1}{k})S(r)$$
$$+ \frac{1}{k}\{m(r,\psi) + log\ \frac{1}{|\psi(0)|}\}. \tag{40}$$

(31) and (40) yield

$$T(r,f) \leqslant - 2N(r,\frac{1}{\varphi'}) + (2 + \frac{1}{k})S(r) + \frac{1}{k}\{m(r,\psi) + log\ \frac{1}{|\psi(0)|}\}$$
$$\leqslant (2 + \frac{1}{k})S(r) + \frac{1}{k}\{m(r,\psi) + log\ \frac{1}{|\psi(0)|}\} \tag{41}$$

By (27) we have

$$m(r,\psi) \leqslant m(r,\frac{\varphi'}{\varphi - 1}) + m(r,\frac{\varphi''}{\varphi'}) + log\,2(k + 1)(k + 2). \tag{42}$$

Finally from (41),(32) and (42), we get (29).

Now let us come back to the proof of Theorem 2. For the sake of simplicity, in what follows we denote by $\alpha, \alpha', \alpha_1, \alpha_2, \alpha_3, \cdots, \beta, \beta', \beta_1, \beta_2, \beta_3, \cdots$ and $\gamma, \gamma', \gamma_1, \gamma_2, \gamma_3$ positive constants depending only on k, which, in particular, may be numerical positive constants.

Consider a meromorphic function f(z) in the circle $|z|<1$ and an integer $k\geqslant 1$ such that each of the equations (8) has no root in the circle $|z|<1$. Assume first that there exists a point z_0 of the circle $|z|<1/8$, such that

$$|f(z_0)|<\alpha_1,\ |\varphi(z_0)|<\alpha_2,\ |\varphi'(z_0)|>\alpha_3,\ |\psi(z_0)|>\alpha_4, \tag{43}$$

where $\varphi(z)$ and $\psi(z)$ are defined by (27). Consider the function

$$F(z)=f(z_0+z) \tag{44}$$

which is meromorphic in the circle $|z|<7/8$ and is such that each of the equations

$$F(z)=0,\ F^{(k)}(z)=1 \tag{45}$$

has no root in the circle $|z|<7/8$. Put

$$\Phi(z)=F^{(k)}(z),\ \Psi(z)=(k+2)\frac{\Phi'(z)}{\Phi(z)-1}-(k+1)\frac{\Phi''(z)}{\Phi'(z)}. \tag{46}$$

Then from (43), we have

$$|F(0)|<\alpha_1,\ |\Phi(0)|<\alpha_2,\ |\Phi'(0)|>\alpha_3,\ |\Psi(0)|>\alpha_4. \tag{47}$$

Hence by Lemma 9, we have for $0<r<7/8$,

$$\begin{aligned}T(r,F)\leqslant{}&(2+\frac{1}{k})\{m(r,\frac{\Phi}{F})+m(r,\frac{\Phi'}{F})\}+(2+\frac{2}{k})m(r,\frac{\Phi'}{\Phi-1})\\&+\frac{1}{k}m(r,\frac{\Phi''}{\Phi'})+(2+\frac{1}{k})\log\left|\frac{F(0)\{\Phi(0)-1\}}{\Phi'(0)}\right|+\frac{1}{k}\log\frac{1}{|\Psi(0)|}+a_k.\end{aligned} \tag{48}$$

Applying Lemmas 2,3,4 to F(z), we see, by (47), that for $0<r<\rho<7/8$, we have

$$T(r,F)<\beta\{\overset{+}{\log}\,T(\rho,F)+\log\frac{1}{\rho-r}+\log\frac{1}{r}+1$$

$$+ \overset{+}{\log}\,\overset{+}{\log}\,\frac{1}{|F(0)|} + \overset{+}{\log}\,\overset{+}{\log}\,\frac{1}{|\Phi(0)-1|}\}$$

$$+ (2+\frac{1}{k})\log|F(0)\{\Phi(0)-1\}| + \beta' . \tag{49}$$

We have

$$T(r,F) + \log\frac{1}{|F(0)|} = T(r,\frac{1}{F}) \tag{50}$$

and by (47),

$$\overset{+}{\log}\,T(\rho,F) \leqslant \overset{+}{\log}\,\{T(\rho,F) + \log\frac{a_1}{|F(0)|}\} \leqslant \overset{+}{\log}\,T(\rho,\frac{1}{F}) + \overset{+}{\log}\,\overset{+}{\log}\,a_1 + \log 2. \tag{51}$$

On the other hand, by Lemma 8 and (47), we have

$$(1+\frac{1}{k})\log|F(0)| + \beta\,\overset{+}{\log}\,\overset{+}{\log}\,\frac{1}{|F(0)|} \leqslant \beta_1, \tag{52}$$

$$(2+\frac{1}{k})\log|\Phi(0)-1| + \beta\,\overset{+}{\log}\,\overset{+}{\log}\,\frac{1}{|\Phi(0)-1|} \leqslant \beta_2. \tag{53}$$

From (49)—(53), we deduce that for $0<r<\rho<7/8$, we have

$$T(r,\frac{1}{F}) < \beta\{\overset{+}{\log}\,T(\rho,\frac{1}{F}) + \log\frac{1}{\rho-r} + \log\frac{1}{r}\} + \beta''. \tag{54}$$

Applying Lemma 7 to $T(r,\frac{1}{F})$ with $r_0=\frac{1}{8}$, we have for $r_0<r<\rho<7/8$,

$$T(r,\frac{1}{F}) < \alpha\log\frac{\rho}{\rho-r} + \alpha' .$$

In particular for $r=1/2$, $\rho=3/4$, we get

$$T(\frac{1}{2},\frac{1}{F}) < \alpha''.$$

Next applying Lemma 5 to $1/F(z)$, we get

$$\overset{+}{\log} M(\frac{1}{4},\frac{1}{F}) < 3\alpha''.$$

Now consider a point z_1 such that $|z_1|<1/8$. Since

$$f(z_1) = F(z_1 - z_0), \quad |z_1 - z_0| < \frac{1}{4},$$

we have

$$\left|\frac{1}{f(z_1)}\right| < e^{3\alpha''}.$$

Consequently in the circle $|z|<1/8$, we have

$$\left|\frac{1}{f(z)}\right| < e^{3\alpha''}. \tag{55}$$

In order to complete the proof of Theorem 2, we distinguish several cases.

A. In the circle $|z|<1/8$ we have

$$\sum_{j=0}^{k+1} |f^{(j)}(z)| \geqslant \frac{1}{4}. \tag{56}$$

Distinguish again two cases:

A_1. In the circle $|z|<1/32$, we have always $|f(z)|\leqslant 1$ or always $|f(z)|\geqslant 1$.

A_2. In the circle $|z|<1/32$ there is a point z_0 such that

$$|f(z_0)| = 1. \tag{57}$$

As above consider the function

$$F(z) = f(z_0 + z)$$

which is meromorphic in the circle $|z|<31/32$. By (56), in the circle $|z|<3/32$ we have

$$\sum_{j=0}^{k+1} |F^{(j)}(z)| \geqslant \frac{1}{4},$$

and hence

$$\frac{1}{|F(z)|} \leqslant 4\sum_{j=0}^{k+1} \left|\frac{F^{(j)}(z)}{F(z)}\right|.$$

Consequently for $0<r<3/32$ we have

$$m(r, \frac{1}{F}) \leqslant \sum_{j=1}^{k+1} m(r, \frac{F^{(j)}}{F}) + log\,4(k+2),$$

and then by Lemma 2 and (57), for $0<r<\rho<3/32$ we have

$$\begin{aligned} T(r, \frac{1}{F}) &< \Gamma\{\overset{+}{log}\, T(\rho, F) + log\,\frac{1}{\rho-r} + log\,\frac{1}{r} + 1\} \\ &= \Gamma\{\overset{+}{log}\, T(\rho, \frac{1}{F}) + log\,\frac{1}{\rho-r} + log\,\frac{1}{r} + 1\}. \end{aligned}$$

Applying Lemma 7 with

$$r_0 = \frac{1}{64}, \qquad r = \frac{5}{64}, \qquad \rho = \frac{11}{128},$$

we get

$$T(\frac{5}{64}, \frac{1}{F}) < \Gamma'$$

and then by Lemma 5, we have

$$\overset{+}{\log}\, M\left(\frac{1}{16}, \frac{1}{F}\right) < 9\Gamma',$$

$$M\left(\frac{1}{16}, \frac{1}{F}\right) < e^{9\Gamma'},$$

which implies that in the circle $|z| < 1/32$ we have

$$\left|\frac{1}{f(z)}\right| < e^{9\Gamma'}. \tag{58}$$

B. In the circle $|z| < 1/8$ there is a point z_0 such that

$$\sum_{j=0}^{k+1} |f^{(j)}(z_0)| < \frac{1}{4}. \tag{59}$$

Distinguish again two cases:

B_1. There is an integer $j(1 \leqslant j \leqslant k+2)$ such that in the circle $|z| < 1/8$ we have

$$|f^{(j)}(z)| \leqslant 1. \tag{60}$$

Then for $|z| < 1/8$ we have successively

$$f^{(j-1)}(z) = f^{(j-1)}(z_0) + \int_{z_0}^{z} f^{(j)}(\zeta) d\zeta,$$

$$|f^{(j-1)}(z)| < \frac{1}{4} + \frac{1}{4} = \frac{1}{2},$$

$$|f^{(j-2)}(z)| < \frac{1}{4} + \frac{1}{2}\,\frac{1}{4} < \frac{1}{2},$$

....

$$|f(z)| < \frac{1}{2}.$$

B_2. There does not exist an integer $j(1 \leqslant j \leqslant k+2)$ satisfying the condition in case B_1. Then there is a point z_1 of the circle $|z| < 1/8$, such that

$$|f^{(k+1)}(z_1)| = \frac{1}{4} \tag{61}$$

and that on the segment $\overline{z_0z_1}$ we have

$$|f^{(k+1)}(z)| \leqslant \frac{1}{4}. \tag{62}$$

As in the above, we find successively that on $\overline{z_0z_1}$ we have

$$|f^{(j)}(z)| < \frac{1}{2} \quad (j = 0, 1, \cdots, k). \tag{63}$$

Distinguish again two cases:

B_2'. $|f^{(k+2)}(z_1)| \geqslant 1/2$. In this case we have

$$|f(z_1)| < \frac{1}{2}, \ |f^{(k)}(z_1)| < \frac{1}{2}, \ |f^{(k+1)}(z_1)| = \frac{1}{4}$$

and

$$\begin{aligned} &|(k+1)f^{(k+2)}(z_1)\{f^{(k)}(z_1)-1\}-(k+2)\{f^{(k+1)}(z_1)\}^2| \\ &\geqslant \frac{k+1}{4} - \frac{k+2}{16} = \frac{3k+2}{16}. \end{aligned}$$

Hence the condition (43) is satisfied and consequently in the circle $|z|<1/8$ we have an inequality of the form (55), namely

$$\left|\frac{1}{f(z)}\right| < e^{3r^*}. \tag{64}$$

B_2''. $|f^{(k+2)}(z_1)|<1/2$. Then there is a point z_2 of the circle $|z|<1/8$, such that

$$|f^{(k+2)}(z_2)| = \frac{1}{2} \tag{65}$$

and that on the segment $\overline{z_1z_2}$ we have

$$|f^{(k+2)}(z)| \leqslant \frac{1}{2}. \tag{66}$$

From (61) and (66), on $\overline{z_1z_2}$ we have

$$\frac{1}{4} - \frac{1}{2}\frac{1}{4} < |f^{(k+1)}(z)| < \frac{1}{4} + \frac{1}{2}\frac{1}{4}. \tag{67}$$

On the other hand, by (59) and (62), we have

$$|f^{(k)}(z_1)| < \frac{1}{4} + \frac{1}{4}\frac{1}{4} = \frac{5}{4^2}$$

and then by (67), on $\overline{z_1z_2}$ we have

$$|f^{(k)}(z)| < \frac{5}{4^2} + \frac{1}{4}\frac{3}{2}\frac{1}{4} = \frac{1}{4^2}\frac{13}{2}. \tag{68}$$

Next from (63) and (68), we find successively that on $\overline{z_1z_2}$ we have

$$|f^{(j)}(z)| < 1 \ (j = 0, 1, \cdots, k-1). \tag{69}$$

Hence we have

$$|f(z_2)| < 1, \ |f^{(k)}(z_2)| < \frac{1}{4^2}\frac{13}{2}, \quad \frac{1}{4}\frac{1}{2} < |f^{(k+1)}(z_2)| < \frac{1}{4}\frac{3}{2}$$

$$|(k+1)f^{(k+2)}(z_2)\{f^{(k)}(z_2)-1\}-(k+2)\{f^{(k+1)}(z_2)\}^2|$$

$$> (k+1)\frac{1}{2}(1 - \frac{1}{4^2}\frac{13}{2}) - (k+2)\frac{1}{4^2}\frac{9}{4} = \frac{1}{4^3}(10k+1).$$

The condition (43) is therefore satisfied and consequently in the circle $|z|<1/8$, we have an inequality of the form (55), namely

$$\left|\frac{1}{f(z)}\right| < e^{3r^*} \tag{70}$$

Finally from (58), (64) and (70), if we put

$$a_k = max(9I', 3I'', 3I'''),$$

then the positive constant $A_k = e^{a_k}$ has the required property in Theorem 2.

REFERENCES

1. W. Blaschke, "Eine Erweiterung des Satzes von Vitali über Folgen analytischer Funktionen", *Leipzig Sitz. ber.*, **67**(1915).
2. A. Bloch, "Les théorèmes de Valiron sur les fonctions entières et la théorie de l' uniformisation". *Annales Fac. sc. Toulouse* **17**(1925)
3. A. Bloch, " Quelques théorèmes sur les fonctions entières et méromorphes d' une variable", *Comptes rendus* **181** (1925) *Comptes rendus* **182** (1926).
4. H. Bohr, "Über einen Satz von Edmund Landau", *Scripta Univ. Hierosolymitanarum* **1** (1923).
5. C. Carathéodory, "Sur quelques généralisations du théorème de M. Picard", *Comptes rendus* **141** (1905).
6. C. Carathéodory, "Sur quelques applications du théorème de Landau-Picard", *Comptes rendus* **144** (1905).
7. C. Carathéodory, "Beitrage zur Konvergenz von Funktionen Folgen", *Sitz ber. Berliner Akad.* (1991).
8. C. T. Chuang,"A generalization of a theorem of Montel", *Science Reports of the National Tsinghua University* **3** (1935) 215-220.
9. C. T. Chuang,"Ètude sur les familles normales et les familles quasi-normales de fonctions méromorphes", *Rendiconti del Circolo Matematico di Palermo* **62** (1938).
10. C. T. Chuang, "Un théorème général sur les fonctions holomorphes dans le cercle unité et ses applications", (I) *Scientia Sinica* **6** (1957) 569-621
11. C. T. Chuang, "Generalizations of the notion of normal families of meromorphic functions", *Contemporary Mathematics*, **25** (1983) 1-18.
12. C. T. Chuang, "On the theory and applications of Q_m-normal families of meromorphic functions", *Complex Analysis I, Proceedings*, University of Maryland (1985-86)44-77.
13. C. T. Chuang and C. C. Yang, *Fix-Points and Factorization of Meromorphic Functions*, World Scientific Publishing Co., Singapore (1990).

14. P. L. Duren, *Univalent Functions*, Springer-Verlag (1935).

15. W. K. Hayman, *Meromorphic Functions*, Clarendon Press, Oxford (1964).

16. G. Julia, *Lecons sur les Fonctions Uniformes à Point Singulier Essentiel Isolé*, Imprimerie Gauthier-Villar, Paris (1924).

17. Y. X. Gu, "A criterion of normlity of families of meromorphic functions", *Scientia Sinica Special Issue* (I) (1979) 267-274.

18. Y. X. Gu, *Normal Families of Meromorphic Functions* (in Chinese), Educational Press of Szechuan (1991).

19. E. Landau, "Ueber eine Verallgemeinerung des Picardschen Satzes", *Sitz. Kg. Akad. Wiss.*, Berlin (1904).

20. E. Landau, *Darstellung und Begrundung einiger neuerer Ergebnisse der Funktionentheorie* (1929).

21. A. J. Macintyre, "Wiman's method and the 'flat regions' of integral functions", *Quart. J. of Math.*, **9** (1938).

22. A. J. Macintyre, "On Bloch's theorem", *Math. Zeit.* **44** (1939).

23. F. Marty, "Recherches sur la répartition des valeurs d'une fonction méromorphe", *Ann. Fac. Sci. Univ. Toulouse* **23** (1931) 183-261.

24. H. Milloux, "Les cercles de remplissage des fonctions méromorphes ou entières et le théorème de Picard-Borel", *Acta Mathematica* **52** (1928).

25. C. Miranda, "Sur un nouveau critère de normalité pour les familles de fonctions holomorphes", *Bulletin Soc. Math.* **63** (1935) 185-196.

26. P. Montel, *Lecons sur les Familles Normales de Fonctions Analytiques et leurs Applications*, Gauthier-Villars, Paris (1927).

27. P. Montel, *Lecons sur les Fonctions Univalentes ou Multivalentes*, Gauthiers-Villars, Paris (1933).

28. R. Nevanlinna, *Le Théorème de Picard-Borel et la Théorie des Fonctions Méromorphes*, Gauthier-Villars, Paris (1929).

29. W. F. Osgood, "Note on the functions defined by infinite series whose terms are analytic functions of a complex variable", *Annals of Math.* **3** (1901).

30. A. Ostrowski, "Ueber Folgen analytischer Funcktionen und einige Verscharfungen des Picardschen Satzes", *Math. Zeit* **24** (1925).

31. F. Schottky, "Ueber den Picardschen Satz und die Borelschen Ungleichungen", *Sitz. Kg. Akad.*, Berlin (1940).

32. W. Schwick, "Normality criteria for families of meromorphic functions", *Journal d' Analyse Math.* **52** (1989) 241-289.

33. T. J. Steiltjes, "Recherches sur les fractions continues", *Ann. Fac. Sci. Univ. Toulouse* **8** (1894).

34. G. Valiron, "Sur les fonctions méromorphes qui admettent des valeurs quasi-exceptionnelles," *Assoc. fr. av. sc.*, (1926).

35. G. Valiron, "Sur un théorème de MM. Koebe et Landau", *Bull. des Sc. math.* **51** (1927) 34-42.

36. G. Valiron, *Familles Normales et Quasi-normales de Fonctions Méromorphes*, Mémorial des Sc. Math. Fasc. 38 (1929).

37. G. Vitali, "Sopra le serie di funzioni analitiche", *Rendiconti del R. Ist. Lombardo* **36** (1903).

38. L. Yang, *Value Distribution Theory and its New Researches* (in Chinese), Science Press, Beijing, 1982.

39. L. Zalcman, "A heuristic principle in complex function theory", *Amer. Math. Monthly* **82** (1975) 813-817.

INDEX

www.ingramcontent.com/pod-product-compliance
Lightning Source LLC
LaVergne TN
LVHW051054210926
848486LV00002B/102
9789810212575